SOFTWARE-ENABLED CONTROL

Perspectives in Control Engineering: Technologies, Applications, and New Directions
Edited by Tariq Samad
2001   Hardcover   532pp   ISBN 0-7803-5356-0

Control Theory: Twenty Five Seminal Papers (1932–1982)
Edited by Tamer Basar
2001   Hardcover   532pp   ISBN 0-7803-6021-4

Linear Time-Invariant Systems
Martin Schetzen
2003   Hardcover   384pp   ISBN 0-47-23145-2

SOFTWARE-ENABLED CONTROL
Information Technology for Dynamical Systems

Edited by

TARIQ SAMAD
Honeywell Automation and Control Solutions
Minneapolis, Minnesota

GARY BALAS
University of Minnesota

IEEE PRESS

A JOHN WILEY & SONS, INC., PUBLICATION

For general information on our other products and services please contact our Customer Care Department within the U.S. at 877-762-2974, outside the U.S. at 317-572-3993 or fax 317-572-4002.

Wiley also publishes its books in a variety of electronic formats. Some content that appears in print, however, may not be available in electronic format.

Library of Congress Cataloging-in-Publication Data is available.

ISBN 978-0-471-23436-4

10 9 8 7 6 5 4 3 2 1

CONTENTS

v

III   ONLINE MODELING AND CONTROL   147

9   ONLINE CONTROL CUSTOMIZATION VIA OPTIMIZATION-BASED CONTROL   149

Richard M. Murray, John Hauser, Ali Jadbabaie, Mark B. Milam, Nicolas Petit, William B. Dunbar, and Ryan Franz

10   MODEL PREDICTIVE NEURAL CONTROL FOR AGGRESSIVE HELICOPTER MANEUVERS   175

Eric A. Wan, Alexander A. Bogdanov, Richard Kieburtz, Antonio Baptista, Magnus Carlsson, Yinglong Zhang, and Mike Zulauf

11   ACTIVE MODEL ESTIMATION FOR COMPLEX AUTONOMOUS SYSTEMS   201

Mark E. Campbell, Eelco Scholte, and Shelby Brunke

12 AN INTELLIGENT METHODOLOGY FOR REAL-TIME ADAPTIVE MODE TRANSITIONING AND LIMIT AVOIDANCE OF UNMANNED AERIAL VEHICLES   225

George Vachtsevanos, Freeman Rufus, J. V. R. Prasad, Ilkay Yavrucuk, Daniel Schrage, Bonnie Heck, and Linda Wills

13 IMPLEMENTATION OF ONLINE CONTROL CUSTOMIZATION WITHIN THE OPEN CONTROL PLATFORM   253

Raktim Bhattacharya and Gary J. Balas

CONTRIBUTORS

MUKUL AGRAWAL, Honeywell Laboratories, Minneapolis, Minnesota, mukul. agrawal@honeywell.com

PANOS ANTSAKLIS, University of Notre Dame, Notre Dame, Indiana, antsaklis@nd.edu

GARY BALAS, Aerospace Engineering and Mechanics, University of Minnesota, Minneapolis, Minnesota, balas@aem.umn.edu

ANTONIO BAPTISTA, Department of Environmental Science and Engineering, OGI School of Science & Engineering, OHSU, Beaverton, Oregon, baptista@ese.ogi.edu

JOHN BAY, Information Exploitation Office, Defense Advanced Research Projects Agency, Arlington, Virginia, jbay@darpa.mil

ALEXANDER BAYEN, Aeronautics and Astronautics, Stanford University, Stanford, California, bayen@stanford.edu

RAKTIM BHATTACHARYA, Aerospace Engineering and Mechanics, University of Minnesota, Minneapolis, Minnesota, raktim@aem.umn.edu

GAUTAM BISWAS, Institute for Software Integrated Systems, Vanderbilt University, Nashville, Tennessee, biswas@vuse.vanderbilt.edu

ALEXANDER A. BOGDANOV, Department of Electrical and Computer Engineering, OGI School of Science & Engineering, OHSU, Beaverton, Oregon, alex@ece.ogi.edu

STEPHEN P. BOYD, Electrical Engineering, Stanford University, Stanford, California, boyd@stanford.edu

MARK CAMPBELL, Mechanical & Aerospace Engineering, Cornell University, Ithaca, New York, me288@cornell.edu

MAGNUS CARLSSON, Department of Computer Science & Engineering, OGI School of Science & Engineering, OHSU, Beaverton, Oregon, magnus@cse.ogi.edu

DARREN COFER, Honeywell Laboratories, Minneapolis, Minnesota, darren. cofer@honeywell.com

DAVID E. CORMAN, The Boeing Company, St. Louis, Missouri, david.e. corman@boeing.com

MUNTHER A. DAHLEH, Electrical Engineering and Computer Science, Massachusetts Institute of Technology, Cambridge, Massachusetts, dahleh@mit.edu

WILLIAM B. DUNBAR, Control and Dynamical Systems, California Institute of Technology, Pasadena, California, dunbar@cds.caltech.edu

JOHAN EKER, Electrical Engineering and Computer Sciences, University of California, Berkeley, johane@eecs.berkeley.edu

ERIC FERON, Aeronautics and Astronautics, Massachusetts Institute of Technology, Cambridge, Massachusetts, feron@mit.edu

RYAN FRANZ, Electrical and Computer Engineering, University of Colorado, Boulder, Colorado, ryan.franz@colorado.edu

EMILIO FRAZZOLI, Aeronautical and Astronautical Engineering, University of Illinois at Urbana-Champaign, frazzoli@uiuc.edu

HELEN GILL, Embedded and Hybrid Systems, National Science Foundation, Arlington, Virginia, hgill@nsf.gov

MURAT GULER, School of Electrical and Computer Engineering, Georgia Institute of Technology, Atlanta, Georgia, gt1452c@mail.gatech.edu

JOHN HAUSER, Electrical and Computer Engineering, University of Colorado, Boulder, Colorado, hauser@colorado.edu

BONNIE HECK, School of Electrical and Computer Engineering, Georgia Institute of Technology, Atlanta, Georgia, bonnie.heck@ece.gatech.edu

THOMAS A. HENZINGER, Electrical Engineering and Computer Sciences, University of California, Berkeley, tah@berkely.edu

BENJAMIN HOROWITZ, Electrical Engineering and Computer Sciences, University of California, Berkeley, California, bhorowit@eecs.berkeley.edu

ALI JADBABAIE, Department of Electrical and Systems Engineering, University of Pennsylvania, Philadelphia, Pennsylvania, jadbabaie@seas.upenn.edu

MIKAEL JOHANSSON, Aeronautics and Astronautics, Stanford University, Stanford, California, mikaelj@stanford.edu

SURESH KANNAN, School of Aerospace Engineering, Georgia Institute of Technology, Atlanta, Georgia, suresh_kannan@ae.gatech.edu

GABOR KARSAI, Institute for Software Integrated Systems, Vanderbilt University, Nashville, Tennessee, gabor@vuse.vanderbilt.edu

RICHARD KIEBURTZ, Department of Computer Science & Engineering, OGI School of Science & Engineering, OHSU, Beaverton, Oregon, dick@cse.ogi.edu

CHRISTOPH M. KIRSCH, Electrical Engineering and Computer Sciences, University of California, Berkeley, Berkeley, California, cm@eecs.berkeley.edu

T. JOHN KOO, Electrical Engineering and Computer Sciences, University of California, Berkeley, Berkeley, California, koo@eecs.berkeley.edu

XENOFON D. KOUTSOUKOS, Palo Alto Research Center, Palo Alto, California, koutsouk@parc.com

TAMAS KOVACSHAZY, Measurement and Information Systems, Technical University of Budapest, Budapest, Hungary, khazy@mit.brne.hu

EDWARD A. LEE, Electrical Engineering and Computer Sciences, University of California, Berkeley, Berkeley, California, eal@berkeley.edu

XIAOJUN LIU, Electrical Engineering and Computer Sciences, University of California, Berkeley, Berkeley, California, liuxj@eecs.berkeley.edu

JIE LIU, Electrical Engineering and Computer Sciences, University of California, Berkeley, Berkeley, California, liuj@berkeley.edu

BRIAN R. MENDEL, The Boeing Company, Berkeley, Missouri, brian.mendel@mw.boeing.com

MARK B. MILAM, California Institute of Technology, Pasadena, California, milam@caltech.edu

RICHARD M, MURRAY, Control and Dynamical Systems, California Institute of Technology, Pasadena, California, murray@cds.caltech.edu

SRIRAM NARASIMHAN, Institute for Software Integrated Systems, Vanderbilt University, Nashville, Tennessee, nsriram@vuse.vanderbilt.edu

GEORGE J. PAPPAS, Electrical Engineering, University of Pennsylvania, Philadelphia, Pennsylvania, pappas@grasp.cis.upenn.edu

TAL PASTERNAK, Institute for Software Integrated Systems, Vanderbilt University, Nashville, Tennessee, pasternak@vuse.vanderbilt.edu

JAMES L. PAUNICKA, The Boeing Company, Berkeley, Missouri, james.paunicka@boeing.com

GABOR PECELI, Measurement and Information Systems, Technical University of Budapest, Budapest, Hungary, peceli@mit.brne.hu

NICOLAS PETIT, Centre Automatique et Systèmes, École Nationale Supérieure des Mines de Paris, Paris, France, petit@cas.ensmp.fr

J. V. R. PRASAD, School of Aerospace Engineering, Georgia Institute of Technology, Atlanta, Georgia, jvr.prasad@ae.gatech.edu

FREEMAN RUFUS, School of Electrical and Computer Engineering, Georgia Institute of Technology, Atlanta, Georgia

TARIQ SAMAD, Honeywell Laboratories, Minneapolis, Minnesota, samad@ieee.org

SAM SANDER, School of Electrical and Computer Engineering, Georgia Institute of Technology, Atlanta, Georgia, sam@sander.com

SHANKER SASTRY, Electrical Engineering and Computer Sciences, University of California, Berkeley, Berkeley, California, sastry@eecs.berkeley.edu

DANIEL SCHRAGE, School of Aerospace Engineering, Georgia Institute of Technology, Atlanta, Georgia, daniel.schrage@aerospace.gatech.edu

GYULA SIMON, Technical University of Budapest, Budapest, Hungary, simon@mit.brne.hu

TIVADAR SZERNETHY, Institute for Software Integrated Systems, Vanderbilt University, Nashville, Tennessee, tivadar@vuse.vanderbilt.edu

CLAIRE TOMLIN, Aeronautics and Astronautics, Stanford University, Stanford, California, tomlin@stanford.edu

GEORGE VACHTSEVANOS, School of Electrical and Computer Engineering, Georgia Institute of Technology, Atlanta, Georgia, gjv@ece.gatech.edu

DALE W. VAN CLEAVE, AFRL/IFSC, Wright-Patterson AFB, Ohio, dale.vancleave@wpafb.af.mil

ERIC A. WAN, Department of Electrical and Computer Engineering, OHSU, Beaverton, Oregon, ericwan@ece.ogi.edu

LINDA WILLS, School of Electrical and Computer Engineering, Georgia Institute of Technology, Atlanta, Georgia, linda.wills@ece.gatech.edu

LIN XIAO, Electrical Engineering, Stanford University, Stanford, California, lxiao@stanford.edu

ILKAY YAVRUCUK, School of Aerospace Engineering, Georgia Institute of Technology, Atlanta, Georgia

YINGLONG ZHANG, Department of Environmental Science and Engineering, OGI School of Science & Engineering, OHSU, Beaverton, Oregon, yinglong@ccalmr.ogi.edu

MIKE ZULAUF, Department of Environmental Science and Engineering, OGI School of Science & Engineering, OHSU, Beaverton, Oregon, mazulauf@ese.ogi.edu

PREFACE

This edited volume is the product of a research initiative undertaken by the U.S. Defense Advanced Research Projects Agency (DARPA) and the U.S. Air Force Research Laboratory (AFRL) to exploit recent developments in software and computing technologies for applications to control systems in general, and autonomous aircraft in particular. Control, in this context, should not be interpreted in some narrow sense. Here it encompasses algorithms for inner-loop regulation as well as supervisory and mission-level optimization, modeling and estimation of vehicle dynamics and environmental influences, real-time computing platforms and software design tools, and much else besides.

The "Software Enabled Control" program is ongoing and the chapters in this volume do not document the culmination of the research. But with some years of effort completed by a number of multidisciplinary teams there is much to report: A number of innovations have resulted and been validated through some combination of theoretical analyses, simulation experiments, and laboratory demonstrations. In the near future, many of the developments detailed in this book are planned to be further evaluated through flight testing.

The SEC program was envisioned and initiated at DARPA by Dr. Helen Gill and Dr. David Tennenhouse and has subsequently benefited from the support and leadership provided by Dr. Shankar Sastry and Dr. John Bay. At the U.S. Air Force Research Laboratory, Ray Bortner, Bill Koenig, Reed Morgan, and Dale Van Cleave have been instrumental in overseeing the program and its constituent projects. Todd Carr, Jessica Greenhalgh,

Nikki Morris, and Carmen Whitson have ably fulfilled a variety of coordination and administration responsibilities. We speak for the SEC research community in expressing our gratitude to these individuals, and to several others who were involved in advisory capacities, for creating and supporting this practically important and intellectually exciting program.

TARIQ SAMAD
GARY BALAS

Minneapolis, Minnesota
October, 2002

PART I

INTRODUCTION

CHAPTER 1

THE SEC VISION

HELEN GILL and JOHN BAY

1.1. THE LEGACY OF CONTROL TECHNIQUES

At the time of conception of the DARPA Software Enabled Control (SEC) program, control research was proceeding down a path determined by old views of the computational and systems context. The assumptions were as follows: highly constrained sensing and actuation, limited processing and communication resources, computational intractability of large or even moderate state spaces, poorly characterized and unpredictable switching effects, and target systems that operated independently and without interaction with other systems.

Control theory and engineering have a remarkably successful history of enabling automation, and information-centric control is by now pervasive. Yet today's controllers are conservative: Being products of overdesign, they often yield underperformance. Their designs are statically optimized for nominal performance, around simplified time-invariant models of system dynamics and a well-defined operational environment. They also fail in unexpected circumstances: control vulnerabilities that arise in extreme environments are frequently ignored. System modification (reconfiguration, damage, failure) may demand large changes in the controller, perhaps online during operation.

Research in adaptive control has sought to accommodate change through the use of online feedback to the governing parameters; robust control research introduced periodic recalculation of the controller, using model-based prediction. Relative to simple PID and set-point control, these model-centric strategies yield improvement. However, there are several problems. Disruptive events occur unexpectedly, not periodically, and the changes required may be dramatic. Principles and support are lacking for reactive but systematic online reconfiguration of models and software. A popular modern

Software-Enabled Control. Edited by Tariq Samad and Gary Balas
ISBN 0-471-23436-2 © 2003 IEEE

strategy is to enlarge models for "full-envelope design," attempting to accommodate an ever-larger span of configurations and environmental conditions in a single control law. This results in monolithic, flattened models with immense state spaces.

1.2. THE LEGACY OF CONTROL SOFTWARE

Mirroring the progress of control engineering, the emerging reality in information technology was one of great promise: exponential growth in processing speeds; new and better-used communications modalities; new storage capacity for capture and online exploitation of information (both sensed and model-generated); and new strategies in software composition that enabled extensible, configurable open software and systems. We had developed device networks, smart sensors, programmable actuators, and systems-on-a-chip, along with distributed objects, real-time operating systems and code generators, and application-specific design tools, but the synthesis of embedded control with these tools was a new problem.

The challenge for the SEC program was how to exploit software and computation to achieve new control capabilities. The program was formulated to create the necessary linkage between physical systems and the software and computation strategies needed to enable next-generation control systems. It was also intended to jump-start new technologies with updated assumptions about distributed embedded systems.

1.3. A NEW PERSPECTIVE ON SOFTWARE AND CONTROL

With the advent of networked sensors and actuators, distributed computing algorithms, and hybrid control, the term "system dynamics" has taken on a whole new meaning. Whereas it used to bring to mind only ordinary differential equations with perhaps some parameter uncertainty, noise, or disturbances, we can now include dynamic tasking, sensor and actuator reconfiguration, fault detection and isolation, and structural changes in plant model and dimensionality. Consequently, the ideas of system identification, estimation, and adaptation must be reconsidered.

This new perspective of the world also requires new models for control software implementations. But we must avoid the temptation to think of software as simply the language of the implementation. Control code—particularly embedded control code—is a dynamic system. It has an internal state, responds to inputs, and produces outputs. It has time scales, transients, and saturation points. It can also be adaptive and distributed. As any control engineer knows, if we take this software dynamic system and couple it to the plant dynamics through the sensor and actuator dynamics, we have a composite system whose properties cannot be decided from the subsystems in isolation.

Thus, when we put an embedded controller on a hardware platform, we have not only a coupled system with significant off-diagonal terms, but a distributed hybrid one at that. To borrow from the computer engineering lexicon, we have a problem in control/software co-design. The control design is evolving through the development of hybrid optimal control, reachability analysis, multiple-model systems, and parameter varying control. The software is being facilitated by distributed computing and messaging services, distributed object models, real-time operating systems, and fault detection algorithms. What we seek in the SEC program is a mutual catalysis of such control and embedded software technologies that will push the boundaries of performance, complexity, and applicability.

1.4. SOFTWARE ENABLED CONTROL FOCUS AREAS

To address these new challenges in software control, four focus areas were originally identified, with a fifth added after the first year's progress.

1.4.1. Managed Models for Predictive Control

Parallel distributed processes, together with time management strategies available in modern real-time operating systems, are particularly effective in facilitating multiple model control approaches. By maintaining system identifiers and parameter estimators in separate processes, more diverse plant models can be generated without the immense time penalties incurred if the estimators share a CPU with the controller. Such models can include previously time-prohibitive components such as nonlinear filters, disturbance predictors, neural network training, and environment dynamics. If the transitions between these models can be properly managed through hybrid controllers, then there are few limits on the number or complexity of plant models and controller configurations that can be used.

1.4.2. Online Control Customization

As these models converge, though, we require a service for identifying the appropriate one, switching it online, and managing this transition in time. The result will be that real-time control will no longer be limited to single-model methods, and we will see improvements over gain-scheduled approaches that assume a fixed controller structure. This represents a revolution in adaptive control, from parameter or gain adaptation in slowly time-varying systems to major structural changes or varying plant dimensionalities.

In addition, the use of smart sensors and programmable actuators adds flexibility to the control configuration and presents numerous options for fault detection, isolation, and reconfiguration. Because these possibilities change the controller's input and output mapping, they also represent control modes and must be managed accordingly. Of course, the challenge in implementing such diverse control configurations in a single system is transition management, both logical and temporal.

1.4.3. Hierarchical, Multimodel, and Multimodal Coordinated Control

The distributed controllers enabled by emerging real-time middleware support can consist of hierarchical systems, integrated subsystems, or independent confederated systems, such as multivehicle systems. All these require coordinated control. Currently, coordination is managed in ad hoc supervisory levels of software control, usually with inadequate vertical integration among hierarchical levels. Today's prototypical control system is one that has one or only a few modes of operation and is implemented (often at the expense of significant manual effort) as a strict and static hierarchy of supervisory levels. These controllers are realized in software as nests of loops. If they coordinate, the strategy is "hard-wired." Such monolithic software is difficult or impossible to change. By contrast, today's trends suggest that control systems increasingly are needed that are much more dynamically configured; that can manage extremely large state spaces; that have open, modular designs; and that enforce a more precise relationship between low-level physical control and supervisory or coordination levels.

1.4.4. Open Control Software Tools and Services

Software services for implementing embedded controllers have remained at a primitive level and have not addressed the embedded software needs of control systems designers. Attention typically has focused on low-level real-time operating system services that are not at a level of abstraction well matched to control software design. These services only poorly support such techniques as adaptive and robust control and hybrid mode transition, all of which may require coherent dynamic reconfiguration of processes and data streams.

Techniques are emerging for open distributed systems that address these needs. They are offered as service layers in real-time operating systems and as portability and interoperability services such as the real-time common object broker, RT CORBA. However, they are not yet exploited in control systems. The monolithic code currently synthesized by control design frameworks lacks suitable levels of abstraction and is neither well-tuned nor amenable to improvement by hand. It is neither portable nor analyzable for such problems as software and communication transients and interference with other critical functionality in the system. Current technology for reasoning and synthesis does not sufficiently address the many interacting issues: the composite of continuous and discrete behavior, software and control, offline design, and online operation.

1.4.5. High-Confidence Systems and Software

Pushing the boundaries of performance and complexity in functionality means that there will be commensurate problems with validation, verification (V & V), and certification for each application. Automation does us little good if we have no confidence in the tools we have created. However, quality assurance and certification analyses generally lag the functional technologies and are too often measured by after-the-fact performance metrics rather than being treated as design-time constraints. An emerging focus area for the DARPA SEC program is the treatment of V & V and certification as design-time criteria, as well as to develop scalable checking and proof technologies that will ultimately result in correct-by-construction or "autocertifiable" code. Although such constraints are application-specific, they are as real as other timing or quality of service constraints. Yet, they have not been addressed in control software.

1.5. THE DARPA SOFTWARE ENABLED CONTROL PROGRAM

It was against this context that the Software Enabled Control program was formulated. The program objective was to unite computer science, systems software, and control technologies, with a primary goal of improving actual control performance in the face of demanding environments and requirements for interactive, cooperative control. However, instead of attacking the problem only at the higher levels of supervisory control, which typically exploit only discrete methods, it was determined that even low-level differential control of reasonably uncomplicated systems could and should be targeted. Thus, the program challenges were seen to lie in three areas: (1) so-called "simple systems," (2) systems built by integrating subsystems at the time of design or configuration, and (3) multisystems, which might form dynamically yet have tight physical constraints on interoperation. Examples include: fuel conservation in a single vehicle; coordinated braking, fuel management, and transmission control within a vehicle; highly automated management of "swarms" of autonomous air vehicles; redundantly instrumented and actuated systems capable of reconfiguration under catastrophic failure; resolution determination for a group of aircraft in a sudden collision avoidance situation or other joint encounter; or coordination of varying, proximity-based enclaves of safe, energy-optimizing vehicles in an intelligent highway system or on a battlefield.

In the pursuit of these goals, the SEC program was started in late fiscal year (FY) 1999 and was fully underway in FY 2000. It focused initially on four technical areas: the active management of state models and online information for predictive control; hierarchical, multimodal coordinated control; online control redesign and software reconfiguration; and software technologies for open control systems. In FY 2001, a fifth emphasis was added in high-confidence software and systems, to build on and guide the technologies developed under the initial four areas. Building on the research begun in this DARPA program, a new base program in Embedded and Hybrid Systems has been created at the National Science Foundation to support and sustain the ongoing effort, both foundational and experimental, that will be required to realize the SEC vision.

As seen in this volume, the vision of software-enabled control is starting to bear fruit: New architectures, algorithms, and models are emerging. At the same time, new directions and opportunities continue to emerge, and the vision appears likely to outlive the program itself!

ACKNOWLEDGMENTS

The superb technical leadership we received from Dr. David Tennenhouse and Dr. Shankar Sastry, in their service as Directors of the DARPA Information Technology Office, has earned our lasting respect and gratitude. They provided wisdom and thoughtful guidance, wit and encouragement, and (perhaps most important) an enthusiastic commitment to the merit of the SEC endeavor. Reed Morgan (now retired), Bill Koenig, and Ray Bortner of the USAF AFRL have been invaluable partners. We owe permanent debt to Jessica Greenhalgh, Nikki Morris, and Carmen Whitson for their cheerful, professional, and dedicated efforts to ensure the smooth operation of the SEC program. We would also like to thank Mark Swinson and Janos Sztipanovits for their encouragement and support.

CHAPTER 2

TRENDS AND TECHNOLOGIES FOR UNMANNED AERIAL VEHICLES

DALE W. VAN CLEAVE

2.1. INTRODUCTION

The possibilities presented by unmanned aerial vehicles (UAVs) have beckoned researchers, military planners and aviation enthusiasts for years. With just a little imagination, one could envision UAVs taking over a host of duties, resulting in numerous benefits. It is thought that UAVs can be used to good advantage in certain situations that are very tedious or present undue danger for pilots. These situations are commonly summed up as the "dull," the "dirty," and the "dangerous." UAV advocates also tout cost savings and other expected benefits.

UAVs are now beginning to realize some of their potential. They are making definite inroads in surveillance and reconnaissance applications and advancing in other areas as well. In this chapter we examine some of the historical background of UAVs and the motivation behind the push to develop UAVs. We review a number of UAV development efforts and their difficulties, and we examine some current UAVs and their successes as well as take a brief look at the future.

2.2. UAV BACKGROUND

2.2.1. Early Days

Curiously enough, like so many other "modern" ideas the concept of UAVs is not a new one. Interest in them goes back almost as far as the origin of

Software-Enabled Control. Edited by Tariq Samad and Gary Balas
ISBN 0-471-23436-2 © 2003 IEEE

Figure 2.1. The Bug pilotless aircraft with engine detail inset. (Photograph courtesy of Dr. Russell Naughton, http://www.ctie.monash.edu.au/hargrave/rpav_usa.html.)

manned flight. During World War I, pilotless planes were conceived as an aerial equivalent to the naval torpedo. A prototype made a number of short test flights in 1916–1917, and from this the Army started a project that resulted in 20 pilotless aircraft called Bugs (Figure 2.1). H. H. "Hap" Arnold was a leader in the project that involved Orville Wright, Charles Kettering, and other industrial-age pioneers [1]. These planes were launched from a dolly on a track and were guided by preset vacuum-pneumatic and electrical controls. After a preset length of time the engine was turned off, the wings folded, and the Bug with its 82-kg warhead would fall to earth (hopefully in the vicinity of a worthwhile enemy target). Successful test flights were accomplished in 1918, but motivation for development of this potential weapon was significantly reduced with the end of the war and the project was discontinued in 1925. These early products were forerunners of drones and guided missiles, and the project was classified until World War II when a similar concept, the German V-1 buzz bomb, was used against England.

In 1937 the first pilotless planes (or drones) were used for antiaircraft target practice; and in the early 1940s, tests were conducted on aircraft converted for remote control use to attack land and sea targets. During World War II, B-17s that were near the end of their useful life were outfitted with remote controls, television cameras, and nine tons of high explosive and used as flying bombs to attack German V-1 missile sites, submarine pens, and deep fortifications that had resisted conventional bombing. A two-man crew would get the aircraft off the ground and to an initial altitude and then the BQ-7 (as it was designated) would be remotely controlled from escort planes flying at higher altitude, allowing the crew to bail out. Many of the planes went out of control and most of the rest missed their targets, although one is reported to have been successful in an attack on an underground submarine pen.

The United States used low-tech UAVs in Vietnam for reconnaissance, surveillance, and some electronic intelligence gathering tasks. However, information gathered was on film, which was ferried to Washington for analysis, and consequently the intelligence gained was often outdated by the time it got back to Vietnam [2], [3]. The Firebee target drone, which actually had its start in the 1950s, has a 4-m wingspan, a gross weight of 1500 kg, and an endurance time of 115 minutes. This aircraft is jet-powered with rocket assistance for launch. The basic target drone was modified to perform autonomous, preprogrammed, long-range reconnaissance missions. It was air-launched, and at the end of a mission a specially equipped helicopter was used to retrieve the plane while it was under parachute descent. The modified UAVs joined the Strategic Air Command reconnaissance units in the late 1960s and were extensively used in Southeast Asia. With one of the longest life spans on record, the basic drone remains in service today.

2.2.2. International Developments

Israel has been a leader in the development and fielding of UAVs. In the early 1970s, the Israeli Air Force's 200 Squadron was formed and today operates three types of UAVs: the Scout and the Searcher, built by Israel Aircraft Industries' Malat Division; and the Hermes 450, built by Silver Arrow. The Malat Division offers four basic types—the Searcher, the Hunter, the Heron, and the Ranger—and has reportedly sold more than 600 UAVs to date to 17 customers worldwide [4]. The Searcher has a wingspan of 7.2 m and a payload capacity of 63 kg. It has the ability to loiter 12 hours at a 100-km range and can provide real-time imagery day or night. The Hunter also has 12 hours endurance, 150-km range, and a maximum speed of 200 km/h and is in service with the U.S. Army and Marines for tactical use. The Heron is a newer design than the Searcher or Hunter and is larger and carries a greater payload. It can loiter for up to 52 hours with a 100-kg payload at 150 km and has a maximum payload capacity of 250 kg. The Hermes 450 is intended for surveillance, reconnaissance, and target acquisition, has a total endurance of 20 hours, can fly up to an altitude of 6100 m and has a range of 200 km. This UAV addresses intelligence requirements for ground forces, supplying real-time battlefield data day or night. It assists in locating enemy targets and adjusting artillery fire.

The Pioneer UAV was begun in the early 1980s, first fielded in 1986, and was produced jointly by American and Israeli firms. A total of 72 vehicles was procured. This UAV was intended for naval gunfire spotting and to provide the Marines with a tactical UAV capability. This pusher-prop aircraft is 4.3 m long, weighs 210 kg, and carries gyrostabilized high-resolution TV or FLIR (forward-looking infrared) payloads for day or night operation [5].

By 1997 a number of countries, in addition to Israel, were producing various UAV models with modest capabilities for use within those countries and also for export to a number of other countries. France produced the Crecerelle and Fox TS drone version and AT version for surveillance, Italy produced the Mirach 26 and Mirach 150 for surveillance and target acquisition, Great Britain was nearing completion of the Phoenix, also for real-time surveillance and target acquisition, Germany produced the Brevel drone, and South Africa produced the Seeker [6]. Today, additional countries have UAVs either in operation or under development, including China, Iran, Iraq, Russia, India, and Japan.

2.2.3. Consolidation in the United States

In 1987 the U.S. Congress consolidated funding for the development of UAVs into a single account instead of separate service accounts. This resulted in the formation of the UAV Joint Program Office overseen by the Defense Airborne Reconnaissance Office within the Office of the Secretary of Defense. The Predator UAV was developed primarily by the Joint Program Office using an advanced concept technology demonstration (ACTD) process. ACTDs are designed to assemble and demonstrate significant, new, and improved military capability that is based upon relatively mature advanced technologies. ACTDs are a critical precursor to formal acquisition. The Predator (Figure 2.2) progressed from a concept to a three-system

Figure 2.2. Predator.

Figure 2.3. Global Hawk.

operational capability in a period of less than 30 months [7, 8]. The goal of the Predator was to provide near-real-time imagery, even in adverse weather. It has a 15-m wingspan and a payload of 204 kg, and it can stay on station for 16 hours at a range of 725 km. It carries an electro-optic/infrared camera system as well as synthetic aperture radar. The Air Force took operational control in 1996 [5].

The Global Hawk (Figure 2.3) is a high-altitude, long-endurance UAV and is also the result of an ACTD program designed to provide extended reconnaissance capability to the Joint Force commander. Extended reconnaissance is defined as data from anywhere within enemy territory, day or night, regardless of weather. It carries both EO/IR (electro-optic/infrared) and SAR (synthetic aperature radar) sensors to generate both wide-area and spot imagery while standing off from high-threat areas. The Global Hawk has a 35-m wingspan, payload capacity of 890 kg, and an endurance of 38 hours (20 hours at 5500-km range). Development of this UAV was begun in 1994. First flight was achieved in February 1998. The Global Hawk has completed the initial phases of its development and has entered the low-rate initial production (LRIP) phase. Initial capability to support limited operations should be achieved by 2003.

The Navy and Marines are beginning low-rate initial production of the RQ-8A Fire Scout, a vertical-takeoff unmanned aerial vehicle (VTUAV) that is scheduled to replace the RQ-2 Pioneer. The vertical-takeoff ability allows it to operate from ships besides carriers but does not require an operator that is experienced in flying helicopters. It is designed for reconnaissance, surveillance, and targeting and has an endurance of 6 hours and a range of 180 km [9].

The Army is developing a new tactical UAV (TUAV) since the Hunter fleet is wearing thin due to normal attrition and combat losses. The Army TUAV program seeks to develop a tactical UAV and procure 44 systems by 2005. Each system consists of three UAVs, ground control stations, a launcher, and support elements. The UAVs will possess forward-looking infrared and video imagery capability and thereby provide battlefield reconnaissance, surveillance, target acquisition, and damage assessment as well as maintain situation awareness [10, 11].

In a joint program, DARPA and the Air Force are sponsoring the development of the unmanned combat air vehicle (UCAV). This craft is intended to take on the mission of attacking enemy air defense sites and high-value targets, particularly in the initial phase of a conflict. The stealthy UAV is designed to carry up to a ton and a half of weapons to points as far as 1600 km away. It will be equipped with video cameras, a global positioning system (GPS), and radar to carry out precision-targeting tasks. The UCAV will have a high degree of autonomy, allowing it to operate in groups of up to four with a single operator. If the program stays on track, UCAVs could attain operational status around 2010. DARPA and the Navy have also initiated a program to develop a carrier-based UCAV that has similar goals to the above.

2.3. THE PROMISE OF UAVs

2.3.1. The Dull, the Dirty, and the Dangerous

In considering the rationale for developing UAVs, it seems apparent that there are situations where they might be employed more easily, safely, cheaply, or efficiently than manned aircraft. The U.S. Department of Defense Roadmap for developing and employing unmanned aerial vehicles labels these situations as the "dull," the "dirty," and the "dangerous" [12]. The "dull" applies to surveillance or sentry duty. Flying in monotonous patterns over and over, watching, listening, and waiting, a UAV will never tire or let its attention wander, as humans naturally will. Simply flying to and from the mission area can create additional hardship for crews when forward bases are not available. With UAVs the operators do not have to spend countless hours in a cramped cockpit, and they may even change shifts when a mission warrants. The "dirty" is described as reconnoitering areas that may be contaminated due to nuclear, biological, or chemical hazards. Although U.S. pilots have flown such missions in the past, it is obviously preferable not to subject humans to such hazards if possible. The "dangerous" pertains to any mission that poses an obvious threat to personnel such as the suppression of enemy air defenses (SEAD). Flying into the teeth of enemy air defenses at the onset of a conflict in order to "suppress" these defenses and protect other friendly aircraft requires excellent equipment and training, great courage, and a good bit of luck and unfortunately can result in short life expectancies. This type of mission is becoming even more dangerous as integrated air defenses are maturing. Enforcing "no fly" zones in Iraq is an example of a mission that is both dull and dangerous (although maybe not simultaneously) and could, therefore, be a candidate for UAVs in the future.

2.3.2. Other Personnel Related Advantages

Removing pilots from dangerous missions where possible is desirable not simply for the immediate physical danger, but also to avoid having them become prisoners during conflicts or political pawns during peacetime events. Conflicts in Iraq, Bosnia, and Yugoslavia all saw downed airmen in need of search and rescue. This type of undesirable situation can also occur during peacetime. In April 2001 a U.S. EP-3 electronic warfare and surveillance aircraft was damaged and forced to land on Chinese soil after a collision with a Chinese naval F-8 fighter. Twenty-four crewmembers were detained for 11 days on China's Hainan Island. Such situations can be avoided in cases where UAVs are employed.

The military aviation accident rate has reached historic lows in recent years. In spite of the reduced number of accidents, there were still 44 fatalities in 1999 and 58 in 2000 for the combined U.S. services. Day-to-day operation and training can be just as deadly to aircrews as actual combat. In the 78-day bombing campaign over Kosovo, only two U.S. Air Force aircraft were lost and no pilots; however, during that same year (1999), nine Air Force personnel lost their lives in aviation-related accidents. (Incidentally, up to two dozen or more UAVs were lost by the NATO forces during the conflict with Serbia, but there was little publicity because friendly forces personnel were not at risk.) Safety considerations in training obviously favor UAVs. Training with UAVs, presumably, does not entail the same risk to personnel as does training with manned aircraft. Also, training can be accomplished for lower cost since simulators can be used to a greater degree.

Another factor arguing for the development of UAVs is the shortage of pilots. Even though the U.S. Air Force has 92% fewer airplanes and 91% fewer pilots today than in World War II, it is still difficult for the services to maintain the desired number of pilots [1]. Frequent and extended deployments place strains on pilots and their families and this coupled with attractive salaries in commercial aviation and other factors have caused a large turnover. We still need manned aircraft but as some missions are transferred to unmanned craft we can reduce the demand for pilots.

2.3.3. Air Vehicle Advantages

UAVs, obviously, do not need a cockpit or any equipment associated with creating an environment required for a pilot. UAVs can be made smaller, more aerodynamic and stealthier than manned aircraft. UAV designs can consider G-forces far in excess of that tolerated by humans and thereby achieve superior maneuverability. If UAVs can be controlled appropriately these factors can add up to survivability and effectiveness of weapons systems and a significant edge in times of conflict.

With the continuing pressures on defense budgets and the rising price of weapon systems, cost is a major consideration. Arguments are put forth that UAVs can be made more cheaply than manned aircraft. One frequently cited comparison is that between the UCAV and Joint Strike Fighter (JSF). Three UCAVs can be purchased for the cost of a single JSF [13].

"Smart" weapons gained considerable favor as a result of their performance in the Gulf War. During the Gulf War, only 2% of the ordinance used were precision-guided weapons; but during the NATO Operation Deliberate Force in Bosnia, over 90% of the bombs dropped were in this category [8]. Since precision-guided weapons are much more accurate than standard bombs, they can be smaller and fewer in number and still accomplish the same goals. Small, smart weapons fit very nicely with the concept of a combat UAV.

2.3.4. Intelligence, Surveillance, and Reconnaissance

UAVs are already employed for a variety of purposes and are under consideration for many more. The most obvious use to date for military UAVs has been for intelligence, surveillance, and reconnaissance (ISR). Accurate and timely information is supremely important for the success of military operations but has historically been a scarce commodity. While satellites are invaluable, the coverage they provide is limited in terms of times and area and they are directed to the highest-priority target areas, thereby inevitably skipping over other areas. Furthermore, they are not under the control of field commanders. UAVs can loiter in an area, use long-range sensors to cover a wide area, deny an enemy an opportunity to move assets without risk of detection, and provide real-time coverage of a battlefield to field commanders. They can provide initial reconnaissance, continuous surveillance, and targeting-quality information for both strike airplanes and cruise missiles and follow the attacks up with immediate bomb damage assessments. UAVs can provide an accurate picture of enemy resources before a conflict, real-time information about enemy positions and movement during a conflict, and monitoring to verify compliance with agreements after a conflict [14].

2.3.5. Weather Research and Monitoring

Aerosonde Robotic UAVs are used to support weather research. They have provided weather data to the Navy and Marines. "We fly them into dangerous weather. Not all come home, but most do," stated the operations manager. "We can fly them into anything except tornados" [15]. NASA plans to use the Altus UAV, by General Atomics, to monitor thunderstorms and better understand how lightning forms and dissipates [16]. Also planned are three new 700-pound payload research UAVs based on the Predator. These UAVs will have 15,000-, 18,000-, and 19,800-m operational altitudes [7].

The U.S. Department of Energy's interest in UAVs stems from the measurement of atmospheric radiation [18].

2.3.6. Other Applications

UAVs with infrared cameras for day or night use to provide live, real-time video have been used by the Marines in a Joint Task Force operation to perform border patrol on the U.S./Mexican border to hunt smugglers and illegal aliens [19].

Arguments have been put forth for using UAVs for search and rescue operations by the Coast Guard and for artillery spotting, target acquisition, decoys, jamming, suppression of enemy defenses and strike missions, and coverage of areas contaminated by biological and chemical agents [2]. UAVs may be used as a communications relay or as a stand-in for a poor man's satellite.

UAVs are receiving growing attention as a way to find and destroy explosive mines on land and at sea. Video and laser-based sensor systems, sensor fusion techniques, and advanced algorithms are technologies that are making it possible to consider using UAVs for this purpose [20]. Miniature UAVs that are launched from howitzers have been proposed for battlefield reconnaissance, target location, and damage assessment [21]. Hand-launched micro-UAVs (6 inches) are under consideration to provide situation awareness at the platoon level, in hostage situations, and for other applications.

An attractive potential use for UAVs is to locate and destroy mobile ballistic or air-defense missile launchers. One obvious difficulty encountered by coalition air forces during Desert Storm was locating and destroying SCUD missiles and their launchers. Locating mobile targets on the ground was difficult in the Kosovo air campaign since manned aircraft were required, for the most part, to fly above 4500 m to eliminate the threat posed by antiaircraft fire. Given these difficulties, potential enemies will surely emphasize these weapons and tactics in the future. While further work is required involving sensors and processing, the UAV is certainly an attractive platform to combat the problem [4, 22].

Advancements in a number of technologies make UAVs a credible solution for many applications. Strong, lightweight materials, advanced, reliable, compact power plants, data links, high-speed embedded processing, navigation aids such as GPS, compact, powerful sensors, and precision-guided munitions all contribute to the practicality of UAVs.

2.4. SUPPORT FOR DEVELOPMENT

With the promise that UAVs represent, they have found proponents both in the services and in the U.S. Congress. A 1997 document, *Global Engagement: A Vision for the 21st Century Air Force*, gave strong support for the development of UAVs across a broad range of combat missions. According to *Global Engagement*,

> The highest payoff applications in the near-term are Intelligence, Surveillance, Reconnaissance (ISR), and communications. A dedicated Air Force UAV squadron will focus on operating the Predator medium-range surveillance UAV, which also will serve as a testbed for developing concepts for operating high altitude, long endurance UAVs. In the mid-term, the Air Force expects that suppression-of-enemy-air-defense (SEAD) missions may be conducted from UAVs, while the migration of additional missions to UAVs will depend upon technology maturation, affordability and the evolution to other forms of warfare.

Senator John Warner, then chairman of the Senate Armed Services Committee, went on record with his belief that in the near future many missions currently accomplished by manned aircraft will be successfully transferred to unmanned vehicles. Senator Warner pushed to have one-third of all deep strike aircraft be unmanned by 2010. His belief was that "... in my judgement this country will never again permit the armed forces to be engaged in conflicts which inflict the level of casualties we have seen historically. So what do you do? You move toward the unmanned type of military vehicle to carry out missions which are high risk in nature...". He noted the "extraordinary record" of Operation Allied Force in which the military conducted a 78-day air campaign that included 38,000 combat sorties but no (allied) casualties; he also noted that, while it may be unreasonable, the expectation of the American people was that future conflicts should also avoid casualties to U.S. military personnel [1, 23, 24]. It is worthwhile to note here that the National Defense Authorization Act for FY2001 stated, "It shall be a goal of the Armed Forces to achieve the fielding of unmanned, remotely controlled technology, such that by 2010, one-third of the aircraft in the operational deep-strike force aircraft fleet are unmanned" [25].

2.5. DIFFICULTIES

In some ways the mid-to-late 1990s were the dark days of UAV development. Several UAV development programs ran into trouble, from one cause or another, and were cancelled or cut back. Congressional lawmakers were running out of patience and beginning to exercise close oversight of some development efforts. The combined investment, across the services, to develop and procure UAVs between 1979 and 1997 totaled approximately $3 billion with little success to show [3]. During the late 1990s, virtually all UAV development programs were experiencing schedule and cost problems and had lost aircraft due to crashes, which generated additional bad publicity. Various factors contributed to this situation. In some cases too much was attempted (or promised) too soon, and in others "requirements creep" was to blame. Attempts to meet joint services needs resulted in conflicting requirements, inevitable cost growth, and program failure.

2.5.1. Aquila

The goal of Aquila was to provide real-time intelligence on enemy troop locations beyond the line of sight of friendly forces. It was abandoned due to schedule and technical difficulties with "requirements creep" playing a role in its demise. While the Israeli drones used to good advantage in 1982 had been developed in a year and cost about $50,000 each, the Aquila's initial estimated cost was $560,000 per plane and before cancellation the estimated cost grew to over $3 million per plane. The difference was that the Aquila was designed to penetrate and survive against Soviet air defenses. By 1987, only 7 of 105 flights were successful in meeting mission requirements, over $1 billion was spent, total cost was projected to be over $2 billion, and the program was cancelled [5, 26].

2.5.2. Pioneer

The Pioneer UAVs purchased by the Navy were successors to the drones used by the Israelis in the Bekka Valley. The UAVs were procured with a relatively modest sum, but they were not designed for carrier operation as desired by the Navy so an additional $50 million was spent developing a recovery system and addressing other problems. They were used successfully during the Gulf War, but had a number of shortcomings that were brought out then and during support of operations in the Balkans. They had a high accident rate, difficulties were encountered with data dissemination, and they could not be used in bad weather due to lack of weatherproofed avionics and synthetic aperature radar. Overpricing of spare parts also was reported (but this problem is not unique to UAVs) [5, 7, 26].

2.5.3. The Medium-Range UAV

The Medium-Range UAV was a joint effort of the Navy and Air Force. This jet-powered UAV was intended to penetrate enemy airspace at high subsonic speed in advance of air strikes and provide reconnaissance using near-real-time video. The Navy was to develop the air vehicle, and the Air Force was to develop the sensor package. Both parts ran into developmental difficulties, resulting in schedule delays and cost overruns. The project was cancelled in 1993 [5, 7, 26]

2.5.4. Hunter

Development of the Hunter UAV was undertaken by the Joint Program Office in 1988. This UAV was intended to perform day or night ISR and target acquisition missions and provide near-real-time video. The program experienced significant cost growth, however, and suffered several crashes that led to program termination in 1996 [5, 7, 26].

2.5.5. Outrider

The Outrider Tactical UAV [TUAV] program was initiated in 1996 with the goal of providing near-real-time reconnaissance, surveillance, and target acquisition information to Marine air/ground task forces, Army brigades, and deployed Navy units. In an attempt to meet both Army and Navy needs, the range requirement was increased by a factor of four, and for ship-board safety reasons an attempt was made to switch from gasoline powered to a diesel-fuel-powered engine. This program also experienced cost growth and schedule delays and was cancelled in 1999 [5, 7, 26].

2.5.6. U.S.A.F. Reluctance

The Air Force was minimally engaged in most UAV developments until the mid-1990s. Of seven developmental programs representing all DOD UAV efforts from 1980 to early 1990, only the Predator had significant USAF involvement. General Ronald R. Fogelman, USAF Chief of Staff, acknowledged in 1995 that the Air Force had been slow to adapt the potential of UAVs as an integral part of the force structure. He wrote, "In the past the Air Force had reservations about UAVs—some technical, some bureaucratic, and some cultural." Subsequent endorsement of UAVs by the Air Force is due, at least in part, to a perception that it would lose UAV-related missions to the other services since they were all active in developing UAVs for various mission areas. This situation has been likened to a similar course taken by the Air Force in the 1950s with regard to Inter-Continental Ballistic Missiles [5].

2.6. ACHIEVING SOME SUCCESS

2.6.1. Israeli UAV Success

The Israeli Air Force used massed drone launches to locate and decoy Syrian surface-to-air missile sites in 1982, before its successful air raids over Lebanon. The drones were used to draw radar-guided missile fire from the Syrians, allowing the Israelis to launch radar-guided missiles in return. With their on-board cameras the UAVs provided real-time video revealing the position and movements of the Syrian troops. In this conflict the Israelis lost only one aircraft while 54 Syrian aircraft and 19 missile batteries were destroyed [26, 27].

2.6.2 U.S. Military Operations

UAVs have been put to good use in military operations in the Gulf War, Bosnia, Somalia, Kosovo, and Afghanistan. Pioneer was used during the Gulf War to target Iraqi artillery, for updated target imagery, for warning of Iraqi attacks, and to identify shore targets for artillery fired from ships. After the war the Vice Chairman of the Joint Chiefs of Staff stated, "The outcome of Operation Desert Storm might not have been as swift or decisive if U.S. and allied forces had not made use of intelligence-gathering and tactical recon-naissance platforms such as the Israeli-designed Pioneer ... " [5].

Predator flew its first flight in July 1994 and deployed to the Bosnia theater in July 1995. It was used in Albania and Bosnia, although still in the ACTD phase of acquisition, to provide humanitarian assistance monitoring, NATO troop protection, target location, search and rescue, before-and-after strike surveillance, peace accord monitoring, and general peacekeeping. Predator UAVs have completed five combat area deployments to the Balkans and logged over 20,000 flight hours, and they are considered quite successful [5, 8, 28, 29].

During the Kosovo air war, UAVs were found to be an effective alternative to the Army Apache helicopters to hunt Serb ground forces. "To have used the Apaches then would have put the pilots at unacceptable risk," then Secretary of Defense Cohen told the Senate Armed Services Committee during a hearing. Defense officials were worried about losing Apaches to shoulder-fired missiles and small arms fire. "Unmanned aerial vehicles were used to an unprecedented degree in Operation Allied Force," defense officials wrote in a Kosovo After-Action Review. "Their ability to loiter over hostile territory enabled them to provide surveillance information unavailable otherwise and avoided the risk of losing aircrews," the review says. Fifteen U.S. UAVs were lost during the air war—most, presumably, shot down by Serbs; by contrast only two manned aircraft were downed. Air Force Preda-tors, Army Hunters, and Navy Pioneers were all used in the air war as were German, British, and French UAVs [30].

2.6.3. Conquering Long Distances

Departing from the continental United States, UAVs have crossed both oceans. In a remarkable accomplishment for a small UAV, an Aerosonde robotic airplane made a successful Atlantic crossing in August 1998. This UAV has a wingspan of only 2.9 m and took 26 hours and 45 minutes to make the crossing from Newfoundland to an island in the Outer Hebrides of Scotland. The undisputed long-distance champion of UAVs is Global Hawk. It has completed missions crossing both the Atlantic and the Pacific oceans. In May 2000, it flew nonstop to Portugal, carried out some radar imagery tasks, and then returned to the United States some 28 hours later. In an operation named Southern Cross II, in April 2001, Global Hawk flew 25,000 km in 23 hours to Australia, was used in joint-force exercises, and then returned in June [31].

2.6.4. Offensive Capability

With the UCAV under development, a commitment has been made to carry weapons aboard an unmanned aircraft. The feasibility of this idea has already been proven by existing UAVs, however. The Predator successfully aimed and launched a "live" Hellfire-C, laser-guided missile that struck an unmanned, stationary Army tank on the ground at Indian Springs Air Force Auxiliary Airfield near Nellis Air Force Base, Nevada. This was the third successful launch in a feasibility demonstration [2]. Another demonstration showed that UAVs could be operated from atypical locations when a Predator was flown by a pilot on board a submarine [33].

2.7. UAV DEVELOPMENT CONSIDERATIONS

Primary among factors to be considered for UAV development is an appropriate, stable set of requirements. As discussed in Section 2.4, "requirements creep" and attempting to force common solutions contributed to cost overruns, schedule delays, and cancelled programs.

The use of commercial-off-the-shelf (COTS) and military-off-the-shelf (MOTS) components and technology can help hold down development and production costs. Use of powerful commercial processing components and systems can provide UAVs with the means to perform on-board processing of sensor data and minimize ground equipment. Available communications equipment can provide the data links for sensor data and command and control instructions. High-resolution TV cameras and infrared sensors, guidance systems, sensor stabilization packages, and laser rangefinder/designators have all benefited from investments in recent years. For example, stabilization packages and optical systems are available for both military and commercial use and are currently used in law enforcement and traffic control [34]. Integrating these and other subsystems into a single system architecture is a challenging endeavor. It is beneficial to use off-the-shelf hardware and software components with standardized interfaces in an open-system architecture. Predator is an example of such a design. It shares common elements with other General Atomics UAVs such as Gnat, Prowler, and Altus [34]. Commonality of payload subsystems and ground support equipment helps reduce costs and the burden associated with logistics. Systems have to be affordable and upgradeable, simple to operate and maintain [22].

2.8. LOOKING FORWARD

Much of the technology needed to develop UAVs is currently available. A number of significant challenges, for many applications, remain to be solved, however. Regardless of the mission, every UAV must, at the topmost level, be under the control of a human being. Questions arise then concerning the level of autonomy desired or achievable, the span of control or number of assets controlled by a single operator, the details of the interface between the operator and UAV(s), how multiple UAVs might communicate among themselves and thereby be controlled in a cooperative fashion, and how to make highly autonomous UAVs robust for operation in dynamic environments.

How do we mix manned and unmanned aircraft? How do we make UAVs an integral part of the overall Joint Forces structure [35]? What tactics should we adopt to best take advantage of these new resources? How do we deal with the vulnerability of the communications and data links and still maintain sufficient bandwidth to retrieve high-quality information? How do we deal with the flood of information made possible by simultaneously using many UAVs? There are difficulties today using data collected by Predator and Global Hawk with other high-value airborne assets such as an airborne warning and control systems (AWACS) aircraft, so what will it be like when the number of UAVs and missions grow [36]?

As control and system integration techniques become more capable, UAVs will find an ever-expanding role to play in both military and civil applications. Figure 2.4 shows the progression of automation levels over time. One study indicates that by replacing 20 support aircraft on board a typical carrier with smaller UAVs, the number of strike aircraft could grow from 36 to 48. If these smaller UAVs are relocatable to other ships in the battle group (e.g., due to vertical takeoff and landing capability), then the number of strike aircraft could even grow to 54 [37]. Potential commercial applications include, among others, fighting crime, power-line or pipe-line surveillance, battling forest fires, drug traffic surveillance, agricultural surveys, maritime patrols, border patrol, and use as a communications link [38].

Advances in control systems, software, and information technologies are playing a critical role in providing the means to achieve the desired capabilities that future UAVs must have. Analog control systems have given way to digital controls, which in turn will benefit from hybrid control techniques.

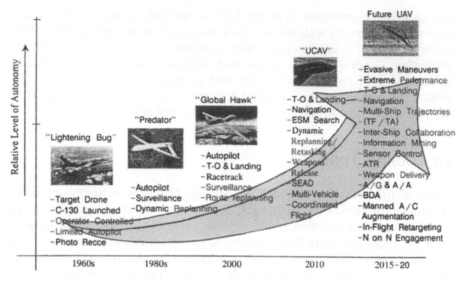

Figure 2.4. UAV automation trends.

Fault detection, identification, and reconfiguration techniques are advancing and will make UAVs more robust in the face of failures or damage. Embedded processing systems with increased throughput, sophisticated real-time software architectures, and integrated design tools will provide the means to implement advanced control algorithms. These control algorithms will make use of accurate models with real-time updates to reflect the current status of the craft and its environment. They will provide for the first time the means to autonomously command an unmanned aerial vehicle to perform a series of high-performance maneuvers that previously only human controlled craft could accomplish. Through advanced, hierarchical control techniques, these craft will establish their own routes using a set of priorities and then replan the route in real-time as necessary to accommodate changing priorities and a dynamic environment. With advancements in design environments and real-time software architectures, less time, effort, and money will be spent to proceed from control concepts to flying software. All of these factors combine to make autonomous UAVs possible, giving the Department of Defense an opportunity to consider a migration from some manned missions to unmanned ones.

Some visionaries see a very heavy dependence on unmanned air vehicles in the future. In 2000, the chief technical advisor to the Secretary of Defense shared his belief that the current production lines for fighter aircraft would be "gone 10 years from now" and a switch would be made to robotic aircraft provided by small companies. He also pointed out what he thought would be the critical technology associated with these craft. "So the innovation that comes will be in robotic control systems and not so much in the aerodynamics of the airplane. But you can imagine a future like this, where it doesn't matter any more that you don't have two or three or four production lines for fighter aircraft with pilots in them. What matters is that you have a competitive base for robotic controls for the airplane" [39].

REFERENCES

1. J. T. Correll, Evolution of the Aerospace Force, *Air Force Magazine*, June 2001.
2. C. Bish, Unmanned aerial vehicles in search and rescue, *U.S. Naval Institute Proceedings* **127**(2):7, February 2001.
3. S. Crock and N. Sandler, Pilotless planes: not cleared for takeoff? *Business Week*, issue 3556, p. 106, December 1997.
4. Israel's UAV makers set sights on new horizons, *Interavia Business & Technology* **56**(354):63, June 2001.
5. M. W. Kennedy, Major, USAF, A Moderate Course for USAF UAV Development, Research Report, Air Command and Staff College, Air University, April 1998. Available: http://www.fas.org/irp/program/collect/docs/98-147.htm.
6. D. Richardson, UAVs come of age, *Interavia Business & Technology* **52**(612):42, September 1997.
7. Statement of Louis J. Rodrigues, Director, Defense Acquisitions Issues, National Security and International Affairs Division to Congressional House Subcommittees on Military Research and Development and Military Procurement, April 1997. Available: http://www.fas.org/irp/gao/nsi97138.htm.
8. Statement of Dr. Paul Kaminski, Undersecretary of Defense for Acquisition and Technology before the Armed Services Subcommittee on Acquisition and Technology of the Senate Committee on Armed Services, 20 March 1996. Available: http://search.cetin.net.cn/internet/DSTI/ACQ/ousda/testimonies/97a_t_prg.doc.
9. R. Burgess, RQ-8A Fire Scout VTUAV receives LRIP approval, *Sea Power* **44**(6):29, June 2001.
10. D. Mulholland, Army flies at the fore of unmanned aerial vehicles, *Army Times* **60**(15):28, November 1999.
11. J. Rhea, Army to select tactical UAV supplier for potential 44 systems, *Military & Aerospace Electronics* **11**(1):6, January 2000.
12. Unmanned Aerial Vehicles Roadmap 2000–2025, U.S. Department of Defense, April 2001.
13. L. C. Bull, Unmanned vehicle will keep U.S. flyers safe, *Journal of Aerospace and Defense Industry News*, August 2000.
14. M. L. McDaniel, High-altitude UAVs should be naval players, *U.S. Naval Institute Proceedings* **125**(2):70, February 1999.
15. FDCH Regulatory Intelligence Database, 12/14/2000.
16. FDCH Regulatory Intelligence Database, 05/08/2001.
17. J. Wilson et al., NASA testing new UAVs, *Popular Mechanics* **177**(5):26, May 2000.

18. G. L. Stephens and R. G. Ellingson, The Department of Energy's atmospheric radiation measurement (ARM) unmanned aerospace vehicle, *Bulletin of the American Meteorological Society* **81**(12):2915, December 2000.

19. G. Fuentes, UAV increasing vision for border patrol missions, *Air Force Times* **60**(12):17, October 1999.

20. Unstable world highlights importance of mine-detection, *Military & Aerospace Electronics* **10**(8):12, August 1999.

21. G. I. Seffers, Small, howitzer-launched UAVs are proposed, *Army Times* **59**(18):25, November 1998.

22. D. A. Fulghum, UAV appetite grows, questions linger, *Aviation Week & Space Technology* **155**(1):66, July 2001.

23. G. C. Wilson, A Chairman pushes unmanned warfare, *National Journal*, p. 718, March 4, 2000.

24. F. Wolfe, Warner: Speed development of unmanned combat, *Defense Daily*, p. 1, February 9, 2000.

25. M. Kauchak, UAVs favored on the hill, *Armed Forces Journal International* **138**(12):12, July 2001.

26. A. Piore, Expensive tastes, *Washington Monthly* **29**(6):16, June 1997.

27. M. Fritz, What me worry about losing my job to a UAV? I'm ready for the future, *Air Force Times* **61**(38):54, April 2001.

28. W. Sweetman, US UAVs stick to their guns, *Interavia Business & Technology* **53**(622):36, July/August 1998.

29. J. Benn, Unmanned aerial vehicles soar to new heights, *San Diego Business Journal* **21**(40):48, October 2000.

30. W. Matthews, Cutting the risk, *Army Times* **60**(14):22, November 1999.

31. FDCH Regulatory Intelligence Database, 06/11/2001 and 05/22/2001.

32. S. Baker, Predator missile launch test totally successful, *Program Manager*, **20**(2):81, March/April 2001.

33. M. Mathews, Robot planes popular, but mission undefined, *Air Force Times* **59**(8):17, September 1998.

34. J. Rhea, Pilotless aircraft proliferate through the military spectrum, *Military & Aerospace Electronics* **9**(5):15, May 1998.

35. 1996 Study, UAV Technologies and Combat Operations.

36. D. Ortiz, A new role for today's UAVs, *Aerospace Power Journal* **14**(3):110, Fall 2000.

37. S. Denny and R. Holzer, UAVs on carriers & destroyers? It could happen, *Navy Times* **47**(39):29, June 1998.

38. S. Denny, Interest grows in commercial UAVs, *Army Times* **57**(41):30, May 1997.

39. Pentagon's Mark sees pilotless future, *Defense Week*, p. 6, May 30, 2000.

40. *Intervia Business & Technology* **56**(651):8, March 2001.

41. C. Mathews, UAVs will maintain air edge, *Navy Times* **49**(18):63, February 2000.

CHAPTER 3

PREVIEWING THE SOFTWARE-ENABLED CONTROL RESEARCH PORTFOLIO

TARIQ SAMAD and GARY BALAS

3.1. INTRODUCTION

Over the last decade or so, the area of information technology has sustained a remarkable level of growth. Processor speed, memory capacity, network bandwidth, and other metrics of progress show few signs of relenting from their exponential trends. To those of us whose first exposure to computing was in the days of mainframe time-sharing, punch cards, and 300-baud modems, the IT revolution has been personally experienced. Recalling these once-ubiquitous artifacts helps place the downturn the IT sector is facing at the time of this writing in its proper context!

Yet even as the virtual world takes over more and more aspects of our lives and work, we are continually reminded of its limitations. We, and the "real world" around us, are not made of bits and bytes, but of matter and energy. The physical world, unlike a virtual one, isn't a malleable entity. Video games—to note an application domain that has been one of the key drivers for the development of faster processors—are free to ignore the laws of physics, but building real systems that reliably operate in the real world transcends software engineering. Ultimately, the engineering and scientific research communities need to address difficult problems related to national defense, antiterrorism, the environment, critical infrastructures, and other domains—and these problems cannot be simulated away.

Software-Enabled Control. Edited by Tariq Samad and Gary Balas
ISBN 0-471-23436-2 © 2003 IEEE

In the context of such technological endeavors, software and information technology should be seen not as an end in itself but as an enabling force. That the computing revolution is relevant to progress in engineering systems goes without saying—it is hardly controversial to assert that advances in computer hardware and software constitute the largest potential enabler of such progress in recent times. To realize this potential, we must seek multi-disciplinary syntheses that integrate developments from a variety of fields.

To exemplify these general remarks, consider the application discussed in the preceding chapter. For national defense, uninhabited aerial vehicles (UAVs) are seen as a warfighting solution that will minimize risk to the nation's military personnel. The first UAVs may have flown in World War I, but it is only now that they are seriously being contemplated as an effective alternative to piloted aircraft. The issue has always been the automation and control required for high-performance missions, and the belief until recently has been that the complexity of the decision and control process performed by a human pilot will defeat any attempt toward a software-based computational substitute.

Four years ago, the vision of "software-enabled control" for systems as complex as military UAVs was turned into a funded program by the U.S. Defense Advanced Research Projects Agency. The transformation of this vision to reality is currently underway and being engineered by a number of leading-edge research teams from industry and academe. Several research successes have been achieved, many of which would likely never have arisen if not for the SEC program; deeper levels of understanding of the problems involved have been gained; and, overall, the research effort is on track to validate the vision over the next few years.

One of the programmatic innovations of the SEC program is that it has brought together two research communities that have, it turns out, much more to contribute to each other than either had previously appreciated. In addition, the program spans efforts ranging from basic theoretical research to proof-of-concept demonstrations of new functionality. The UAV domain, by providing a common target for research, has served as a nexus among the diversity of disciplines and efforts, and it has been instrumental in ensuring a coherent overall program.

Most of this volume contains reports from most of the key contributors to the SEC research. We have organized the individual chapters into three categories that are central themes of the research:

- Software architectures for real-time control
- Online modeling and control
- Hybrid dynamical systems

Below we briefly summarize the contents of each of these parts.

3.2. PART II: SOFTWARE ARCHITECTURES FOR REAL-TIME CONTROL

The control systems for all modern aircraft are computer-based, with flight control laws, onboard diagnostics, state estimators, and flight management systems all implemented in software. Some of these systems (e.g., flight management systems) are seen as non-mission-critical—if the function failed, the safety of the aircraft and the pilot is not immediately at risk—whereas others (such as the flight controls) are mission-critical. It is the latter that are of particular interest in the SEC program, and it is here that the limitations of the state-of-the-practice in avionics are most apparent.

The software architectures that are employed today have fundamental limitations that preclude the implementation and real-time execution of advanced algorithms. Until new architectures are devised that can overcome these limitations, we will not be able to take full advantage of advances in processing power. The five chapters in this part of the book present new real-time architectures and approaches, or elements thereof.

Chapter 4 describes a new architecture for real-time systems that is being developed to serve as a standard resource for the SEC program and other related efforts. The authors are from Boeing Phantom Works, and the new development is an *open control platform* (OCP) that provides support for all levels of vehicle control, including inner-loop flight control, dynamic reconfiguration of control laws, and communication among vehicles in a formation. It is envisioned that ultimately several of the technology components being developed by other SEC researchers will be incorporated within the OCP, either as infrastructure enhancements or as new control algorithms. Support for simulation demonstrations is also included. The OCP extends the RT CORBA (the real-time extension of the Common Object Request Broker Architecture) programming interface in its middleware implementation; the extensions enable the higher performance required for the execution of sophisticated control algorithms and include new resource management methods that permit dynamic reconfiguration of processing resources.

Chapter 5, by Wills et al. at Georgia Institute of Technology, also focuses on the open control platform. The specific challenges addressed by the Georgia Tech research are the integration of a variety of software components in a distributed computational environment and dynamic reconfiguration based on mission and operational tasks. A three-layer hierarchical structure is proposed. The bottommost core layer allows distributed, heterogeneous components to communicate asynchronously. The middle "reconfigurable control" layer provides an abstract interface based on familiar control engineering concepts (e.g., block diagram components and I/O ports) enabling the controls engineer to specify designs at a higher conceptual level. The "hybrid controls" layer focuses on reconfiguration management support with specific features for mode transitioning. The Georgia Tech team is working in close coordination with the Boeing OCP team, and transfers of research results have already taken place. Initial flight tests have been conducted using a commercially available remote-control helicopter testbed that has been outfitted with new avionics.

One of the advances of the OCP over previous real-time control platforms is the incorporation of techniques to reliably execute algorithms that can trade quality of service for computing time. Such "anytime" algorithms introduce significant complications for real-time resource management. Chapter 6 describes this research contribution, from Honeywell. An "anytime CPU assignment controller" (ACAC) has been developed and ported to the OCP. This work is based on a real-time adaptive resource management technology developed in a previous DARPA-sponsored project at Honeywell. This chapter also illustrates the application of anytime algorithms for onboard trajectory optimization in UAVs, indicating how different considerations such as threat avoidance, terrain-following, and target overflight can be addressed in a flexible, situation-specific manner. A simulation framework that combines real-time scheduling and execution of control algorithms with aircraft and environmental models has also been developed.

Another aspect of the complexity of real-time control systems is addressed in Chapter 7, authored by Lee and colleagues at the University of California, Berkeley. Lee et al. introduce a componentized, system-level design architecture, Ptolemy II, that has been designed for applications consisting of a variety of models of computation. The authors argue that managing this variety is the key question to be answered when designing complex control software. Ptolemy II represents a structured approach to heterogeneity management by providing an environment within which hierarchies of components can be constructed. A hierarchical Ptolemy II model specifies execution orderings and intercomponent communication interfaces, aspects that define a model of computation. Examples relevant to control systems include the numerical simulation of ordinary differential equations, discrete event dynamical systems, and dataflow models. Different models of computation can be integrated within one Ptolemy II model, thereby allowing the representation of modal systems.

The authors of Chapter 8 are also from UC Berkeley. Henzinger and colleagues focus on one computational model, the time-triggered paradigm. They present a design methodology for embedded control systems, called Giotto, that is based on time-triggered concepts. Time triggering implies that all communication among components of an embedded system occurs at clock ticks, thereby providing timing predictability. Giotto offers this predictability, which can be crucial for safety as well as for platform independence. The latter can be realized, for example, by layering a clock synchronization service atop any real-time communication protocol. A Giotto program specifies a set of modes, each mode defining a set of tasks and a set of mode switches. The Giotto compiler is used to generate embedded software for a given platform, relying on information about task execution times. A task and communication schedule is generated that guarantees timing requirements. For distributed platforms the user can annotate a Giotto program; the annotations serve as compiler directives that can facilitate generation of a feasible schedule.

3.3. PART III: ONLINE MODELING AND CONTROL

The research efforts summarized above are focusing on developing the software-related technology for enabling the online application of high-performance control algorithms. In Part III of the volume we turn our attention to the complementary control-algorithmic activities that are also underway in the SEC program. New methods for feedback and supervisory control, state estimation, and fault-tolerant control are included, and in all cases some discussion of software implementations and issues related to such implementations are provided.

The first two chapters adopt a model predictive control (MPC) framework. MPC is a computationally intensive control approach that, by virtue of solving an optimization problem at each sample instant, can realize high-performance control. Another key property is the explicit handling of constraints on input and state variables. MPC is widespread in the process industries, where the slow response times of processes and their open-loop stability considerably simplifies the computational challenge that it poses. How to implement MPC for aircraft control is still an open issue, which the SEC project is directly addressing.

In Chapter 9, Murray and colleagues from Caltech and the University of Colorado extend the usual MPC formulation with a novel terminal cost factor that allows a proof of stability—a property typically hard to ensure with MPC. The authors also show that when a system is differentially flat the problem of generating an optimal trajectory of the system output can be formulated as a nonlinear program over a decision space of basis function coefficients. A software package has been developed to solve this problem. The implementation of the MPC-based trajectory optimization approach on a real-time flight control testbed, the Caltech ducted fan, is described in detail, documenting computing times and performance metrics for several parametric variations of the approach. As the authors note, MPC represents a migration of control design from the offline to the online, real-time realm; as a result, it becomes possible to implement more versatile controllers that can respond to changes in system dynamics, mission intent, and environmental constraints.

An MPC-like methodology is described by Wan et al. from the Oregon Graduate Institute in Chapter 10. In this case, the controller does not explicitly involve an online optimization at every sample time. Instead, an optimized neural network feedback controller is designed through extensive offline simulations. The neural network can then be used online (with minimal computational cost). For improved stability, a state-dependent Riccati equation controller is also employed in parallel with the neural network. The SDRE controller requires an approximate analytical model of the system under control and involves the solution of Riccati equations at every sample time. The application being targeted in this project is aggressive maneuvers by autonomous helicopters, and several simulation experiments with a high-fidelity simulator are presented in the chapter.

Regardless of the sophistication of a model-based control algorithm, control performance is crucially dependent on model fidelity. The practice today is to develop models through offline analysis; the models are then held fixed over the course of the system's operation. But systems change over time, and in any case the offline model will always be of limited fidelity. Ideally, we would like to be able to adapt models online in response to the observed discrepancies between the model predictions and the system behavior. In Chapter 11, a new approach for the online adaptation of models is presented by Campbell and co-workers from Cornell University. The approach relies on a recent development in nonlinear state estimation, the *unscented* Kalman filter. A distinctive aspect of this approach is that it also generates uncertainty bounds on the derived model. Model adaptation is particularly important for fault management, and the simulation results presented—for an F-15-like aircraft and UAVs—show how component faults are handled by the scheme.

Next, in Chapter 12, Vachtsevanos and collaborators from Georgia Tech present two algorithms based on fuzzy control and neural networks that relate to mode behavior problems in UAVs. The first problem considered is the reliable and stable transition from one mode to another, and a fuzzy neural network (FNN) structure is adopted to solve it. The FNN implements blending gains that gradually transition from one local mode controller to another. An adaptation scheme has also been developed that uses a model prediction error to adjust a system model that is then used to adapt the blending controller. Simulation results are presented for the transition from hover to forward flight for a simplified helicopter model. The second part of this chapter focuses on flight envelope limit avoidance and detection. The approach presented is an integration of a neural network and a recently developed extension to adaptive control architectures called pseudo-control hedging. The neural network is trained to predict steady-state values of vehicle parameters, thereby anticipating the evolution of the actual value. When a prediction indicates that the flight envelope is in danger of being compromised, the command to the controller is "hedged" accordingly. Simulation results for rotorcraft applications are included.

Balas and Bhattacharya from the University of Minnesota conclude Part III by bringing together the topics of control design and software architecture in Chapter 13. Three control implementations are discussed, all based on a textbook F-16 model: a linear quadratic regulator, a linear parameter varying controller, and a waypoint tracking controller. In addition to describing the control algorithms, the authors discuss how they integrated their Matlab/Simulink implementations within the Open Control Platform framework. The *sim2ocp* tool, provided by the OCP and which is intended to help control engineers create OCP components from their Simulink diagrams, is described and enhancements to it are proposed. *sim2ocp* currently only handles the I/O interface among the components. The enhanced version will incorporate a code generator as well. A limitation of the Simulink-based OCP implementation approach is also noted, relating to a potential timing mismatch for asynchronous systems—the order of execution may be different in the Simulink and OCP implementations.

3.4. PART IV: HYBRID DYNAMICAL SYSTEMS

The area of hybrid dynamical systems—"hybrid systems" for short—is perhaps the most significant product of cross-disciplinary research between control engineering and computer science over the last decade. The relevance of this area to systems such as uninhabited vehicles is, or should be, obvious. Aircraft dynamics (as with the dynamics of virtually all physical systems) is inherently a continuous-time phenomenon, extensively explored within the traditional scope of control science and engineering. The symbolic realm doesn't intrude just by our wrapping a control loop around this dynamics, but as soon as our interest expands from regulatory objectives (e.g., maintaining constant altitude cruise flight under nominal conditions) to the realization through automation of greater autonomy and multifunctional capabilities, software ceases to be just an implementation vehicle for a control law. The control system must now incorporate multiple, qualitatively different "modes" of operation, each with its own dynamics and its own control criteria. The notion of control now encompasses both discrete-time symbolic behavior and continuous-time dynamics. This multimodal perspective is proving especially productive for handling safety constraints, achieving fault-tolerant control, and multivehicle formation management. Part IV of this volume presents new results in all these areas.

We begin Part IV, however, with a tutorial-level introduction by Antsaklis (University of Notre Dame) and Koutsoukos (Palo Alto Research Center) in Chapter 14. The authors review hybrid system concepts and discuss several areas of current research interest. Hybrid models arise in numerous ways in control applications: embedded controllers for multimode systems, the integration of planning and regulatory control, sequential supervisory control (e.g., with PLCs) for continuous processes, and hierarchical organizations of complex nonlinear dynamics, among others. Much of the research effort today is expended toward developing software tools for the simulation, analysis, and design of hybrid systems; the chapter lists about a dozen, briefly noting their scope and utility. Three specific research directions within the hybrid systems field are reviewed. The first is hybrid automata, and the important (because tractable) special cases of linear hybrid automata and rectangular automata are described. Second, stability analysis approaches for hybrid systems are outlined, with particular attention to multiple Lyapunov functions. Finally, supervisory control of hybrid systems—where a continuous plant is approximated as a discrete-event system—is discussed. In this case, the discretization is realized by partitioning the state space, an approach that is attractive because it allows control design to be undertaken in the discrete domain (but that can result in nondeterministic behavior). Examples from a variety of problem areas illustrate the concepts and techniques presented in this chapter.

A novel approach to motion planning for autonomous vehicles is introduced by Frazzoli et al., from the University of Illinois at Urbana-Champaign and the Massachusetts Institute of Technology, in Chapter 15. This approach is based on a precomputed library of maneuvers or trajectory primitives. A maneuver is defined as a finite-time transition between two relative equilibria or trim trajectories and is characterized by a duration for the maneuver and an open-loop control signal. Motion planning then consists of composing maneuvers from the library that take the system through a series of trim trajectories. The resulting architecture is labeled a *hybrid maneuver automaton*, and its definition ensures that all generated trajectories satisfy all constraints on the vehicle and its control. The complexity of motion planning is dramatically reduced at the expense of suboptimality of the generated trajectories—but given the computational intractability of the general motion planning problem, global optimality is a quixotic pursuit in any case. Experimental results and real-time system architecture details are described for a three-degree-of-freedom helicopter testbed.

The general problem addressed in Chapter 16, by Koo and co-authors from UC Berkeley and the University of Pennsylvania, is the following: Assuming that low-level continuous-time controllers have been satisfactorily designed, can we realize higher-level tasks by automatically switching between these controllers? The operation of the system under a low-level controller is referred to as a control mode of the system. The mode switching problem then is concerned with finding a sequence of the control modes as well as the switching conditions to achieve the higher-level objectives. Within a mode, a reference trajectory may also need to be defined, although by assumption its tracking is guaranteed. The authors show how the mode switching problem can be transformed into the problem of searching a discrete graph, the nodes of which are submodes of the closed-loop system and the edges define the allowable transitions. Simulation results with an autonomous helicopter are presented; modes in this case include hover, cruise, ascend, and descend. Finally, the authors highlight the close connection between multi-mode control and embedded system implementations.

Karsai et al. from Vanderbilt University and the Technical University of Budapest adopt a hybrid systems framework for fault-adaptive control in Chapter 17. Component failures naturally result in qualitative changes in plant dynamics, and the authors are interested not only in the hybrid models that arise as a consequence of these changes but also in the problem of reconfiguring a controller to maintain control in the face of faults and failures. The modeling formulation for this research is hybrid bond graphs, an extension of bond graphs that includes controlled junctions to model discrete mode transitions. The chapter presents a fault-adaptive control architecture that uses hybrid bond graphs as the underlying formalism and includes a number of algorithmic components. These include a control reconfigurator that uses a language for defining controllers with both regulatory and supervisory controller modeling features. A hybrid observer is also included with a mode-switched Kalman filter. Fault isolation is a particular challenge for hybrid systems; the authors have developed a combination of qualitative reasoning and parameter estimation methods for this purpose. A controller reconfiguration approach is proposed and is work-in-progress. The application domain discussed by the authors is a two-tank system, analogs of which arise in aircraft fuel management.

In Chapter 18, Tomlin and colleagues from Stanford University ask the question, *Is a potentially unsafe system state reachable from an initial state?* They also present computational methods that can help give an answer. Their approach is based on a hybrid system model that is not limited to linear hybrid automata but incorporates accurate nonlinear models of the continuous dynamics of the system under control. The central notion is to consider the evolution of the boundaries of the reachable states given the hybrid dynamics. The problem can be transformed into a Hamilton–Jacobi partial differential equation formulation, an efficient numerical solution method for which has been implemented as a software tool. Examples of aircraft autolanding and multiaircraft collision avoidance are presented. For larger-scale problems, the computational complexity of this method is likely to be a limiting factor. However, the authors show that the problem of computing polyhedral overapproximations to the reachable set of states can be formulated as a convex optimization problem, for which fast algorithms are available even for high-dimensional instances.

3.5. CONCLUSION

It is not an undue exaggeration to assert that automation and control have so far derived only limited benefits from the revolution in information technologies. This lack of impact is not because new software, communication, and network technologies are less relevant to physical systems such as aircraft, chemical plants, and automobiles than they are to the domains of finance, office automation, and computer games. But it is the case that substantial advances in, say, the state-of-the-art in flight control will never be *solely* a matter of software technology. Dynamical systems and control engineering are equally crucial disciplines. By bringing together the real and the virtual worlds (so to speak), the projects under the software-enabled control umbrella are exemplifying how digital architectures, models, and algorithms can lead to more capable and higher-performing complex engineering systems.

SOFTWARE ARCHITECTURES FOR REAL-TIME CONTROL

CHAPTER 4

OPEN CONTROL PLATFORM: A SOFTWARE PLATFORM SUPPORTING ADVANCES IN UAV CONTROL TECHNOLOGY

JAMES L. PAUNICKA, BRIAN R. MENDEL, and DAVID E. CORMAN

Editors' Notes

It is only by better exploiting the processing horsepower, memory capacities, and communication bandwidths that are readily available today that our plans for tomorrow's autonomous systems will be realized. The Software Enabled Control program is addressing this challenge through the development of an *Open Control Platform* for autonomous uninhabited vehicles (UAVs). This term refers to software that permits a variety of application services to be readily deployed on real-time systems. Among other features, the OCP provides abstractions so that applications need not be tailored for specific operating systems, it provides support for resource management so that sophisticated computational tasks can be reliably executed, and it enables plug-and-play composionality of software components that may be developed by different organizations at different times.

The OCP development effort is being led by The Boeing Company with support from several of the other SEC program participants. The OCP builds upon a previous system developed at Boeing, Bold Stroke, as well as on middleware developed at Washington University. It is intended that the OCP will be usable by control engineers directly, and therefore an OCP Controls API has been defined that allows the OCP to be linked with familiar control development tools such as Matlab/Simulink. A simulation environment allows control designs to be validated within the OCP environment.

A parallel and collaborative OCP development is also underway at the Georgia Institute of Technology and this is described in Chapter 5. More detailed discussions of specific OCP aspects also appear in Chapter 6, which describes the adaptive resource management functionality developed at Honeywell and incorporated within the OCP, and Chapter 7, which describes the Ptolemy II tool from the University of

Software-Enabled Control. Edited by Tariq Samad and Gary Balas
ISBN 0-471-23436-2 © 2003 IEEE

California at Berkeley for modeling heterogeneous real-time systems that has been interfaced with the OCP.

4.1. INTRODUCTION

The open control platform (OCP) is an important software technology component of the U.S. Defense Advanced Research Projects Agency (DARPA) Software Enabled Control (SEC) research program. The OCP provides an open, middleware-enabled software framework and development platform for controls technology researchers who want to demonstrate their technology in simulated or actual embedded system platforms. The middleware layer of the OCP provides the software layer isolating the application from the underlying target platform. It provides services for controlling the execution and scheduling of components, inter-component communication, and distribution and deployment of application components onto a target system. The embedded system domain of particular interest to the SEC program is that of uninhabited aerial vehicles (UAVs).

This chapter describes the OCP approach to handling the challenges of UAV control, including support for all levels of vehicle control (including low-level flight control), interaction between a UAV and other platforms (e.g., another UAV, ground station, piloted aircraft), innovative scheduling techniques, adaptive resource management, and support for dynamic reconfiguration.

The OCP is being developed by the Embedded Systems Research Team within the Boeing Phantom Works Open Systems Architecture organization. Assisting Boeing in these efforts are the Georgia Institute of Technology, Honeywell Labs, and the University of California at Berkeley. The OCP is being delivered to a host of university and industrial researchers who are participating in the SEC program.

The OCP is an object-oriented (OO) software infrastructure designed to enhance the ability to analyze, develop, and test UAV control algorithms and embedded software. Figure 4.1 shows the role of the OCP in pulling together technologies into real demonstrations and ultimately transitions to future programs. The OCP includes a middleware framework for integrating embedded application software components with useful services important to the UAV control domain, support for UAV simulation with a simulation framework, and integration with embedded software design and analysis tools. The OCP is also planned for use in a variety of live flight demonstrations on a variety of autonomous flight platforms.

The OCP presents many benefits to the embedded system developer. In particular, the OCP provides a path for quick transition of controls design to desktop and embedded targets. It allows a controls designer to focus on controls design instead of software design by handling the issues of integration, communication, distribution, portability, execution, scheduling, system configuration, and resource management.

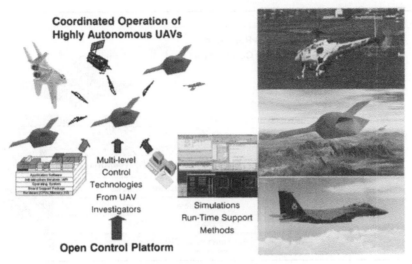

Figure 4.1. OCP overview.

4.2. OCP GOALS AND BACKGROUND

The OCP was developed based on the goals for the SEC program [1], which included development of innovations in software and controls technology that could leverage the dramatically expanding embedded processor execution speeds, memory sizes, and communication bandwidths. The goal of the program was to combine control theory and software design to allow for utilization of new control technology on embedded computing platforms.

The goals of the OCP [2] are to:

- Provide an open platform for enabling control research and technology transition.
- Support dynamic configuration of components and services.
- Provide a mechanism enabling the transition between execution and fault management modes while maintaining control of the UAV.
- Allow for coordinated control of multiple vehicles.
- Provide a software system infrastructure that is isolated from a particular hardware platform or operating system.

The goals of the SEC program were strongly aligned with the goals of an internal Boeing-funded software initiative called Bold Stroke. The Bold Stroke initiative was responsible for the creation of a CORBA-based middleware platform based on open standards targeted for real-time embedded flight platforms in the mission avionics domain. Due to this close alignment between the Bold Stroke software platform and the DARPA OCP goals, Boeing proposed to DARPA and won a contract on the SEC program to

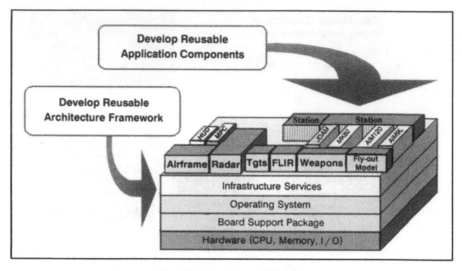

Figure 4.2. Bold Stroke layered architecture.

extend the Bold Stroke middleware to support the challenging requirements of UAV flight control.

The CORBA-based middleware from Bold Stroke (see Figure 4.2) was initially developed at Boeing to support a product line approach for avionics software development. The product lines targeted for support at that time included fighter jets produced by the then McDonnell Douglas Corporation. The focus of Bold Stroke was additionally limited to the class of embedded software applications functioning in what was known as *mission computing*. Mission Computing handles on-board tasks such as navigation, pilot-vehicle interfaces (PVIs), stores management, and weapon control. In contrast, CORBA-based middleware was not used at Boeing in Vehicle Management processing, where safety-critical flight controls are implemented.

Figure 4.2 highlights two of the major benefits of the OCP layered architecture. The lower layers isolate the upper-layer application components from the underlying operating system and the hardware platform. This allows for flexible usage of application components, which will be contributed by numerous controls technology researchers on the SEC program. These application components execute on a variety of simulation and flight demonstration platforms. Secondly, the Infrastructure Services layer, made up of OCP middleware services, provides a reusable framework that operates on a variety of operating systems and hardware platforms.

As stated earlier, the open middleware framework is being extended on the SEC program to bring middleware-enabled software development to vehicle-management-type processing. In this regard, the OCP is being designed to support all levels of UAV flight control with specific features to be described later.

4.3. OCP OVERVIEW

The major components of the OCP software are as follows:

- *A middleware framework based on RT CORBA* [3]. Provides the mechanism for connecting application components together to control their execution.
- *A simulation environment.* Allows the embedded application to execute and be tested in a virtual world, reading simulated sensors and driving simulated actuators on vehicle models.
- *Tool integration support.* Provides linkages to useful design and analysis tools, such as Matlab/Simulink [4] and Ptolemy II [5], allowing controller designs realized in these tools to more easily transition to embedded targets.
- *Controls API (application programmer interface).* Provides a controls-domain-friendly look and feel to the OCP. This is accomplished by using familiar terminology and simplified programming interfaces.

A primary motivating factor in implementing a middleware-based architecture was the promise of isolating the application components from the underlying platforms. This allows for a more cost-effective path for implementing common software components that (1) could be used across different product lines and (2) could be rehosted onto evolving embedded computing platforms. Support for rapid reimplementation of existing, tested designs onto evolving computing platforms is important for maintaining an effectiveness advantage in currently fielded embedded systems.

These evolving computing platforms are starting to be dominated by commercial hardware and software components, which have more dynamic life cycles than previous military components and which create more opportunities for incorporation of computing advances into existing systems. Furthermore, commercial and military flying platforms are experiencing longer lifetimes, and the ability to inexpensively rehost existing functional and tested software on updated computing platforms is very attractive.

The embedded software framework of the OCP inherits the RT CORBA-based middleware of Bold Stroke. The inherent advantages of a middleware-based solution within OCP should prove to be an enabler for current and future UAV software developments.

Use of the CORBA-based middleware helps isolate application software components from the underlying hardware and operating systems in the embedded computing platform. The middleware also facilitates distributed processing and intercomponent communications by supporting CORBA event-based communication.

The ability of the OCP to isolate application components from the underlying system also can be used to enable efficient embedded system software development. For initial design and development, application code using the OCP middleware can be executed on familiar desktop systems, such as those using the Windows NT or Linux operating systems. Certain levels of software testing can take place more conveniently on the desktop. Then, the operating system and hardware isolation features of the middleware would be instrumental in implementing the smooth transition of the developed embedded application code to the embedded hardware target—perhaps a target executing a POSIX-compliant real-time operating system (RTOS), such as VxWorks or QNX. Isolation of the application would result in minimal code changes during transition to the embedded target. *Note:* The Institute of Electrical and Electronic Engineers (IEEE) Portable Operating System Interface (POSIX) standard was first published in 1986 to document a desired standard interface for UNIX operating systems.

4.4. OCP FEATURES

4.4.1. RT CORBA Core

The OCP middleware is written in C++. It includes an RT CORBA component [3], which leverages the ACE and TAO products [6] developed by the distributed object computing (DOC) research team at Washington University. TAO provides real-time performance extensions to CORBA. These extensions are discussed briefly in Section 4.5.

4.4.2. CORBA Services (Real-Time Publish / Subscribe)

Figure 4.3 illustrates the use of the CORBA event channel (EC) to route information between components. The power of the EC can be seen from its

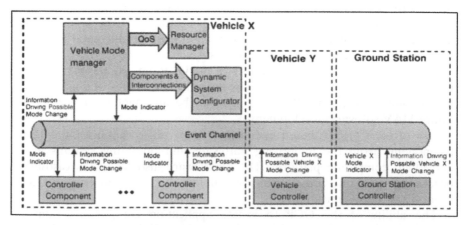

Figure 4.3. Distributed processing and intercomponent communications.

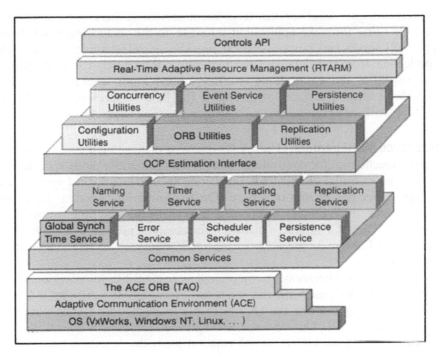

Figure 4.4. OCP layered architecture.

data distribution capabilities. The components transferring data, including data suppliers that publish data and data consumers that subscribe to data, can either (a) be placed in the same computer or process or (b) distributed onto other processes or computers—all without a change to the components' source code. Only configuration data, specifying where the components reside, needs to be updated to implement a working redistributed architecture.

4.4.3. CORBA Services (Naming Service)

The naming service provides a mechanism for getting a reference to a component. This allows an application to store and retrieve references to components that are independent of address space and are therefore portable across processes [7].

4.4.4. Middleware Services

The OCP provides a set of services in addition to the standard services provided by the CORBA specification. Figure 4.4 shows the services that make up the OCP, including lower-level services (e.g., naming service, timer service, etc.), which provide additional real-time performance enhancements, as well as higher-level services (e.g., event service utilities, replication utilities, etc.) that form the application interface to the underlying middleware. This API serves two purposes: (1) It provides an interface that isolates the application from the details of the underlying RT CORBA implementation and simplifies the applications interface to the lower level services; (2) it provides a clean interface for extending the base features provided by the controls API.

Resource Management. The OCP's resource management component provides a mechanism for controlling resource utilization in a mode-specific way. This is an essential element for supporting modes in a hybrid system. The designer specifies quality of service (QoS) information, which is input into the resource management component to control the run-time execution of the OCP. The OCP resource management component is an extension of the Honeywell Labs real-time adaptive resource management (RT-ARM) capability [8].

The resource management component is responsible for partitioning the system resources based on the mode of execution. It performs a sort of meta-scheduling task for the OCP with the assistance of TAO's scheduling component, [9, 10]. The resource management component adjusts rates of execution based on utilization information from the scheduling component and notifies application components when rates have been adapted. These components that are scheduled with the adapted rates can then modify their behavior based on their assigned rate (e.g., they can adjust controller gains).

Hybrid Systems Support. A hybrid system is a system that combines both continuous and discrete elements. For example, in a typical flight controls system, the lower levels of the architecture tend to be designed as continuous-time controllers. When moving to higher levels of the architecture, controllers tend to be of the discrete supervisory type. These controllers are designed to function in one or more modes. Adding mode support to the OCP provides a set of new challenges especially when controlling a UAV, such as the following:

1. The OCP must support stable operation of the continuous (physical) system during mode changes.
2. Mode changes require that the system be reconfigured at run time. Reconfiguration imposes definite real-time constraints on the system.
3. To better utilize limited computing resources, the OCP must support adaptation of resources in a mode-specific way.

The OCP includes a transition service extension to RT CORBA to provide the foundation infrastructure for seamless mode transitions. The transition service provides a distributed mode manager that enforces structured mode

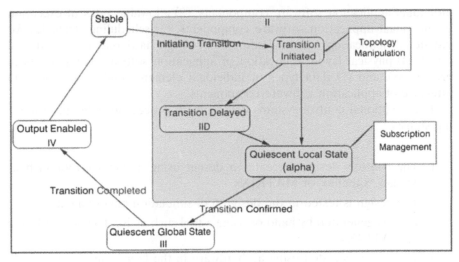

Figure 4.5. OCP transition service.

switching. Characteristics include: (1) allowing components to identify the current mode of the system, (2) allowing components to initiate transition to a new mode, and (3) managing the transition between system modes. Figure 4.5 provides a state transition diagram showing the transition service flow. Although a complete description of transition management is beyond the scope of this chapter, it is useful to note from the figure that transition management enforces a set sequence of coordinated actions. The states shown in the diagram are not as important as the concept of enforcing predefined states of the system prior to transition to another mode.

OCP Controls API. As described earlier, the OCP provides several advanced mechanisms such as dynamic scheduling and resource management. To help hide the complexity from the controls designer, the OCP includes a control designer abstraction layer above the RT CORBA implementation. This API allows the designer to focus on familiar tools and terminology while enabling the use of RT CORBA extensions. This helps provide a consistent view of the system that is meaningful to the controls designer. This abstract layer was a collaborative Boeing–Georgia Institute of Technology effort.

The controls API provides an interface for managing OCP components, setting system information, and controlling system execution. This provides the user with a higher-level interface to the OCP and represents the "out of the box" view of the OCP.

The controls API is more than just a programming interface, as the name would suggest. It is a combination of a descriptive language, Extensible Markup Language (XML), and a simple programming interface. The XML description encodes information such as: (1) worst-case execution times of individual application software components, (2) allowable rates of execution of individual application software components, (3) scheduling strategies for individual application software components (e.g., hard real time, soft real time), (4) physical layout (allocation of application software components to processors), and (5) data types of individual elements making up the I/O interface of application software components.

The traditional control systems development process includes the following steps:

1. The controls designer creates a design using a familiar tool such as Matlab/Simulink or MATRIXx.
2. The design is tested in the design tool simulation environment.
3. Code is generated by hand or using a tool such as Real-Time Workshop or MATRIXx.
4. The design is tested using a "hardware in the loop" simulation.
5. The design is flown.

This traditional approach has several limitations addressed by the OCP. For example, the generated code and its execution model are static, so advanced features such as resource management and dynamic scheduling are difficult to implement. This approach also does not adequately address the complex reconfiguration issues associated with hybrid systems.

The OCP helps minimize these drawbacks by providing the tools to map the design onto the OCP's component architecture. The traditional steps are recast as follows when developing embedded software with the OCP:

1. The controls designer generates a software system design with Simulink (see Figure 4.6), specifying the meaningful application software components and their interconnections, and uses the controls API front-end tool to add additional QoS information. Examples of QoS information are worst-case execution times, allowable rates of execution, scheduling type (hard or soft), physical layout, and data types.
2. The controls API front-end tool generates an XML representation of the system.
3. The XML description of the system is provided as an input to the controls API back-end tool, which auto-generates C++ code implementing the OCP component framework.
4. Finally the designer populates the auto-generated framework with code implementing the application components using source code that can come from a variety of sources, such as heritage modules, new modules hand-coded for the application at hand, or code that is auto-generated from tools such as Mathworks's Real-Time Workshop.

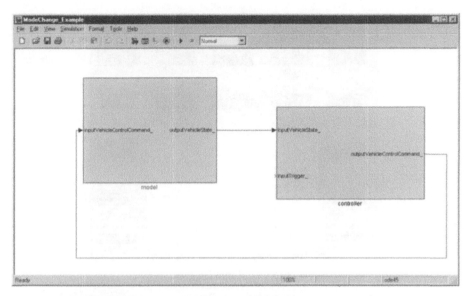

Figure 4.6. User-entered diagram of software component layout.

OCP Simulation Environment. The OCP simulation environment presents
a host of useful modeling and visualization functions to the SEC controls
researcher. Included in the OCP are generic UAV models for researcher
experimentation. Models exist for both fixed-wing and rotorcraft UAVs.
Some models are delivered in both higher-order language (HOL) realizations
and Simulink realizations. Delivered along with the models in the OCP are
controllers for those modeled vehicles. The controllers are also delivered in
HOL and Simulink realizations. The University of Minnesota contributed
heavily to the fixed-wing models and controllers delivered with the OCP, and
Georgia Tech was instrumental in the development of the rotorcraft UAV
models and controllers.

Simulation visualization interfaces are also supplied with the OCP. For
example, interfaces to Simulink scope displays allow for strip-chart-like
display of dynamics parameters during simulation execution. Network interfaces to the AVDS (aviator visual design simulator) display from Rasmussen
Simulation Technologies, Inc., are supplied to allow three-dimensional view-
ing of one or more vehicles in a simulated environment (see Figure 4.7).

The simulation models, controller, and displays are interfaced and controlled with the middleware-based OCP infrastructure. The middleware
allows flexible distribution of the components across multiple platforms and
facilitates rehost of controller code from initial desktop development and
testing to embedded processing boards. Timer services within the middleware
simplify the task of synchronizing periodic simulation and control processes
to time-based triggers.

Simulation Parameter Display

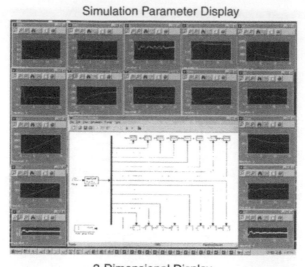

3-Dimensional Display

Figure 4.7. Simulation environment displays.

The models, controllers, and displays are all used in buildable and runnable examples that are supplied with the OCP. SEC controls developers can use, modify, and extend these examples to implement new controllers, or plug in different models or display functions.

OCP Tool Integration. The OCP provides integration with popular and useful controls and software design, development, and testing tools. Commercial tools like Microsoft Visual C++, Microsoft Visual Debugger, the VxWorks real-time operating system, and Matlab/Simulink have wide acceptance in the software and controls communities. Buildable and runnable examples delivered with the OCP to illustrate OCP features and generic vehicle model capabilities make use of these well-supported commercial products. The debugger can be used, for example, during a running vehicle simulation to set breakpoints, single-step, and perform other useful testing functions while simulation displays show the dynamic state of the vehicle.

Matlab and Simulink are used in the examples for data display as well as for expressing vehicle models and vehicle controller designs. Simulink is also used in the OCP as a schematic editor for laying out OCP software components that will be interconnected through an RT CORBA-like interface with the automated functions of the Controls API.

Ptolemy II from the University of California at Berkeley is another system design and analysis tool that has been interfaced with the OCP. Its use is illustrated in a delivered executable OCP example that uses a Ptolemy II realization of a nonlinear actuator interfaced through the OCP to a Simulink vehicle model.

4.5. OPTIMIZATIONS IN SUPPORT OF REAL-TIME PERFORMANCE

To leverage the useful features of CORBA in the domain of hard real-time fighter plane applications, the Boeing Bold Stroke software initiative made use of the real-time performance extensions to CORBA found in TAO. Further real-time performance optimizations were also added including: (1) events that are allowed to bypass the CORBA Event Service ("lightweight events"), (2) client-side caching, and (3) pluggable protocol support. This section discusses the features that are provided by the OCP in support of real-time performance.

4.5.1. Optimizations in ACE / TAO

ACE and TAO provide many optimizations in support of real-time operation. Although describing the numerous optimizations is beyond the scope of this chapter, it is worth mentioning just a few of the optimizations important to the real-time performance of the OCP. In particular, optimization of the ORB core, including minimized data copying, minimized dynamic memory allocation, delayered demarshaling, no-copy octet sequences, and direct co-location support, is crucial for achieving real-time performance for distributed communication. All of these optimizations are covered in detail via a wealth of technical publications accessible from the DOC web site [11–15].

4.5.2. Additional OCP Optimizations (Lightweight Events)

TAO's Real-Time Event Service provides a mechanism for decoupling publishers and subscribers. This publish/subscribe mechanism is a fundamental capability inherent in the communication approach of the OCP. Although the TAO Event Service is a generic solution for removing the undesirable tight coupling between suppliers and consumers, it can add unacceptable overhead in certain situations such as the high rates required for low-level control. It can also add unnecessary overhead when the full capabilities of the Event Service are not required such as a publish/subscribe connection that does

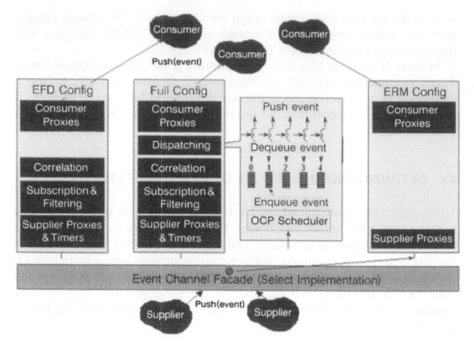

Figure 4.8. OCP event channel facade.

not require scheduling intervention to determine rate of execution (e.g., the supplier and consumer run at the same rate). To maintain the decoupling provided by the Event Service without adding unnecessary overhead when the full capabilities are not required, the OCP provides a mechanism for receiving lighter-weight events. This is accomplished by maintaining the interface to the EC through a higher-level API and switching the actual implementation underneath the API. Use of lighter-weight events reduces the number of messages compared to the number required for events sent through the EC. In our UAV applications, we have found that this can be a key contributor to achieving real-time performance and meeting required process timelines. Figure 4.8 illustrates the differences between the different implementations supported by the OCP.

This interface is an implementation of the Façade pattern [16]. This allows the application to select the event implementation while maintaining a consistent event service interface. The figure shows three variants of events (EFD, full, and ERM) characterized by the paths that are taken through the EC. Each of these event types has varying degrees of overhead due to additional functionality needed to send or receive the event. The following describes the distinguishing characteristics of each.

- *Event Forwarding Discriminator* (*EFD*). The event type provides all of the capabilities of the EC except for dispatching or queuing of the event. Events of this type are delivered in the publisher's thread. These events are used when no queuing of events is needed (i.e., publisher and subscriber run at the same rate).
- *Full Channel Events* (*Full Configuration*). The event type provides all of the capabilities of the EC including dispatching and therefore scheduling. This is the most resource-intensive form of an event, and it allows multirate communication between publishers and subscribers.
- *Event Registration Manager* (*ERM*). The event type provides all of the capabilities of the EC except for dispatching or queuing of the event and correlation. This is the most optimized form of event delivery. Events of this type are delivered in the publisher's thread and are sometimes referred to as *synchronous events*. These events are used when no queuing of events is needed and no correlation between events is necessary. Correlation allows a subscriber to wait for the occurrence of multiple events before being notified by the EC. Correlation is the distinguishing feature between EFD and ERM events.

The OCP supports all three variants of events.

4.5.3. Additional OCP Optimizations (Client-Side Caching)

Client-side caching provides a mechanism for preserving the logical data flow model at the application level. In the UAV domain, client-side caching is important to reduce the need for data passing via the EC in a distributed application. Specific scenarios include ground-station–vehicle communications, swarm communications, and so on. Normal request/reply semantics are at odds with real-time communication in a distributed environment. In the normal request/reply scheme the caller makes a request and blocks the caller's thread until a response is received from the called object. This results in an indeterminate amount of time that the calling process could be performing some other task while it is waiting for a response from the distributed object. Figure 4.9 illustrates the concept of client-side caching. The top portion of the figure shows the desired logical behavior of the system. In the figure, the AF object services requests from multiple clients in the same or different processes. This is a simple programming model that is preserved by the OCP and implemented underneath as illustrated in the lower portion of the figure. As illustrated, clients have a local cache of the AF object that looks and behaves like the actual object. The OCP is responsible for updating the caches for the original object when the actual object's state changes. This results in a call that does not block for an indeterminate amount of time.

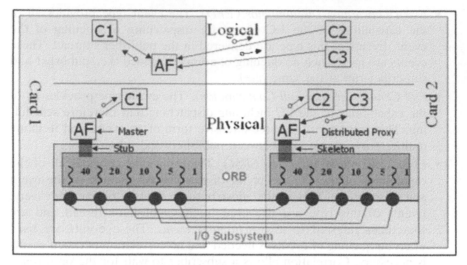

Figure 4.9. Client-side caching in the OCP.

4.5.4. Additional OCP Optimizations (Pluggable Protocol Support)

The CORBA specification provides a protocol model for ensuring interoperability between object request broker (ORB) end systems. These models are referred to as Inter-ORB Protocols (IOPs) in the specification. They include a standard messaging protocol, called General Inter-ORB Protocol (GIOP), and a standard transport mapping over Transmission Control Protocol (TCP), called Internet Inter-ORB Protocol (IIOP) [7].

Standard Transmission Control Protocol/Internet Protocol (TCP/IP)-based IIOP is not well-suited for the stringent high-performance hard real-time requirements of the SEC community. In this environment, the added overhead of IIOP can result in missed deadlines, excessive jitter, and unacceptable request latencies. TAO provides a mechanism for creating user-specific protocols that can be plugged into the ORB core as an alternative messaging and transport mechanism [17].

The OCP provides an optimized mapping, called GIOP-Lite, over shared memory for Versa Module Eurocard (VME)-based systems. This protocol is essential to maintaining the real-time characteristics of the ORB end system. GIOP-Lite is a slightly optimized variant of GIOP which takes advantage of known items from the target environment such as communication between homogeneous processes. This allows the protocol to bypass portions of the standard protocol (GIOP) such as byte swapping. GIOP-Lite is a TAO optimization to GIOP.

4.5.5. Additional OCP Optimizations (Accurate Timing Triggering at the Application Level)

The OCP has its heritage in the mission avionics domain, which is a hard real-time domain with rates typically up to 40 Hz. In contrast, the flight controls domain has much higher rates especially in the area of low-level inner-loop control. This imposes more stringent requirements on the OCP middleware.

A typical sequence for a low-level control loop would be to sample sensor data at rates in the range of 50–100 Hz, followed by processing and actuation. It is important to not only run at higher rates, but to support accurate and jitter-free sampling of sensor data. This requires the creation of highly accurate timers, which typically means the creation of timers that are implemented in hardware. The challenge is to provide a framework for encapsulating these hardware-specific tasks and exposing them to the application through a clean, portable interface.

The OCP provides this capability through the timer service. The timer service provides a generic interface for creating both hardware and software timers and uses the Strategy pattern to determine the implementation that allows use of the timer [16]. This is an important feature that allows the application to select the implementation based on the timing accuracy requirements for the application. Hardware timers are the most accurate timers available through the controls API for component triggering, and they are triggered by one of the finite number of hardware timers resident on the processing hardware. Software timers, also available through the controls API, are less accurate than hardware timers, but are used to emulate hardware timers in software. Figure 4.10 displays an overview of the OCP timer service.

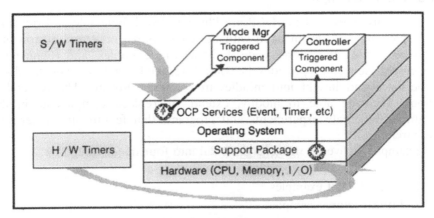

Figure 4.10. OCP timer service.

4.6. CURRENT STATE OF THE OCP

The OCP is being designed to be an open system. DARPA requirements for the software included the notion of a low-cost system "leave-behind" for the current SEC researchers and later researchers. A significant portion of the middleware consists of the open-source ACE (adaptive communication environment) and TAO (The ACE ORB) software developed by research teams at the Department of Computer Science at Washington University in St. Louis, Missouri. The ACE and TAO software is freely available from the Washington University web site [6].

The OCP source code, libraries, and documentation are freely available on a DARPA-sponsored web site hosted at Boeing Phantom Works. Researchers working on the SEC program have access to this web site for downloading. Select researchers from other, related DARPA programs have also been given free access to the software OCP Builds.

4.6.1. OCP Builds

The OCP is being generated with a spiral development process. After significant capabilities have been developed, the OCP middleware, simulation features, illustrative executable examples, and documentation are packaged into a "build" of the OCP for distribution. Over the life of the DARPA SEC program, three major builds of the OCP are planned. Major Build 1.0 was released in December of 2000, Build 2.0 was released in the spring of 2002, and Build 3.0 is slated for 2004. In between major builds, minor builds are generated and distributed to ensure that needed capabilities are made available to SEC researchers.

4.6.2. OCP Examples

The various aspects of the OCP are illustrated with buildable and runnable examples that are downloaded with the OCP source code and documentation. Inspecting the source code, building the examples, and running the generated executables provides the controls researcher with valuable hands-on experience and insight into middleware-enabled software. The researcher should then be able to modify, extend, or replace components in the examples to create experiments and programs that feature his or her own research.

Examples delivered with the OCP fall into four major categories:

1. Controls API examples
2. Infrastructure examples
3. Simulation examples
4. Tool integration examples

The controls API examples give the controls researcher insight and experience with the controls API. These examples let the controls engineer use Simulink to lay out components, use the controls API front-end tool to decorate the component model with various input/output and QoS attributes, and use the controls API back-end tool to auto-generate an RT CORBA C++ interface framework.

Controls API examples include a variety of sample systems with a range of complexities. Included is a "hello world" example to illustrate two simple processes that are communicating. Other examples include use of a mode transitioning controller and the interconnection of multiple software components, which contain behaviors that execute at different rates.

The infrastructure examples illustrate how to use the OCP to reach down to infrastructure capabilities that are not available when using the controls API. These examples demonstrate such things as the use of the CORBA Interface Definition Language (IDL) for creating distributable interfaces for server applications, reserving event types, creating and manipulating event sets, registering suppliers and consumers, receiving periodic timeouts, receiving local and remote events, pushing events, correlating event types, late registration of event suppliers and consumers, and replication of object state between different processes with various update rates.

The simulation examples show the operation of the generic vehicle models and controllers, with state data being sent to displays.

The tools examples show integration with modeling and analysis tools, including Matlab/Simulink and Ptolemy II.

It is a major goal of the DARPA SEC program to develop middleware that supports multiple levels of vehicle control [2]. With this in mind, the OCP team has been developing features, tools, and interfaces to bring middleware to the domain of vehicle control. A list of significant innovations in the development of the OCP are as follows:

- Creation of a controls API, bringing the benefits of CORBA-based middleware (e.g., platform isolation, distribution) to control theorists in a way that does not require a computer science background
- Support for high-rate, low-jitter control loop execution with timer functions
- Making advances in computational and execution flexibility available to the embedded system developer with a usable API—dynamic scheduling, adaptive resource management
- Supporting mode changes, an SEC Program goal, with the OCP Transition Service

The OCP middleware is designed, with its layered architecture, to execute on a variety of desktop and embedded hardware platforms. Various operating systems are also supported, including the desktop computing Windows and Linux operating systems, as well as a number of real-time operating systems.

The controls API tools execute on Windows desktop platforms, as do the OCP simulation displays. The controls API and the reusable simulation scope displays make use of Simulink from the Mathworks. An interface to the AVDS three-dimensional vehicle flight display is supplied with the OCP.

As was stated in the BAA for the SEC program, the ability to utilize advances in computer science and control theory in embedded systems is enabled by advances in embedded processor speeds and memory sizes. These advances in computer science, coming from the OCP, include dynamic scheduling and adaptive resource management. These advances, along with the further advantages of hardware and operating system software isolation, are made possible with a middleware product with a nontrivial memory footprint—in the neighborhood of 2 megabytes. This memory size is much larger than early-generation flight control systems, but fits comfortably within modern embedded computers.

4.6.3. OCP Flying Platforms

The OCP middleware is used on many platforms, and it is planned for use in further vehicles and demonstrations as the SEC program progresses. Georgia Tech and Stanford University are using the OCP in hardware demonstrations of their controls technology. Specifically, Georgia Tech used an OCP-based embedded flight control application on a Yamaha commercial rotorcraft UAV in the spring of 2002.

Boeing uses a closely related middleware infrastructure in real-time embedded applications on a variety of manned and unmanned military vehicles, including the F-15, F/A-18, AV-8B, T-45, and the DARPA/USAF unmanned combat air vehicle (UCAV) (see Figure 4.11).

The OCP development team at Boeing is executing and benchmarking the OCP middleware on PowerPC and Pentium processors within an embedded system laboratory. The VxWorks real-time operating system is being used in these research studies.

Boeing intends to host the OCP on an F-15 strike fighter and a UAV test vehicle as part of a flying demonstration of SEC technology. In this demonstration, the OCP will be used, in part, to distribute control of a UAV to a Weapon Systems Operator sitting in the back seat of an F-15.

4.7. OCP PERFORMANCE

The OCP is taking a middleware platform, which has been proven in the mission computing domain of fighter jets, to the flight controls domain— characterized by higher processing rates and more stringent real-time constraints. Laboratory experiments are ongoing to determine the performance

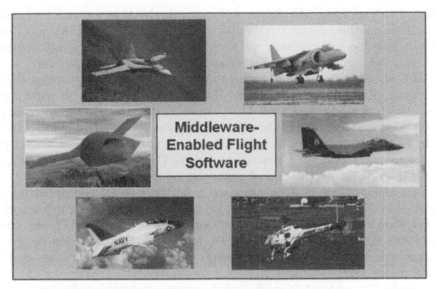

Figure 4.11. Embedded flight platforms using OCP and OCP-related middleware.

of OCP-based flight control designs and to identify areas of needed optimizations. Work recently done at Boeing Phantom Works demonstrated a complex, fixed-wing vehicle controller and a full 6DOF model of the vehicle executing at 100-Hz on a 300-MHz Pentium II processor. The results are encouraging, but middleware optimizations are still being pursued to increase operating margins and to support more complex controllers and other processing that might be done on the vehicle.

4.8. FUTURE OCP DIRECTIONS

4.8.1. Flight Demonstrations

Transition to operational, flying platforms is an important component of the SEC OCP effort. Flight demonstrations exercise the full development and testing cycle for embedded software. The OCP has features that can be exploited in various phases of the development and test cycle, including aiding in the transition from initial designs to code, developer-friendly testing on desktop computers, support for flexible software component interconnections and distribution onto on-board computing assets, and allowing the smooth transition from desktop to embedded platforms.

As was stated before, as part of the OCP effort, Boeing will be architecting a multiple-vehicle SEC flight demonstration involving a piloted strike fighter exercising loose control of a UAV flying a coordinated mission. In the demonstration scenario, the strike fighter will collaborate with the UAV in a mock attack scenario. The fighter will escort the UAV to a search area, and it will dynamically update UAV mission data in the air. The UAV will then depart from the strike fighter, will fly a simulated search and attack maneuver, and will rendezvous with the strike fighter for flight back to base.

SEC research teams with access to air vehicles are also planning flight demonstrations involving the OCP. The Georgia Tech team, in particular, has already demonstrated some OCP capability in the air. In the summer of 2000, Georgia Tech demonstrated OCP capabilities on a rotorcraft UAV. The flight demonstration included (1) communications between the UAV and a ground station, (2) remote control of the UAV with the ground station, and (3) online reconfiguration of processing on the UAV.

In addition, Georgia Tech exercised a full OCP-based embedded rotorcraft UAV flight program in 2002. In this embedded flight program, the application software system was laid out with the controls API, the controls API also auto-generated the C++ framework for population by application code, and application components were triggered in real time by the OCP during flight.

The OCP team is additionally looking at other potential flight demonstration vehicles for hosting and demonstrating controls technology being developed by SEC control research teams.

4.8.2. Middleware Enhancements

Enhancements to the middleware infrastructure are being added. Included in these enhancements are mechanisms for multiple-processor/network resource management. Fault tolerant services are also getting attention, as are services for collecting performance metrics online.

4.8.3. Technology Transition

A possible transition path of the OCP and SEC controls research to the DARPA/USAF-funded UCAV is enabled by the use of Bold Stroke middleware in the UCAV avionics software. Other unmanned vehicles will also be a natural transition possibility for both the OCP software technology and the controls technology coming from the SEC Program. The array of Boeing flying platforms already containing the object-oriented, CORBA-based Bold Stroke middleware will also be able to benefit from advances made to the architecturally similar OCP.

4.9. SUMMARY

The OCP provides a feature-rich execution framework for embedded software. The OCP middleware, built around an RT CORBA interface, includes extensions needed for the challenges of real-time, low-level control of multimodal hybrid systems while living within the resource constraints typically found in flying embedded systems. Included among these extensions are dynamic scheduling and resource management to optimize utilization of limited on-board resources. In addition, the open OCP interfaces to a variety of tools enable cost-effective translation of control designs into embedded code.

The authors wish to acknowledge the contributions from those in the SEC controls technology community for helping guide OCP requirements; the significant contributions of Georgia Tech, Honeywell Labs, and the University of California at Berkeley in developing the OCP and its key concepts; and the technical guidance of Dr. John Bay (DARPA/IXO), Mr. William Koenig and Mr. Raymond Bortner (Air Force Research Laboratory), and Dr. Helen Gill (NSF).

REFERENCES

1. *BAA #99-08 Software-Enabled Control CDB Reference*, DARPA Information Technology Office, 1999.
2. *BAA #99-08 Software-Enabled Control Proposer Information Pamphlet*, DARPA Information Technology Office, 1999.
3. Object Management Group, *Realtime CORBA Joint Revised Submission*, OMG Document orbos/99-02-12 ed., March 1999.
4. Matlab/Simulink, http://www.mathworks.com/.
5. Ptolemy II, http://ptolemy.eecs.berkeley.edu/ptolemyII.
6. ACE/TAO, http://www.cs.wustl.edu/~schmidt.
7. Object Management Group, http://www.omg.org.
8. D. Rosu, K. Schwan, S. Yalmanchili, R. Jha, On adaptive resource allocation for complex real-time applications, in *Proceedings of the IEEE Real-Time Systems Symposium*, December 1997.
9. Christopher D. Gill, Douglas C. Schmidt, and Ron Cytron, Multi-paradigm scheduling for distributed real-time embedded computing, Submitted to the *IEEE Proceedings Special Issue on Embedded Systems*. 2002.
10. Bryan S. Doerr, Thomas Venturella, Rakesh Jha, Christopher D. Gill, and Douglas C. Schmidt, Adaptive Scheduling for Real-time, Embedded Information Systems, in *Proceedings of the 18th IEEE / AIAA Digital Avionics Systems Conference (DASC)*, St. Louis, Missouri, October 24–29, 1999.
11. Irfan Pyarali, Douglas C. Schmidt, and Ron Cytron, Achieving end-to-end predictability of the TAO real-time CORBA ORB, Submitted to the *8th IEEE Real-Time Technology and Applications Symposium*, San Jose, CA, September 2002.
12. Yamuna Krishnamurthy, Vishal Kachroo, David A. Karr, Craig Rodrigues, Joseph P. Loyall, Richard Schantz, and Douglas C. Schmidt, Integration of QoS-enabled distributed object computing middleware for developing next-generation dis-

tributed applications, in *Proceedings of the ACM SIGPLAN Workshop on Optimization of Middleware and Distributed Systems* (OM 2001), Snowbird, Utah, June 18, 2001.

13. Carlos O'Ryan, Douglas C. Schmidt, Fred Kuhns, Marina Spivak, Jeff Parsons, Irfan Pyarali, and David L. Levine, *Evaluating Policies and Mechanisms to Support Distributed Real-Time Applications with CORBA*, Special Issue on Distributed Objects and Applications, Vol. 13, No. 2, John Wiley & Sons, February, 2001.

14. Joseph P. Loyall, Alia K. Atlas, Richard Schantz, Christopher D. Gill, David L. Levine, Carlos O'Ryan, and Douglas C. Schmidt, Flexible and adaptive control of real-time distributed object middleware, submitted to *The International Journal of Time-Critical Computing Systems*, 2000.

15. Christopher D. Gill, Fred Kuhns, David Levine, Douglas C. Schmidt, Bryan S. Doerr, Richard E. Schantz, and Alia K. Atlas, Applying adaptive real-time middleware to address grand challenges of COTS-based mission-critical real-time systems, in *Proceedings of the 1st International Workshop on Real-Time Mission-Critical Systems: Grand Challenge Problems*, IEEE, Phoenix, Arizona, November 30, 1999.

16. E. Gamma, R. Helm, R. Johnson, and J. Vlissides, *Design Patterns, Elements of Reusable Object-Oriented Software*, Addison-Wesley Longman, Reading, MA, 1995.

17. *The Design and Performance of a Pluggable Protocols Framework for Real-time Distributed Object Computing Middleware*, IFIP/ACM Middleware 2000 Conference, Pallisades, New York, April 3–7, 2000.

18. M. Henning and S. Vinoski, *Advanced CORBA Programming with C++*, Addison-Wesley Longman, Reading, MA, 1999.

CHAPTER 5

A PROTOTYPE OPEN-CONTROL PLATFORM FOR RECONFIGURABLE CONTROL SYSTEMS

L. WILLS, S. KANNAN, S. SANDER, M.GULER, B. HECK,
J. V. R. PRASAD, D. SCHRAGE, and G. VACHTSEVANOS

Editors' Notes

This chapter can be read as a companion to Chapter 4. Like the latter, it discusses the SEC open control platform (OCP) for uninhabited autonomous vehicles. Limitations of the current design practice and of the commercial tools currently available are discussed by way of motivation and a layered view of the OCP is presented. The roots of OCP lie in the Common Object Request Broker Architecture (CORBA); important extensions include distributed real-time communication support, a real-time event service that facilitates online reconfiguration of the OCP, and adaptive resource management (see Chapter 6).

Most of this chapter focuses on the topic of reconfigurable control, showing how the OCP can be used for real-time reconfiguration—creating or deleting components, changing intercomponent connections and changing the quality of service for connections. A reconfigurable controls application programmer interface (API) has been developed to provide familiar abstractions to control engineers, who may not have the extensive computer science background needed for unassisted use of the OCP. Through the API, control engineers can design reconfiguration strategies with localized changes to the control system architecture. An API at a further level of abstraction, called the Generic Hybrid Controls API, is under development that is intended to ensure that configuration changes (whether initiated by a person or through automation) do not compromise system integrity.

Georgia Tech has conducted initial flight tests of the OCP on an X-cell helicopter. In these tests, ground station commands triggered online reconfiguration of the onboard software. A simulation example is also presented showing a reconfiguration scenario in which a main rotor collective controller for a helicopter (which uses the blade pitch of the rotor as the actuation variable) is replaced with a rotor rotation

Software-Enabled Control. Edited by Tariq Samad and Gary Balas
ISBN 0-471-23436-2 © 2003 IEEE

speed controller in the event of a failure of the collective actuators. This strategy is not typically viable for large helicopters but it may be for small uninhabited rotorcraft.

5.1. INTRODUCTION

Complex dynamic systems such as aircraft, power systems, and telecommunications networks present major challenges to control systems designers. Both the military and civilian sectors of our economy are demanding new and highly sophisticated capabilities from these systems that traditional controls technology is not offering. Among these capabilities are the following:

- *Adaptability/Dynamic Reconfigurability.* Large-scale engineered systems must have the ability to quickly and gracefully react to a changing environment or to changes in their own configuration without compromising their operational integrity. Extreme performance dynamic systems must be able to support online switching of algorithmic components and rapid redirection of the interconnections among them, as well as changing the priorities at which information is flowing. This ability is particularly useful for hybrid control systems.

- *Plug-and-Play Extensibility.* New advances are continually being made in control algorithm design, communications and sensor technology, and high-performance hardware platforms. To take full advantage of these innovations as they become available, we must be able to insert new technology into the system architecture without redesigning the components already in the system.

- *Interoperability.* Today's control systems operate in distributed, heterogeneous environments; that is, the software components may be running on different processors, using different programming languages, hardware platforms, and network protocols, often over wireless links. Real-time communication must be provided among these distributed components while satisfying stringent constraints on bandwidth, response time, and reliability.

- *Openness.* Reconfigurability and component interchangeability require software architectures that are flexible and that support tools and algorithms from a variety of sources and domains. This requires a shift away from traditional control system implementation, which tends to be built with a particular application in mind and which makes rigid, limiting assumptions about the types of technology that will be used. The development of control systems across applications involves much duplication of effort; there are tremendous opportunities for reuse that are currently not being exploited. A shift to open architectures is essential for addressing this problem.

Meeting these challenges will require a fundamental change in the way control systems are composed, integrated, changed, and reused. Recent advances in software technology have the potential to revolutionize control system design. In particular, new component-based architectures [1, 2] encourage flexible "plug-and-play" extensibility and evolution of systems. Distributed object computing allows heterogeneous components to interoperate across diverse platforms and network protocols [3]. New advances are being made to enable dynamic reconfiguration and evolution of systems while they are still running [4, 5]. New communication technologies are being developed to allow networked, embedded devices to connect to each other and self-organize [6].

This chapter describes a new software infrastructure for complex control systems that exploits these new and emerging software technologies. More specifically, it presents a prototype *open-control platform* (OCP) for complex systems that coordinates distributed interaction among diverse components and supports dynamic reconfiguration and customization of the components in real time. Its primary goals are to accommodate rapidly changing application requirements, easily incorporate new technology (such as new hardware platforms or sensor technology), interoperate in heterogeneous environments, and maintain viability in unpredictable and changing environments. The next section describes the current practice in control system implementation and discusses features of a complex control system architecture. It is followed by a description of the desired features a software infrastructure must have to promote new advances in control system design. It then describes an open-control software infrastructure to support these desired features, followed by a brief overview of a first-generation prototype of this infrastructure that has been developed for an autonomous aerial vehicle control application. This prototype was released in June 2000 and since then has been transitioned to Boeing to take advantage of the technology described in Chapter 4 in supporting its core real-time distributed computing features. The chapter concludes with a discussion of ongoing work and open issues.

5.2. CURRENT PRACTICE IN CONTROL SYSTEM CONFIGURATIONS

Control systems for highly complex systems (such as processing plants, manufacturing processes, aerospace vehicles, and power plants) are themselves very complex. Notions of "control" are expanding from the traditional loop-control concept to include such other functionalities as supervision, coordination and planning, situation awareness, diagnostics, and optimization [7–9].

Consider, for example, the control system for an uninhabited aerial vehicle (UAV) such as a vertical takeoff and landing (VTOL) UAV of the helicopter

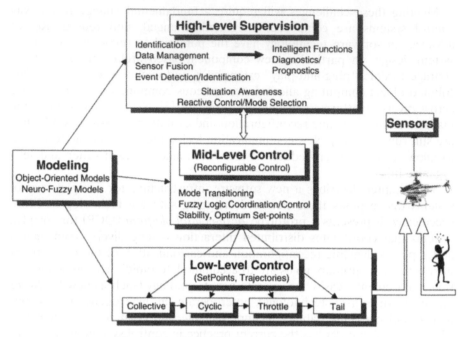

Figure 5.1. Hierarchical control architecture.

type. A hierarchical/intelligent control architecture for the UAV application is depicted in Figure 5.1. It integrates high-level situation awareness and mode selection functionalities with middle-level coordination routines, aimed at facilitating mode transitioning and reconfigurable control, as well as low-level control activities. An intelligent operator interface complements the basic algorithmic modules.

The three-level hierarchy is intended to respond to varying time frames and degrees of intelligence imposed by the behaviors the vehicle must exhibit in the course of a mission. Thus techniques used to respond to changes may range from rapid, reactive approaches to more deliberative ones. Situation awareness, reactive control/mode selection, and mode transitioning/reconfiguration [10] typically require fast response, yet the processing required for communications and computations in these algorithms severely taxes the limits of available hardware/software platforms. Other evolutionary functions such as mission-oriented customization of the onboard systems and transferring generic control design modules to new applications may be more deliberative since they occur over longer time frames. The degree of autonomy also varies across this spectrum from "complete" autonomy in the top half of the hierarchy to greater human interaction during mission-oriented customization and component reuse.

Although the control architecture depicted in Figure 5.1 is specific to a UAV, it possesses generic features that may be found in many complex engineered systems. A similar framework may be realized for the integrated monitoring and control of the powertrain modules of an automotive system [11, 12] or the control of major components of a papermaking machine that aims to accommodate grade changes efficiently and robustly [13]. Thus modern control system configurations entail a series of "modules" (or "components") that perform a variety of functions needed to carry out a mission, a production plan, or a specific operational task. The challenge addressed by the OCP is to deal with the complexity of integrating these components in a distributed environment while still providing the flexibility to reconfigure them dynamically at run time.

5.2.1. Limitations of Current Design Practice

Implementations of complex control systems tend to suffer from a common set of problems:

- *Complex, Brittle Data Interchange*. The interaction of the different components requires a highly sophisticated and reliable communication channel that can provide the needed data transfer in a timely manner with possibly different update rates for different processes and a hierarchy of priorities among the processes. With current methods of designing distributed control systems, several different people may have written code for various controllers or sensors to be run on different types of processors and in different languages. Significant effort is required to design an efficient and stable data transfer protocol between these different components. Changes to the system, such as using a new hardware platform or adding new sensors, may necessitate modifications in the entire data transfer protocol (timings, priorities, interconnections, etc.). This would not be necessary if the control system were built on a software infrastructure with a standardized interface so that new components, even on different types of hardware platforms, could be added easily without having to explicitly change the data transfer properties manually.
- *Tight Coupling*. The modules in a typical modern control system tend to be tightly coupled; that is, there are direct communication links between the various components that share information. This can make the system faster but may make changes to the system harder to achieve, particularly with respect to the constraints imposed by the data transfer protocol, as mentioned above. With the advent of ever-faster processors, tight coupling is not necessary to gain efficiency, particularly at the expense of reduced ability to update the control system quickly and

easily. It would be much easier to make changes to the system if its components were less coupled and if functionally similar components could be easily interchanged in a "plug-and-play" fashion. This requires an open software infrastructure that supports a component-based control system design.

- *Computational Limitations.* Many extreme-performance control systems, such as that of a small helicopter UAV, are operating under stringent technological constraints that limit their flexibility in responding to change. For example, they must be small enough to be carried on board, and they must operate with low power requirements. They must also communicate with external systems such as a groundstation, other UAVs, or a mothership, often using mediums of limited bandwidth such as a wireless link. Yet traditional hardware/software systems conforming to these constraints are typically not able to meet the computationally intensive requirements of system monitoring and reconfiguration needed for autonomous operation and distributed communication. Recent advances in high-performance, low-power embedded processors, compact packaging technology, and high-bandwidth wireless communications are starting to provide the resources needed for autonomous, extreme-performance control.

- *Closed/Proprietary Systems.* The current state of the practice in control implementation does not offer the flexibility required to allow open architectures for communications, computing, and control. Traditionally, control firms have provided customized or proprietary hardware and software platforms to their customers, severely limiting component interchangeability, reconfigurability, and distributed and concurrent processing. More recently, though, market drivers are dictating open-control architectures and flexible programming tools. Control firms are beginning to adopt this paradigm shift by making available new hybrid, distributed control systems with interchangeability and reconfigurability features.

5.2.2. Limitations of Current Commercial Tool Support

Developing hierarchical control systems such as depicted in Figure 5.1 involves integrating many different types of components, including software algorithms, device drivers for hardware components, sensor data processing code, and mathematical flight dynamics models written in more than one programming language. The components may also originate from a variety of sources: newly developed using MATLAB/Simulink, off-the-shelf commercial products, or legacy components. These systems are further complicated in that the components are often distributed across different hardware platforms and multiple types of networks within the same system.

Integrating these diverse components so that they seamlessly communicate with one another both locally and remotely is challenging. Even more challenging is trying to *change* these systems—either offline (e.g., to incorporate new types of sensor technology) or at run time (e.g., to switch to a new control algorithm in response to a vehicle failure). Yet the ability to flexibly and easily adapt the system architecture is critical to the ability to reuse and to the "plug-and-play" extensibility of complex control systems. Furthermore, rapid adaptation is critical to enabling innovative online customizable controls technology for extreme-performance applications such as aerial robotics.

Recently, a variety of commercial tools have appeared on the market for implementing real-time control systems, including ControlShell/NDDS from Real-Time Innovations, Inc. [14], Real-Time Workshop from The Math-Works, Inc. [15], and ARCSware from Advanced Realtime Control Systems, Inc. [16]. Most provide simulation and code-generation capabilities that support primarily the *development* of control systems. They make available to the developer repositories of prebuilt and preverified reusable components (such as common types of controllers (PID), filters, and matrix computations) for efficient construction of systems. However, they do not support rapid adaptation or dynamic reconfiguration of these systems online. They do not facilitate changing algorithms or inserting and removing components at run-time while maintaining adherence to real-time constraints.

Some commercial tools, most notably ARCSware [16] and ControlShell/NDDS [14], address to a limited extent the issue of integrating distributed components by supporting remote communication and hiding many low-level network programming tasks from the controls engineer. However, these tools typically provide "best-effort" communication methods where all communication events have equal priority. Since all components compete for communication resources on an equal basis, critical messages may have to wait for low-priority messages to process. These tools rely on the real-time features of the target system, which typically provide priority only at the process level and a fixed process schedule. This causes a lack of adaptability and often necessitates an underutilization of available resources, relative to what could be provided with the use of advanced real-time communication services such as dynamic scheduling and adaptive resource management [17, 18]. (These are described in the next section.)

Furthermore, while these tools facilitate reuse at the component level, there are tremendous opportunities for also reusing common ways of integrating and reconfiguring components across similar control applications. These opportunities should be exploited.

5.3. OPEN CONTROL PLATFORM DESIGN

A new software infrastructure called the open control platform (OCP) is being developed by Georgia Tech, in collaboration with Boeing, Honeywell,

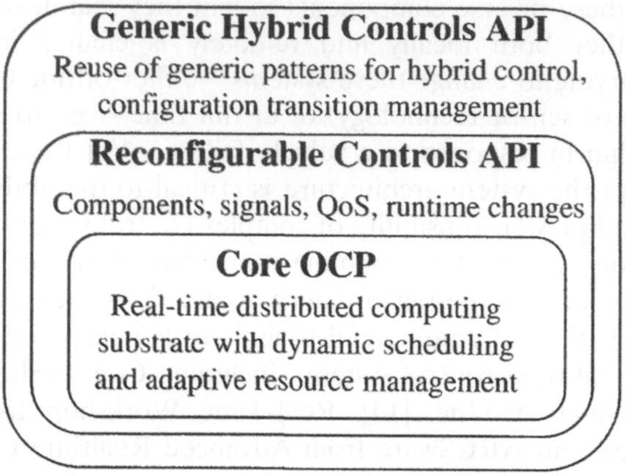

Generic Hybrid Controls API

Reuse of generic patterns for hybrid control, configuration transition management

Reconfigurable Controls API

Components, signals, QoS, runtime changes

Core OCP

Real-time distributed computing substrate with dynamic scheduling and adaptive resource management

Figure 5.2. Open control platform.

University of California–Berkeley, and the University of Minnesota, to serve as a substrate for integrating innovative control technologies. The OCP specifically provides more comprehensive support than existing approaches for integrating distributed, heterogeneous components while hiding details of distributed computing from the control system developer. It moves beyond development-only support to enabling the rapid runtime adaptation and dynamic reconfiguration of control systems. It also complements the component-level reuse of commercial products with reuse of generic integration patterns and reconfiguration strategies.

A prototype version has been successfully demonstrated [19–21] and is being applied to the autonomous control of UAVs. As illustrated in Figure 5.2, the OCP consists of multiple layers of application programmer interfaces (APIs) that increase in abstraction and become more domain-specific at the higher layers. The layers of the OCP are intended to form a bridge from the controls domain to distributed computing and reconfigurability technology so that controls engineers can exploit these technologies without being experts in computer science or computer engineering.

In the bottommost "core" layer, the OCP leverages from and extends new advances in real-time distributed object computing [3, 22], which allows distributed, heterogeneous components to communicate asynchronously in real time. It also supports highly decoupled interaction between the distributed components of the system, which tends to localize architectural or configuration changes so that they can be made quickly and with high reliability.

The middle "reconfigurable controls" layer provides abstractions for integrating and reconfiguring control system components; the abstractions bridge the gap between the controls domain and the core distribution substrate. The abstract interface is based on familiar controls engineering concepts, such as block diagram components, input and output ports, and measurement and command signals. It allows real-time properties to be specified on signals that translate to quality-of-service constraints in the core real-time distribution substrate. It also allows run-time changes to be made to these signal properties, which are then handled by lower-level dynamic scheduling and resource management mechanisms. This layer raises the conceptual level at which the controls engineer integrates and reconfigures complex, distributed control systems.

The third "hybrid controls" layer supports reconfiguration management by making reconfiguration strategies and rationale for reconfiguration decisions explicit and reusable. It contains generic patterns [23] for integration and reconfiguration that are found in hybrid, reconfigurable control systems. It provides generic algorithmic patterns that can be specialized with logic for choosing reconfigurations as well as blending strategies for smoothly transitioning from one configuration to another. This is critical to hybrid systems in which continuous dynamics must be maintained between discrete reconfiguration events and where multiple control and blending strategies are applicable.

These layers can guide online adaptation of the software architecture, ensuring that system integrity is continually maintained so that all changes and reconfigurations to the control system are valid, safe, consistent, and coordinated at all levels of granularity. The layers of the OCP are described in more detail in this section. The core and reconfigurable controls layers are the most mature. The highest ("hybrid controls") layer is in the early stages of development.

5.3.1. Core OCP: Real-Time Distributed Computing Substrate

Todays complex control systems are increasingly made up of distributed, heterogeneous hardware and software components. The underlying network technology in place to allow communication between these components may also vary both within and across the applications. For example, consider the various components used during the design and implementation phases of a UAV application (see Figure 5.3). The initial design work involves a great deal of simulation, possibly across different languages (such as a high-fidelity vehicle model written in C and a control module written in MATLAB). The next step is to perform hardware-in-the loop simulation where part of the simulation is performed in software and part is performed in the actual hardware that is to be used on board the vehicle. In this step, there may be control algorithm modules and sensor-processing modules running on a desktop PC, actual sensors and actuators, and a high-fidelity math model of the vehicle running on a multiprocessor computer. Also, a high-fidelity simulation capable of providing realistic graphics and visualization may be running on specialized hardware in order to visualize the aircraft dynamics

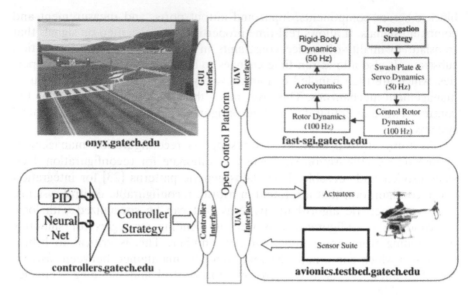

Figure 5.3. Distributed communication.

and provide mission information. The control and sensor processing algorithms will eventually be moved from the desktop PC to a flight computer. During flight, these algorithms will communicate with the actual hardware sensor and actuation suites on board the vehicle. Some of these components may need to communicate over a wireless Ethernet link, whereas others may be connected through a high-speed, high-bandwidth backplane bus on board the vehicle. Moreover, the distribution topology is likely to evolve as the system develops, as more functionality is moved on board, and as the UAV interacts with other vehicles and other sources of mission guidance, such as a mothership.

An important challenge is providing real-time communication among these distributed, heterogeneous components while dealing with tight constraints on bandwidth, response time, and reliability. In addition, the different steps of the design (as mentioned above) should not require redesign of the communications protocol as the design evolves and as various validation procedures are performed. Currently, building systems like these requires detailed, tedious activities: Complex timing relationships have to be worked out, data interchange protocols have to be created and maintained, detailed network programming (e.g., using sockets and remote procedure calls) has to be performed, and careful prioritization and scheduling of tasks have to be accomplished. The OCP is designed to provide reconfigurable, adaptive real-time distributed communication while hiding most of these tedious network programming details from the control system developer.

To meet these challenges, the OCP leverages from the latest middleware technology for distributed object communication. Middleware is software that provides a substrate through which components can communicate with each other regardless of whether they are co-located on the same processor (local) or remote and regardless of whether they are written in different programming languages (within certain supported sets, typically Ada, C, COBOL, C++, Smalltalk, and Java) or running on different operating systems or hardware platforms. Middleware also allows components to communicate regardless of the communication protocol or type of network used to connect the components.

The middleware used by the OCP is the Common Object Request Broker Architecture (CORBA) standard set by the Object Management Group (OMG) [24]. This hides many details of remote communication by automating several common network programming tasks, including registering, locating, and activating objects that may not be local to a process, dispatching operations, error handling, and marshalling/demarshalling parameters [24] (i.e., translating data values from their local representations to a common network protocol format). Thus, interoperability is enabled between applications, and multiple components are seamlessly interconnected within and across control systems.

Extensions have been made to CORBA technology by Washington University and by Boeing Phantom Works to allow distributed communication to occur in real time [3, 22]. A particularly powerful mechanism used in the core OCP to provide high-performance distributed communication is *replication*. This is a common technique in distributed computing that caches local copies of remote data objects so that they can be efficiently accessed frequently by several local client objects. Two other important extensions to the underlying CORBA technology are a real-time event service and dynamic event scheduling capabilities, as described below. Additional details on these real-time performance extensions are given in Chapter 4.

Prioritized Event-Based Communication for Dynamic Reconfiguration.
One of the key extensions provided by Washington University is a new real-time event service [22], which enables components to interact efficiently without being tightly coupled. This eases architectural evolution and facilitates online reconfiguration by localizing structural changes. The event service provides a communication abstraction called an *event channel*, similar to a bus. Components that generate data ("suppliers") or use data ("consumers") connect to the event channel. The suppliers "publish" certain data event types, whereas the consumers "subscribe" to certain event types. (An event is any data communicated between suppliers and consumers; it may be, for example, a message, command, control flag, or method invocation.)

The event channel acts as a mediator [23] between components so that their interconnections are flexible. When a component subscribes to some

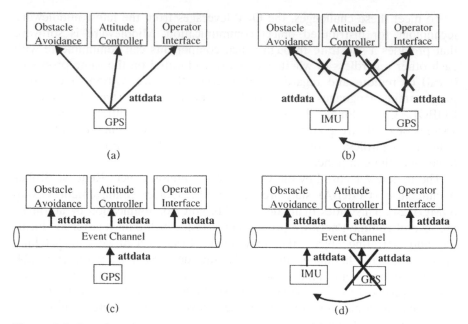

Figure 5.4. Benefits of event channel communication: (a) direct coupling between components, (b) replacement requires many changes, (c) decoupled communication through an event channel, (d) replacement requires local change.

type of event (e.g., navigational state information), it does not necessarily care where that information is coming from. It would like to receive that information from whatever sensors or sensor processing modules are active. The software should not make rigid commitments as to which components are providing this information by hard-coding a call to a specific source of data. The event channel provides the level of abstraction needed by mediating information flow between suppliers and consumers. When a new type of sensor is added to the system or replaces another type of sensor, for example, it can be connected to the event channel to publish its type of data, and all consumers subscribed to that data will receive it without changing their implementation or interface. (See Figure 5.4.)

Dynamic Scheduling and Adaptive Resource Management. The core OCP allows control system developers to specify the real-time properties of the system in the form of quality-of-service (QoS) requirements for the components (such as worst-case execution time, rate, and priority). The event service used by the OCP provides facilities for scheduling of events in real time [22], based on assigned priorities and desired QoS constraints. This is useful during an extreme performance maneuver—for instance, when attitude data may not only be needed at a higher rate and accuracy, but also at a higher priority than other types of events. In particular, when the event channel receives a set of supplier events, it dispatches them to the targeted consumers according to the priorities of the consumers, either by queuing them for later dispatch or interrupting (preempting) a running thread for immediate dispatch.

Supporting online adaptation means that the mechanisms used to support high-performance distributed communication and decoupled communication themselves be customizable at run time. For example, run-time changes to QoS parameters must be supported so that tasks that might have previously been running at a lower priority than others can be placed at a higher priority to respond to a failure or some unexpected condition. Similarly, replication update strategies must be customizable. For example, replicated information being received on the ground may need to be updated more often when a critical mission replanning task is being performed to avoid a threat.

To support these needs, Boeing has extended the core OCP to incorporate recent advances in dynamic scheduling from Washington University [17] and adaptive resource management from Honeywell Laboratories [18] (see Chapter 6). By using dynamic scheduling algorithms in the event channel scheduler, the OCP can accommodate run-time changes in QoS requirements for events by allowing QoS specification changes to be specified and enforced at run time. Dynamic scheduling allows run-time changes in the control system architecture while maintaining real-time operation. And since a fixed schedule is not used, higher utilization of resources can be achieved for nonperiodic tasks [17].

5.3.2. Reconfigurable Controls API

The core OCP hides details of dealing with remote objects, network protocols, and the like, but it still requires extensive computer science background (e.g., details about event channels, scheduling, and replication) to use it effectively. The controls API raises the level of abstraction to provide a convenient interface familiar to controls engineers, bridging the gap between the controls domain and the core OCP.

In the controls API, a control system consists of an interconnected system of components that communicate via ports. The controls API provides a component interface for sending and receiving data (e.g., measurement signals and command signals), sending and receiving notifications, and performing system reconfiguration. Notifications include timeout events, signal update events, and correlation events (which are user-specified unions of events).

Components consist of user-defined application code, input ports, and output ports. An example is shown in Figure 5.5, where quality-of-service parameters (e.g., execution time, criticality, and rate) can be associated with ports. When data are available on an input port or at a periodic interval, the application code is executed. This code can run algorithms, send data using

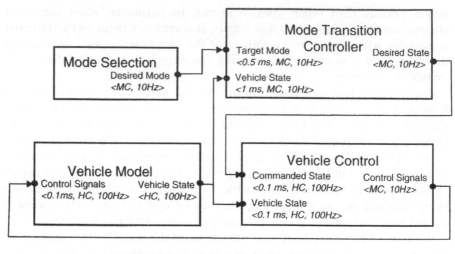

QoS parameters: <Execution Time, Criticality, Rate>
MC = Medium_Criticality
HC = High_Criticality

Figure 5.5. Sample control system with associated quality-of-service parameters.

an output port, and perform system reconfiguration. System reconfiguration can consist of creating components, changing the interconnections between components, and changing the quality of service for connections.

5.3.3. Generic Hybrid Controls API: Reconfiguration Management and Reuse

The software infrastructure supports a loose coupling between control system components to provide flexibility, extensibility, and reuse. The reconfigurable controls API provides a layer of abstraction for easily specifying configurations and reconfigurations with localized changes to the control system architecture. However, configuration management is needed to ensure that the configurations are valid and consistent with overall system requirements. Moreover, it is critical that changes to the configuration (either through human-directed evolution or dynamic, online reconfiguration) maintain overall system integrity by being globally coordinated and consistent. The generic hybrid controls API, which is currently under development as the top layer of the OCP (shown in Figure 5.2), will capture general mechanisms for managing configurations and reconfigurations.

Reconfigurations of control systems follow standard strategies, analogous to what Oreizy et al. call "change application policies" [4]. These are strategies for making changes without violating reliability, safety, and consistency constraints. For example, they may dictate how quickly one algorithm can be switched for another, or whether a redundant component needs to work concurrently with the component it is replacing before the swap occurs to allow the new component to "come up to speed." For instance, a fault occurring in a system may necessitate a change from the nominal algorithms to the fault control algorithms. The fault may be abrupt, requiring a fast switch in the control, or it may be more gradual, which may allow for a blending of the outputs from the nominal controller and the fault controller to smooth the transients during the transition. Control reconfiguration is also a necessary component for hybrid controllers.

Reconfiguring (or switching) from one sensor to the other may need to be done in a discrete manner or in a gradual manner, for which there is a transition period while the outputs of the two sensors are blended together. All these methods for reconfiguration of components may be encapsulated within pattern constructs that hold user-defined, application-specific policies on how to replace one component with another.

5.4. A PROTOTYPE OPEN CONTROL PLATFORM

An initial prototype of the OCP has been constructed by Georgia Tech in collaboration with Boeing Phantom Works using real time distributed object technology pioneered by Washington University [3]. The design and implementation of the OCP was driven by a test reconfiguration scenario where a helicopter main rotor collective controller is replaced with an RPM controller when the collective actuators fail and become stuck. Altitude control on a helicopter is normally achieved by controlling the magnitude of lift produced by the main rotor. The lift produced by the main rotor may be changed either by varying the collective pitch setting of the blades or by varying the main rotor's rotational speed (RPM). In practice, the collective is used as the primary control variable to change the overall lift produced by the main rotor. This is because the RPM degree of freedom is slower to respond and the margin of feasible RPM change on full-size helicopters is small. However, this margin is much larger in small unmanned helicopters and may thus be exploited as a secondary control variable in case the collective actuators fail and become stuck.

The functional setup of this scenario is shown in Figure 5.6. The control system consists of a helicopter attitude controller (inner-loop), a trajectory controller (outer-loop), a pilot stick to provide trajectory commands, a mathematical model of the unmanned helicopter, a multimedia GUI providing visualization of the flight dynamics, and a fault detection and reconfiguration module.

The trajectory controller [26] takes commanded trajectories X_C, Y_C, Z_C, ψ_C as inputs and generates the attitude ϕ_C, θ_C, ψ_C, necessary to track the desired command. Within the trajectory controller, either the collective $\delta_{COLLECTIVE}$ or the RPM controller's output Ω_C may be used to control altitude. The

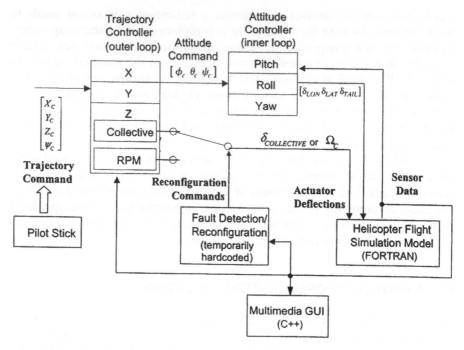

Figure 5.6. Flight control configuration in OCP demonstration prototype.

attitude controller [27] takes as input the commanded attitude and generates the actuator deflections, δ, necessary to achieve and maintain the desired attitude. The actuator signals are then fed into the simulation model (or, in the future, the actual helicopter in flight). A preprogrammed fault event is input to the simulation and is picked up by the fault detection and identification (FDI) module, which then generates reconfiguration commands. The functionality of the FDI module and the decision logic to generate the necessary reconfiguration are assumed to be known. The simulation model will eventually be replaced with the actual vehicle hardware—a Yamaha R-50/Rmax remotely piloted helicopter.

Figure 5.7 shows the behavior of the simulated helicopter's altitude rate response over the course of the test reconfiguration scenario. A video clip of this reconfiguration demonstration, as well as a demonstration involving mode transitioning [10] from hover to forward flight, can be viewed in reference 28. For purposes of this test, which is a vertical descent scenario, altitude rate (h) was commanded rather than altitude itself. Figure 5.7 shows the commanded altitude rate (h_C) and the response (h) of the UAV over time. At point A, a vertical descent rate of approximately 3 ft/s is commanded and the UAV responds. Sometime between B and D, the actuator becomes stuck at point C. At D, a positive altitude rate is commanded in an attempt to recover the aircraft from its continuing descent; however, the

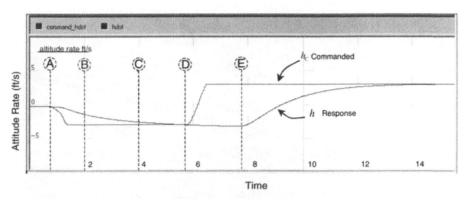

Figure 5.7. Results of an online reconfiguration performed using the OCP.

helicopter fails to respond. A reconfiguration event occurs at E where the RPM controller is substituted for the nominal RPM controller, at the same time switching off the collective control, if necessary. The reconfiguration commands necessary for this discrete switch were preprogrammed in the fault detection and controller reconfiguration module for this example. This allows the helicopter to successfully recover altitude and continue its mission. Without this reconfiguration, the helicopter would have continued to descend until it crashed into the ground.

The components integrated in the OCP demonstration system are able to communicate with each other even though they are running in different processes and are written in different programming languages (e.g., Fortran and C++). In the demonstration system, a simulated collective actuator fault triggers a reconfiguration event, which replaces the main rotor collective controller with the redundant main rotor RPM controller. This online switch is made gracefully through a simple change in the controller connections to the event channel.

The initial OCP prototype also demonstrates the ability to reuse legacy components and plug in new replacement components as the system evolves. A legacy flight dynamics model was used in the initial demonstration system. Since it was written in Fortran, it was wrapped in C++ and a standard CORBA IDL interface was defined to integrate it with the rest of the system. This allowed us to reuse a trusted, existing dynamics model in the original system. The OCP made it easy to replace this legacy component with a higher-fidelity dynamics model (written in C++) when it became available in the next version of the demonstration system a few months later.

5.5. ONGOING WORK AND OPEN ISSUES

Initial flight trials of the OCP were conducted at Georgia Tech using an X-cell helicopter testbed [21], shown in Figure 5.8. The flight trials tested the

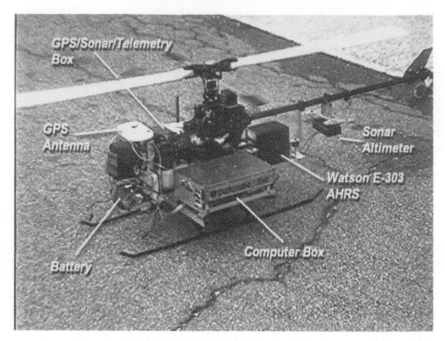

Figure 5.8. Flight test aircraft.

dynamic reconfiguration and distributed communication features of the OCP. The basic mission of the X-cell was to obtain telemetry data and communicate it to the ground computer in real time at rates that could be varied by the user on the ground; that is, commands to switch to a different rate were given from the ground, which caused an online reconfiguration of the software running on board the vehicle. The avionics system used a 486 flight control computer (100 MHz with 128 MB RAM) running the Linux operating system. (This is currently being ported to run on VxWorks and QNX, where it will require much less memory.) The sensor modules consisted of an attitude and heading reference system, a sonar altitude sensor, and a Novatel differential GPS. Flight trials of the OCP running on VxWorks and QNX took place during the summer of 2002, using a Yamaha R-50/Rmax helicopter testbed.

We have also employed the OCP's integration and dynamic reconfiguration capabilities in support of innovative midlevel control algorithms for mode transitioning and for fault detection and identification [19, 29]. We have successfully tested these in simulation (see reference 28 for video clips of these tests). We have performed hardware-in-the-loop simulation tests, in preparation for actual flight tests of these control algorithms as well.

New midlevel control algorithms, primarily for fault-tolerant control, are being developed and will be integrated into the OCP in the near term. Higher-level mission planning and mode-selection algorithms will also be incorporated in the future. Additional areas in which we are focusing our attention include the following.

5.5.1. Real-Time Performance

Several design ideas are being explored to balance tradeoffs that arise in providing system flexibility and modularity while maintaining real-time performance. An important extension developed by Boeing to the real-time CORBA technology is a replication service that essentially caches local copies of remote data objects so that they can be efficiently accessed without the overhead of interacting through the ORB. Extensions to this replication service are being considered to allow efficient distributed data transfer while still maintaining mutual anonymity (for loose coupling) between components. Optimizations to the replica update mechanism are also being designed to minimize unnecessary bandwidth. Another important issue is how to balance real-time performance with reasoning about reconfiguration options and strategies.

5.5.2. Reconfigurable Hybrid Controls

An important area of ongoing work is developing the generic hybrid controls API. In addition to the task of populating it with generic patterns and reconfiguration strategies discussed earlier, several challenging issues are arising. One is that the patterns are applicable at multiple levels of the controls hierarchy, and support must be given for the hierarchical composition of these patterns. This composition may be greatly facilitated by leveraging from existing hybrid modeling and simulation techniques such as those developed in the Ptolemy project [30]. Ptolemy integrates multiple, diverse models of computation, exploiting hierarchical relationships to efficiently reason about model interactions. By using Ptolemy to formally define interaction semantics for generic integration and reconfiguration patterns, it may be possible to validate the composition and use of these generic patterns for hybrid controls.

This chapter has described the use of the latest software technology to enhance the capabilities of emerging control algorithms. A beneficial synergy exists in the concurrent development of the new control algorithms and of the software infrastructure to support it. Although the controls technology is enabled by the open software infrastructure, it is also driving the underlying component-based software technology forward by demanding new capabilities to support online customization, hybrid controls, and extreme performance.

ACKNOWLEDGMENTS

This work is supported under the DARPA Software Enabled Control Program under contracts #33615-98-C-1341 and #33615-99-C-1500, managed by the Air Force Research Laboratory (AFRL). We gratefully acknowledge DARPA's Software Enabled Control Program, AFRL, and Boeing Phantom Works for their continued support. We have benefited greatly from the guidance of John Bay, Helen Gill (DARPA), and Bill Koenig (AFRL). We would also like to acknowledge the contributions of Brian Mendel, James Paunicka, and Bryan Doerr of the Boeing Phantom Works, as well as those of Scott Clements, Eric Johnson, Aaron Kahn, Carlos Restrepo, Freeman Rufus, Gideon Saroufiem, and Ilkay Yavrucuk of Georgia Tech. We are also grateful for the helpful suggestions from the anonymous reviewers of this chapter.

REFERENCES

1. G. T. Heineman and W. T. Councill, *Component-Based Software Engineering: Putting the Pieces Together*, Addison-Wesley, Reading, MA,2001.
2.. C. Szyperski, *Component Software: Beyond Object-Oriented Programming*, Addison-Wesley, Reading, MA, 1998.
3. D. Schmidt and F. Kuhns, An overview of the real-time CORBA specification, *IEEE Computer* 33(6):56–63, June 2000.
4. P. Oreizy, N. Medvidovic, and R. N. Taylor, Architecture-based runtime software evolution, in *Proceedings of the International Conference on Software Engineering* (ICSE 98), pp. 117–186, 1998.
5. P. Oreizy, M. Gorlick, R. N. Taylor, D. Heimbigner, G. Johnson, N. Medvidovic, A. Quilici, D. Rosenblum, and A. Wolf, An architecture-based approach to self-adaptive software, *IEEE Intelligent Systems* 14(3):54–62, May/June 1999.
6. J. Waldo, The Jini architecture for network-centric computing, *Communications of the ACM* 42(7):76–82, July 1999.
7. P. P. Bonissone, V. Badami, K. H. Chiang, P. S. Khedkar, and K. W. Marcelle, Industrial applications of fuzzy logic at General Electric, *Proceedings of the IEEE* 83(3):450–465, 1995.
8. G. Vachtsevanos, Hierarchical control, in *Handbook of Fuzzy Computation*, E. Ruspini, P. Bonissone, and W. Pedrycz, Editors in Chief, Institute of Physics Publishing Philadelphia, pp. F 2.2:42–53, 1998.
9. G. Vachtsevanos, W. Kim, S. Al-Hasan, F. Rufus, M. Simon, D. Schrage, and J. V. R. Prasad, Mission planning and flight control: Meeting the challenge with intelligent techniques, *Journal of Advanced Computational Intelligence* 1(1):62–70, October 1997.
10. F. Rufus and G. Vachtsevanos, Design of mode-to-mode fuzzy controllers, *International Journal of Intelligent Systems* 15(7):657–685, July 2000.
11. R. Jones, A. Cherry, S. Farrall, Application of intelligent control in automotive vehicles, in *Proceedings of CONTROL '94*, pp. 159–164, March 1994.
12. L. Qiao, M. Sato, K. Abe, H. Takeda, T. Mogi, N. Tomisawa, and S. Watanabe, Environment recognition in powertrain control, in *Proceedings of the IEEE*, pp. 730–734, 1995.
13. T. F. Murphy, S. Yurkovich, and S. C. Chen, Intelligent control for paper machine moisture control, in *Proceedings of the IEEE Conference on Control Applications* (Dearborn, MI), pp. 826–833, September 1996.

14. G. Pardo-Castellote and S. Schneider, The network data delivery service: Real-time data connectivity for distributed control applications, in *Proceedings of the 1994 IEEE International Conference on Robotics and Automation*, Vol. 4, pp. 2870–2876.

15. The MathWorks: Real-Time Workshop [Online], September 29, 2000. Available at http://www.mathworks.com/products/rtw/.

16. Advanced Realtime Control Systems, Inc. [Online], September 29, 2000. Available at http://www.arcsinc.com/progref-manual.pdf.

17. D. Levine, C. Gill, and D. Schmidt, Dynamic scheduling strategies for avionics mission computing, in *Proceedings of the 17th DASC. AIAA / IEEE / SAE Digital Avionics Systems Conference*, Vol. 1, pp. C15/1–8, 1998.

18. M. Cardei, I. Cardei, R. Jha, and A. Pavan, Hierarchical feedback adaptation for real time sensor-based distributed applications, in *Proceedings 3rd IEEE International Symposium on Object-Oriented Real-Time Distributed Computing, 2000 (ISORC 2000)*, pp. 181–188.

19. S. Kannan, C. Restrepo, I. Yavrucuk, L. Wills, D. Schrage, and J. V. R. Prasad, Control algorithm and flight simulation integration using the open control platform for unmanned aerial vehicles, in *Proceedings of the 18th Digital Avionics Systems Conference*, St. Louis, MO., pp. 6.A.3-1 to 6.A.3-10, October 1999.

20. L. Wills, S. Kannan, B. Heck, G. Vachtsevanos, C. Restrepo, S. Sander, D. Schrage, and J. V. R. Prasad, An open software infrastructure for reconfigurable control systems, in *Proceedings of the 19th American Control Conference (ACC-2000)*, Chicago, Illinois, June 2000.

21. L. Wills, S. Sander, S. Kannan, A. Kahn, J. V. R. Prasad, and D. Schrage, An open control platform for reconfigurable, distributed, hierarchical control systems, in *Proceedings of the 19th Digital Avionics Systems Conference (DASC-2000)*, Philadelphia, PA, October 2000.

22. D. Schmidt, D. Levine, and T. Harrison, The design and performance of a real-time CORBA event service, in *Proceedings of OOPSLA '97*, pp. 184–200, Atlanta, GA, 1997.

23. R. Johnson E. Gamma, R. Helm, and J. Vlissides, *Design Patterns, Elements of Reusable Object-Oriented Software*, Addison-Wesley, Reading, MA, October 1998.

24. Object Management Group, CORBA 2.2 Common Object Services specification, [Online], December 1998. Available at http://www.omg.org.

25. B. R. Woodley, H. L. Jones, E. A. LeMaster, E. W. Frew, and S. M. Rock, Carrier phase GPS and computer vision for the control of an autonomous helicopter, in *Proceedings of the Institute of Navigation GPS-96 Conference*, pp. 461–465, Kansas City, MO, September 1996.

26. J. V. R. Prasad and A. M. Lipp, Synthesis of a helicopter nonlinear flight controller using approximate model inversion, *Mathematical and Computer Modeling* **18**(3–4):89–100, August 1993.

27. R. Rysdyk and A. Calise, Adaptive model inversion flight control for tiltrotor aircraft, *Journal of Guidance, Control, and Dynamics* **22**(3):402–407, May–June 1999.

28. Demonstration AVIs of Georgia Techs Prototype OCP [Online], October 1999. Available at http://controls.ae.gatech.edu/projects/sec/archive/oct99pimeeting/reconfiguration.avi and http://controls.ae.gatech.edu/projects/sec/archive/oct99pimeeting/modeTransition.avi.

29. F. Rufus, S. Clements, S. Sander, B. Heck, L. Wills, and G. Vachtsevanos, Software-enabled control technologies for autonomous aerial vehicles, in *Proceedings of the 18th Digital Avionics Systems Conference*, St. Louis, MO, pp. 6.A.5-1 to 6.A.5-8, October 1999.

30. J. Liu, L. Xiaojun, T. Koo, B. Sinopoli, S. Sastry, and E. Lee, A hierarchical hybrid system model and its simulation, in *Proceedings of the 38th IEEE Conference on Decision and Control*, Vol. 4, pp. 3508–3513, 1999.

CHAPTER 6

REAL-TIME ADAPTIVE RESOURCE MANAGEMENT FOR MULTIMODEL CONTROL

MUKUL AGRAWAL, DARREN COFER, and TARIQ SAMAD

Editors' Notes

An important issue in real-time systems is resource management: How can processor time (and other computing resources) be allocated among the multiple tasks that have to be executed on the computing platform? If a control calculation fails to complete within a predetermined sampling interval the consequences can be catastrophic.

Today, resource management is done offline and the methods used are designed for computationally simple tasks: processing requirements are assumed fixed and known. But these assumptions are violated for implementations of the complex models and model-based algorithms that are required for high-performance control and autonomous operation. In order to realize these objectives, real-time system middleware must be developed that can accomodate tasks with complex (e.g., nondeterministic) execution properties. Examples of such tasks include multiresolution models and anytime algorithms; in either case the accuracy of the calculation can vary as more or less processing cycles are devoted to the task.

This chapter describes a software mechanism, adaptive resource management, that can enable computationally complex models and algorithms to reliably execute on real-time platforms. Precisely which processes would be invoked in a specific situation, and the relative importance accorded to each, can vary with the mission objectives, the state of the environment, and other factors that can change unpredictably at any time.

The example considered as an illustration in this chapter is online route optimization for a UAV. A route optimizer that is based on a multiscale trajectory representation and an evolutionary computing algorithm is discussed. A simulation framework has also been developed that allows the effectiveness of advanced control algorithms

Note: Research supported in part by DARPA (Defense Advanced Research Projects Agency) as part of the Software Enabled Control program, contract No. F33615-98-C-1340.

Software-Enabled Control. Edited by Tariq Samad and Gary Balas
ISBN 0-471-23436-2 © 2003 IEEE

to be studied and data gathered on their real-time behavior with respect to the adaptive utilization of computing resources.

The adaptive resource management technology described in this chapter is being incorporated within the SEC OCP, as noted in Chapters 4 and 5.

6.1. INTRODUCTION

Most control systems built today are resource-limited. This is especially true for embedded control systems in mobile platforms (whether military or commercial) due to constraints of size, weight, space, or power. Great effort is expended in engineering solutions that provide jitter-free periodic execution while meeting hard real-time deadlines in systems with high CPU (central processing unit) utilization.

The goal of our research under DARPA's Software Enabled Control program has been to develop the software mechanisms that are needed to allow higher-performance automation and control solutions to be deployed in embedded, onboard systems. It has been a continuing complaint of the controls community that sophisticated algorithms that result from research are usually relegated to the archival literature; what ultimately gets implemented is a simplified version of the real thing. The exponential growth in computing capabilities should, in principle, permit implementation of high-performance compute-intensive algorithms for time-critical applications. But the middleware and other infrastructural aspects of real-time systems cannot at present accommodate the novel execution requirements of these algorithms, and until they can do so the full benefits of Moore's Law will not be realized for real-time control.

In this environment, we maintain that a shift from a tightly integrated, highly coupled design approach to one oriented more toward quality of service guarantees and tradeoffs is appropriate. For this shift to occur, real-time platforms must incorporate methods to adapt resource utilization for computational tasks under unpredictable situations. Models and model-based algorithms will be central to this capability. A variety of models— not only representing the physical systems that are being controlled and the physical environments they are operating in, but also characterizing the computational performance of processing tasks—must be available for online interrogation. For example, the *active multimodel* real-time systems we envision will perform complex optimizations based on nondeterministic, incremental processing with variable completion times. The models and algorithms supporting these optimizations must be *resource-aware* and adapt their performance based on the computing time made available to them. The resource allocation parameters can be treated as additional independent control variables in the overall optimization problem to achieve the mission objectives.

Section 6.2 presents some motivating problems for the development and use of adaptive resource management technology. Sections 6.3 and 6.4 discuss computational issues and present mechanisms for adaptive allocation of the on-board computing resources. Section 6.5 introduces the UAV route adaptation problem that serves to identify and test the computational requirements for our work on software infrastructure. In Sections 6.6 and 6.7 we apply our computational models to the multiresolution multimodel optimization of UAV trajectories. Section 6.8 discusses the simulation framework we have developed to evaluate the performance of our control and computing models.

6.2. THE PROBLEM SPACE

Mission-critical command and control is a resource-constrained yet multifunctional enterprise, requiring simultaneous consideration of a variety of activities such as closed-loop control, measurement and estimation, planning, communication, and fault management. On the same computational platform, a large number of tasks must execute. Each may tolerate a range of computing and communication resource availability (*quality of service*) as long as a minimum lower limit is provided. The performance of a task may depend on the allocated resources, and it may be capable of trading off some application service quality to better satisfy the overall mission objectives.

Anytime or *incremental* algorithms are particularly well suited for implementing tasks that must adapt their resource usage and quality of service. In an anytime algorithm, the quality of the result produced degrades gracefully as the computation time is reduced. In particular, such algorithms may be interrupted anytime and will always have a valid result available. If more computation time is provided, the quality of the result will improve. Applications can take advantage of the flexibility that anytime algorithms offer if a mechanism is developed that can regulate their dynamic behavior.

Our objective is to enable available computing resources to be maximally employed toward achieving the highest overall mission performance. We describe dynamic resource management mechanisms that enable computing resources to be dynamically optimized in a context-aware manner. These mechanisms are intended to be used in situations where a variety of complex, scalable control algorithms (in addition to more conventional deterministic algorithms) must be coordinated given the available computing resources, the external situation, and mission objectives, all of which may change unpredictably and at any time. The adaptive resource allocation mechanisms that we are developing therefore assume that the control optimization methods provide some means of adapting their behavior.

Examples of applications that could benefit from dynamic resource management include:

* Automatic target recognition (ATR) systems of the type studied in reference 1, which rely on distributed pipelined processing.

- Intelligent mission planning software (such as the CIRCA architecture of reference 2) in which additional processing time is used to synthesize an improved control strategy.
- Integrated vehicle health monitoring software based on computationally intensive system identification methods to detect and recover from fault conditions.
- Real-time trajectory optimizers that can dynamically replan routes and trajectories (and the underlying control laws) of a fleet of UAVs [3].

6.3. RESOURCE OPTIMIZATION

The set of currently executing models and tasks will have to adapt to both internal and external triggers to make optimal use of available computational resources. Mode changes and environmental changes (new targets, bad weather) will cause changes in task criticality and computing loads. We have adopted the Real-Time Adaptive Resource Management (RTARM) technology developed under the DARPA Quorum program as a starting point for providing the required adaptation capabilities [4]. The key to adaptation and optimization using RTARM is the ability of the application itself to respond to changes in the environment, mission and system state.

Adaptation among the executing control tasks occurs according to the following steps.

1. Based on the computed or observed state, the criticality, completion deadlines, and computing requirements for the control tasks are determined. These values may be statically determined based on the mission mode (e.g., more CPU time allocated for a higher-fidelity weather model during cruise mode) or computed online by a higher-level planning system [2]. The deadlines we refer to here generally correspond to response times derived from mission-level requirements, such as the need to compute a new trajectory prior to reaching a weather system or a pop-up threat.
2. The task scheduler makes CPU computing resources available to tasks based on their criticality, computing requirements, and a schedulability analysis. Resources are measured in terms of CPU utilization, computed as execution rate × execution time per period.
3. Control tasks execute within their allotted time and are subject to preemption if they attempt to consume more than their allowed resources. Tasks adapt to meet application constraints. RTARM provides tasks with the information necessary to adapt their computation to the resources available. The application tasks may need to balance the competing demands of deadlines and accuracy, given the resources made available to them. These tasks can adapt their computation time

to meet the deadlines, or adapt the deadlines to meet the assigned computation times, by including more or fewer algorithm iterations or sensor data sources, adjusting model fidelity, and using longer or shorter planning horizons. However, the analysis performed in step 1 ensures that the most critical tasks for the current operational scenario have the highest claim on resources. One of the fundamental differences between our application adaptation model and RTARM is the responsibility for allocation of the CPU time to each task.

In principle, anytime algorithms can make effective use of any amount of processing time that is available. Execution of multiple anytime tasks on a shared computing resource where the tasks have real-time commitments to satisfy requires a nontraditional scheduling model. We have implemented a special anytime server to manage resource allocation among competing tasks. This server permits anytime tasks with temporarily high computation time (such as a terrain model used during terrain-following mode) or severe time constraints (such as a threat-tracking model used for generating an evasive maneuver) to access more of the available computing time when necessary.

The resource allocation parameters are additional independent control variables that can be changed to achieve the overall control objectives. The resource allocation for different tasks is directly controlled within the application. Algorithms are designed to achieve their results based on available computation resources, thus enabling better results for more critical control problems. This flexibility allows the system to achieve the best solution possible, given the current environmental constraints and the computing resource constraints.

6.4. ANYTIME TASK SCHEDULING

Based on the requirements identified in our multimodel control effort, we have developed a set of mechanisms for scheduling anytime control tasks. Service requirements for the anytime tasks are specified by an application-level policy task called the anytime CPU assignment controller (ACAC). The ACAC is responsible for assigning a weight to each anytime task that indicates its relative CPU assignment. This can be based on deadlines, mission scenario, or other factors. Selection of the appropriate weight is essentially a control activity that can be used to optimize overall performance.

Anytime tasks are modeled as follows:

1. They are continually executing iterative algorithms that are not periodic. Examples include algorithms that continually refine their result (imprecise computation) and that produce new outputs based on new inputs.

2. Computation times and deadlines for each iteration of the algorithm are an order of magnitude larger than the base periodic rate—for example, 1 second on a 50-Hz system.
3. The computation time for each iteration is variable and data-dependent. Furthermore, it is possible for the algorithm to adapt its computation time based on the resources allocated.

Scheduling anytime tasks opens up some interesting issues. For example, since an anytime task may be continually executing, it cannot properly be modeled as a periodic task with properties as desired by rate-monotonic analysis (RMA). In particular, such tasks will all have to be modeled as having a worst-case utilization of 1.0, thus rendering any RMA analysis meaningless. In addition, as mentioned earlier, the allocation of the CPU time is now a control function. However, the problem of designing control algorithms to assign computation time for various functional algorithms is poorly understood. In the short term, these "control" algorithms will necessarily be heuristic in nature—perhaps determining CPU allocation based on mission states (or modes).

From the perspective of the control designer, the scheduling model for allocation of CPU time is a set of control variables allocating the CPU time for each anytime task. Consider a system composed of n anytime tasks. Let P_i indicate the fractional CPU time allocated to task $i \in \{1 \ldots n\}$. Each task will get P_i fraction of the total CPU execution time allocated to all the anytime tasks. The following condition holds:

$$\sum_{i=1}^{n} P_i = 1 \qquad \forall i, P_i \geq 0$$

Alternatively, the computation time for each task may be assigned in some standard units (such as the number of floating point operations) and P_i would then be the normalized values.

Each anytime task is admitted to the system via a negotiator that can determine whether there is sufficient time in the schedule to accommodate the new task. The anytime scheduler then runs the anytime tasks for an amount of time proportional to its assigned weighting. All anytime tasks run within a fixed periodic time block allocated by the system.

It is important that anytime algorithms coexist with the periodic tasks in the control system. In order to achieve this coexistence, the anytime task scheduler executes as a periodic task within the overall control system with period T and execution time C. This periodic task is assumed to run at the system clock rate and can be modeled as a periodic task for rate monotonic analysis. The scheduler allocates a fixed fraction C/T of the overall CPU time for the use of anytime tasks, and this allocation is then subdivided based on the allocation of individual anytime tasks.

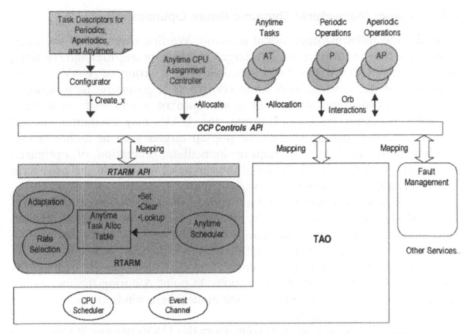

Figure 6.1. Integration of RTARM and anytime scheduling into OCP.

In the original RTARM work, task adaptation was accomplished only at the time of task admission. However, our new anytime scheduler performs continuous task adaptation during each cycle based on the current ACAC parameters. Each assignment of ACAC parameters corresponds to a "mode" on an essentially continuous spectrum of possible modes.

The anytime scheduling functionality of this software is being integrated into the RTARM code to provide the additional anytime support for the open-control platform (OCP) being developed by Boeing under the DARPA Software Enabled Control program [5] (Figure 6.1). We are using adaptive communication environment (ACE) services [6] as part of the integration, and the application programming interface (API) we are specifying for anytime scheduling is included in the OCP design document.

6.5. UAV ROUTE OPTIMIZATION

To demonstrate and evaluate our approach to dynamic resource management, we now discuss an application to in-flight UAV trajectory optimization. We first describe a "vision" for our research and then present an overview of our current architecture and technical approach.

6.5.1. Vision: Multivehicle Dynamic Route Optimization

Consider the following operational scenario. We first assume that an initial route (one that avoids threats, hits targets, and minimizes fuel and/or time, among other criteria) has been obtained for a group of UAVs at the beginning of the mission. This route consists of a geometric path between origin and destination, along with a sequence of modes such as altitude following or target tracking. Enroute, the UAVs may encounter sudden weather changes, pop-up threats, or pop-up targets such as mobile SAMs (surface-to-air missiles) that require immediate generation of optimized maneuvers in such a way as to avoid the threat or weather in the near-term and eventually get back onto the original route.

Figure 6.2a shows a route plan for a wing of three aircraft (one piloted and two UAVs) generated as part of the advance planning for a military mission. The aircraft leave a base and head for a target along a cruise segment of flight where winds-aloft data can influence the route plan. All vehicles fly in close formation, each relying on sensor data that feeds models of its peers and collective dynamics of the formation. At point A during the ingress, all three aircraft drop into terrain following mode where winds are generally not a factor, but weather can be.

At point B the piloted aircraft splits from the UAVs because it only needs to stand off from the target and identify and designate the target for the UAVs' guided weapons. The UAVs themselves then may split at point C to conduct independent bombing runs across the target. The aircraft must coordinate their timing to put the two UAVs across the target at a predetermined time spacing, and the piloted aircraft at its closest point at about the same time as the UAVs are making their bomb runs. Each vehicle uses models for its peers to estimate their flight paths and coordinate the maneuvers. Later there is a rendezvous (in space and time) at point D, followed by a return to the start point along another cruise segment.

As the aircraft execute the mission, they need to constantly reoptimize the plan to take into account variations in actual weather conditions or the appearance of disturbances not predicted during the initial planning. Figure 6.2b shows how our multimodeling technology would be used for dynamic replanning of the trajectories. The piloted aircraft encounters a small weather system after the split of the UAVs but still has to make its time mark for target identification. Additionally, a pop-up target-of-opportunity presents itself after rendezvous. At point E the pilot dispatches a UAV to destroy the target but still requires it to catch up and arrive with the wing. Finally, the winds may have changed between the times of the two cruise segments making a route replan advantageous.

Other unforeseen situations can require maneuver and route reoptimization at much shorter time scales: Avoiding an incoming missile requires extreme evasive maneuvers while damage to the vehicle may limit its dynamic capabilities for high-performance flight.

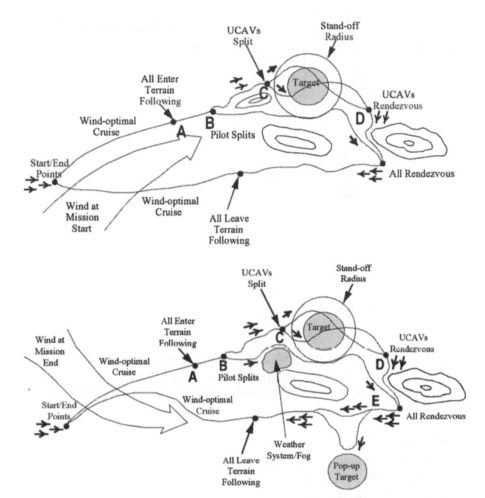

Figure 6.2. (a) Original UAV mission plan. (b) Resulting mission after dynamic replanning.

6.5.2. Dynamic Route Optimization for a Single UAV

Our work so far has focused on in-flight route optimization for a single UAV. Within this single vehicle context, we are addressing the trajectory optimization problem in two dimensions:

1. Optimization of the control solution for maneuver generation, given the available CPU resources.
2. Optimization of the allocation of CPU resources to improve the control solution produced.

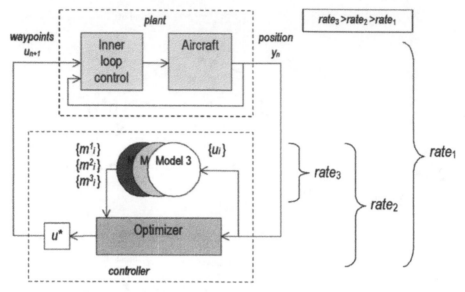

Figure 6.3. Model predictive control of aircraft trajectory.

Our emphasis is on the interaction of these two dimensions. Improved allocation of the available computing resources in the UAV allows the execution of the trajectory optimization algorithms to be adapted to include variations such as:

- More iterations
- Greater communication throughput
- A longer time horizon
- Higher-fidelity models
- More samples (higher resolution)

We are using a model predictive control framework to solve the UAV trajectory optimization problem. Figure 6.3 provides a high-level overview of the approach. The plant to be controlled consists of a stabilized aircraft with an autopilot capable of following a sequence of waypoints. Our immediate objective is to provide new waypoints as they are needed by the autopilot. In the short term, the sequence of waypoints generated must be flyable by the aircraft. Over the longer term, the trajectory defined by this sequence of waypoints must satisfy mission and safety objectives.

The control dynamics in our experimental aircraft model and autopilot require that the waypoints be provided at approximately 30-s intervals. The optimizer is responsible for updating and maintaining the current optimal trajectory u^*. To accomplish this, it is continuously evaluating many candidate trajectories $\{u_i\}$ and must therefore execute much faster than the rate at which waypoints are "consumed" as they are reached by the aircraft. The optimizer, in turn, relies on a set of models to perform its evaluations. Each model describes a different aspect of the mission (aircraft limits, target movements, terrain, etc.). In our framework, these models are themselves dynamic entities that must be executed at a rate faster than the optimizer to determine the effect $\{m_i^j\}$ of a candidate trajectory $\{u_i\}$ on the constraint or cost that they model.

The control optimization algorithms that we are using provide two main mechanisms for adaptation:

1. Multiple distinct models with different performance characteristics and resource requirements are used in the evaluation of candidate trajectories in an *active multimodel control architecture*.
2. Our main trajectory control algorithm supports *multiresolution trajectory optimization* over time and space.

6.6. APPLICATION OF ACTIVE MULTIMODEL ARCHITECTURE

The route control optimization algorithm depends on data from a variety of complex models that are continuously executed and updated during the mission. These models are *active* in the sense that each is executed as an independently schedulable task that communicates asynchronously with the consumers of its data. An active model maintains and updates state information pertaining to the data for which it is responsible. It may be able to perform its required computation using different algorithms, depending on the circumstances. Examples include predictive weather models that forecast wind conditions and threat tracking/sensor fusion models.

Traditionally, avionics systems are restricted to a fixed set of discrete operational modes, each of which specifies the tasks to be performed and the resources to be used. By contrast, we allow computational load to shift (more or less) continuously between tasks at any time. Modes can blend smoothly from one to another, and some tasks are gradually allocated less time while others are allocated more time. (Blending of state and output data during transitions must still be handled by other techniques.)

Active models are configured as schedulable tasks with a variety of service requirements that must be satisfied by the execution environment. Models specify their requirements as ranges of values within which the execution environment may optimize overall vehicle performance. For example:

- A model may be able to run at a range of rates, within a given minimum and maximum, and adapt its execution time correspondingly. *Example:* servicing a message queue.

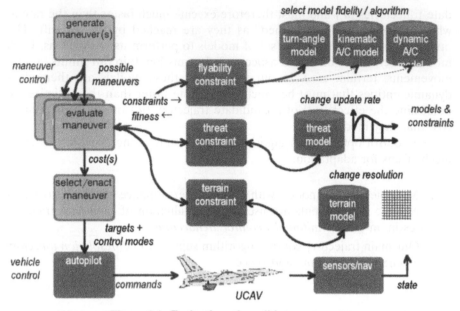

Figure 6.4. Evaluation of candidate maneuvers.

- A model may be able to run at several discrete rates. *Example:* a control loop or state estimator for target tracking that can use different tables of values for gains and time constants.
- A collection of models may provide equivalent data but by using algorithms of varying fidelity and execution time. *Example:* aircraft models for maneuver flyability may include a simple turn angle limit, a kinematic model, and a full dynamic model for simulation.

The decoupling of active models through their execution as independent asynchronous tasks has been applied in several new schemes that we are using for UAV route optimization. For example, the evolutionary optimization algorithm (see later) perturbs the current route to generate many new candidate maneuvers in its search to improve over the current route. These maneuvers are buffered for evaluation against the mission and vehicle constraints by a set of active models (Figure 6.4). The result of each evaluation is saved with the maneuver. Any maneuver found to be infeasible or in violation of one of the constraints can be eliminated immediately and not evaluated by other active models.

In different mission phases or scenarios, different constraint models may be more restrictive than others and therefore more efficient at eliminating infeasible routes. If the more restrictive model is given a larger share of the available computing resources, the optimization can proceed faster. For example, when flying in a low-altitude terrain-following mode, more time can be given to the terrain constraint model. Alternatively, a higher-fidelity model can be used to support the evaluation and enactment of more complex maneuvers.

Computational load will also tend to shift dynamically between optimization and predictive models. Another method we are using to control the shift in scheduling priority is to define a mechanism whereby the route optimization algorithm specifies a set of constraints on the data it is using for its current evaluations. As the predictive models update, they compare their data against the specified constraints and trigger a reoptimization when the constraints are violated. We have used a simple version of this technique to reduce the time allocated to the evolutionary optimization algorithm when it is no longer making significant progress.

The new model-based algorithms that we have developed for generating and evaluating UAV trajectories are anytime algorithms, processing more candidate trajectories with greater detail if more computing time is available. Rather than having periodic time constraints associated with each iteration as in traditional control applications, the deadlines for completion come from higher-level sources. As each algorithm is run, it will be informed of its deadline and time allocation and adapt its internal behavior by using different model resolution parameters.

6.7. MULTIRESOLUTION OPTIMIZATION

The need for *multiresolution trajectory optimization* can be appreciated from the following considerations. Near-term trajectories need to be computed with high accuracy using detailed models. However, it is not sufficient to address only the near term; the overall objectives of the mission—including return to base—must be considered. More distant segments of the route should be considered in the solution with progressively less fidelity. Since the environment and mission needs can change over time in ways that cannot be anticipated, we should not rely on fine-resolution details for future route projections. Computational effort expended on long-term highly detailed predictions is likely to be wasted. In a sense, the route generated should be more robust to perturbations for further out segments than for the nearer-term ones. Computing resources should be expended accordingly, with relatively more processing dedicated to near-term trajectory segments. Furthermore, one unified formulation and approach is needed so that nearer and longer-term considerations can mesh with each other.

To meet these requirements, we have developed a trajectory optimization algorithm based on a wavelet-like decomposition of candidate routes [7, 8]. Wavelets are an inherently multiresolution signal composition and decomposition tool [9]. Higher-frequency and lower-frequency components of a signal are represented separately, not just in frequency but also in the time domain. Thus high-frequency components at different times can be distinguished.

For a two-dimensional trajectory (we have not so far considered route altitude as an independent variable to be optimized) the wavelet-like representation can be expressed as

$$x(u) = \sum_{m=0}^{M_m} \sum_{n=0}^{N_m} a_{m,n} 2^{-m/2} \psi(2^{-m}u - n) + x_0(u)$$

$$y(u) = \sum_{m=0}^{M_m} \sum_{n=0}^{N_m} b_{m,n} 2^{-m/2} \psi(2^{-m}u - n) + y_0(u)$$

(6.1)

where u is a path length parameter, x and y represent the trajectory in the horizontal plane, x_0 and y_0 express the nominal trajectory (possibly computed offline or from a previous optimization), $\psi(.)$ is a basis function (a second-order B-spline function), m and n cover the range of dilations and translations (respectively) of ψ that can be accommodated over the trajectory horizon, and $a_{m,n}$ and $b_{m,n}$ are associated coefficients. For more details, including feedback aspects, see reference 8.

The important point about Eq. (6.1) is that the time/frequency components can be selectively suppressed. Thus both low- and high-frequency factors can be permitted for the short term; and for progressively further out segments of the signal (the route trajectory), more and more higher frequency coefficients can be excluded. The result is a unified representation that automatically structures a route so that it is more complex in its initial stages and increasingly smoother in its later ones.

The level of detail in the trajectory optimization over the course of the route can be adjusted for current and projected future conditions and can be updated dynamically as the situation changes. Satellite surveillance could result in new threats being identified that would affect the UAV at some future point, necessitating low-altitude terrain following in the threat region. The vehicle could dynamically update the relevant part of the trajectory, further optimizing it at higher resolution for terrain following.

We exploit the wavelet-based multimodel representation for trajectory optimization in an evolutionary computing algorithm [8] that performs a directed random search over the space of wavelet coefficients ($a_{m,n}$ and $b_{m,n}$ in Eq. (6.1)). We constrain the optimization to some subset of the basis functions in accordance with the detail required for different look-ahead horizons and the computing resources available. Typically, details of the trajectory (high-frequency components) are only considered for the immediate route duration; high-frequency coefficients associated with later route segments are fixed at zero. As each waypoint on the route is passed, the route is reoptimized starting from the current position with coefficients capturing more detail in the near term. In this way a window of detailed optimization is moved ahead of the aircraft, with a horizon of progressively less detail projecting into the future.

The objectives of route optimization for UAVs can be broadly classified into obstacle avoidance (threat, weather, and terrain avoidance) and constraint satisfaction (stay within aircraft dynamic limits, achieve mission objectives). The obstacle avoidance objective requires longer preview and can tolerate approximate vehicle dynamic models. On the other hand, the generated trajectory must be flyable. This is especially an issue in the near term where the route is represented with greater detail and is likely to be more challenging. We use the evolutionary computing algorithm to optimize the long-range behavior of the system while taking into account obstacle avoidance constraints. Different models of the vehicle performance limits are then used to verify that the route is flyable by the aircraft (i.e., it does not violate dynamic constraints imposed by aerodynamics). The resulting output of the route optimizer is then fed to the autopilot for trajectory tracking.

Multiple models, at different degrees of resolution and/or fidelity, are also needed for the vehicle (to determine what maneuvers are feasible for the vehicle to execute) and for the terrain. Similarly, multiresolution models for threats, targets, and weather also need to be incorporated. The multiresolution optimization scheme provides a seamless integration of both obstacle avoidance and flyability objectives over a range of resolutions appropriate for satisfying both short-term and long-term constraints.

Figure 6.5 shows a UAV route optimized by the evolutionary computing algorithm. Particular choices of optimizer iterations, trajectory resolution

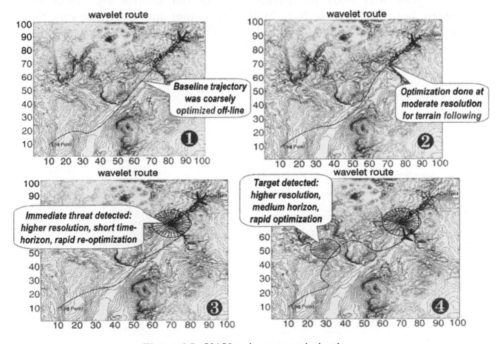

Figure 6.5. UAV trajectory optimization.

profile (basis-function coefficients to be adjusted), and model fidelity are made based on the distance to the target and the importance of other criteria (such as evading radar through low-elevation flight). Flyability constraints were represented as turning restrictions (limited lateral acceleration) for the evolutionary computing algorithm. Elevation is represented as contours in the figure. The terrain data are from north of Albuquerque, New Mexico. The resulting route is shown for four different points in the mission:

1. The UAV starts with an initial route that has been optimized offline before the start of the mission.
2. Once the mission begins, the route is reoptimized with slightly higher resolution.
3. Shortly after the start of the mission, a new threat is detected. The route is rapidly reoptimized over a short time horizon to avoid the threat.
4. Later in the mission, a target is detected. The route is again reoptimized at a higher resolution so that the target can be attacked.

6.8. SIMULATION FRAMEWORK

To integrate and study the performance of our control algorithms and software infrastructure, we have built a simulation and test framework that combines real-time scheduling and execution of control algorithms with aircraft and environmental models running in simulated time (Figure 6.6).

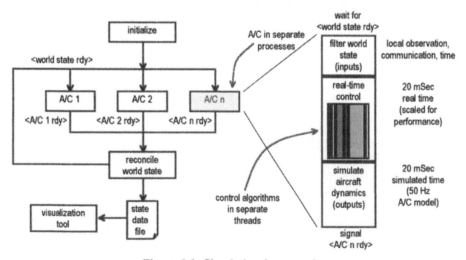

Figure 6.6. Simulation framework.

Our goal is to simultaneously study the effectiveness of advanced control algorithms in high-fidelity simulations while gathering data on their real-time behavior with respect to the adaptive utilization of computing resources.

Our simulation framework is built on the Boeing OCP and runs on an NT platform for ease of demonstration. The following capabilities are provided:

- The framework can be populated with a variety of control and computational models and algorithms.
- Control tasks (active models) run as real executable code with real-time deadlines.
- Control tasks can be structured as true, iterative, anytime algorithms and do not have to be restructured as deterministic periodic tasks.
- Control tasks are executed in real-time "slices" so that we can observe the effectiveness of adaptation.
- High-fidelity aircraft models can be used. Their computations are performed "off the clock" and interleaved with the real-time control tasks.
- Real-time performance data for the control tasks and scheduling infrastructure can be tracked independently of the (non-real-time) simulation.
- Time is scalable to simulate future hardware platforms.
- Multiple aircraft can be modeled as separate processes, using CORBA communication services for coordination. These can later be transferred to separate hardware to more closely approximate a real multiaircraft application.

In our demonstration the simulation framework contains two types of entities: a world state server and multiple aircraft servers. The demonstration shows how to run anytime tasks on a simulated aircraft CPU. In this example, the anytime tasks are route generators for an aircraft along with a task to assign their workload. An aircraft model controls and simulates the aircraft and exchanges its state with a model that tracks the "world state."

The aircraft register with the world state server, and then both will exchange their states periodically. More aircraft can be registered at any time.

The world state server repeatedly performs the following functions:

- Determine which aircraft are active.
- Wait for each aircraft to report its current state.
- Reconcile the state of all the aircraft within the world model and update the world state. This includes communications between aircraft and conflicts or collisions (between aircraft, weapons, or the ground).
- Add/remove aircraft as needed. This will eventually include the addition of weapons launched from the ground or other aircraft.

- Output the world state to a file for visualization.
- Send a world state message to each active aircraft.

Each aircraft server repeatedly performs the following functions:

- Wait for the world state message.
- Filter the world state to account for sensor limitations, communication delays, and so on.
- Execute the aircraft control software for one step in real time (currently set to 20 ms). This includes the execution of our control tasks and scheduling mechanisms. In particular, the scheduler will:
  - Determine the next task to be executed based on the ACAC parameters.
  - Resume the selected task thread.
  - Wait for the next periodic scheduler interrupt.
  - Quit when the current time step ends.
- Simulate the effect of latest control task commands by executing the aircraft dynamics model for one time step (20 ms).
- Send the aircraft state messages to the world state server.

Output data gathered by the world state server can be played back by a 3-D visualization tool.

6.9. CONCLUSION

As part of our work on multimodel control for UAVs, we are investigating new methods for trajectory optimization. Our objective is to develop advanced control and computational models along with the necessary software infrastructure to fully exploit available onboard computing resources for optimizing the performance of UAV missions. We have developed several new model-based algorithms for generating and evaluating aircraft trajectories. These are anytime algorithms that can be adapted to run at varying levels of fidelity, with varying requirements for computing resources. We have produced a nontraditional task scheduling mechanism to manage competing anytime tasks. In our control model, the resource allocation parameters used by the scheduler are additional independent control variables that can be changed to achieve the overall control objectives.

We have implemented these concepts in a simulation framework that allows us to simultaneously study the effectiveness of advanced control algorithms in high fidelity simulations while gathering data on their real-time behavior with respect to the adaptive utilization of computing resources. Our current demonstration includes multiple aircraft models, interleaving of real-time control tasks with aircraft simulation, and multiresolution trajectory optimization. The aircraft software includes inner and outer loop controllers and a waypoint-following algorithm. Results of the simulation include scaled performance data and a 3-D animation of the aircraft undertaking real-time trajectory reoptimization.

REFERENCES

1. R. Jha, M. Muhammad, S. Yalamanchili, K. Schwan, D. Rosu, and C. deCastro, Adaptive resource allocation for embedded parallel applications, in *Proceedings of the 3rd International Conference on High Performance Computing*, December 1996.
2. D. Musliner, R. Goldman, M. Pelican, and K. Krebsbach, SA-CIRCA: Self-adaptive software for hard real-time environments, *IEEE Intelligent Systems* **14**(4):23–29, 1999.
3. W. Koenig, D. Cofer, D. Godbole, and T. Samad, Active multi-models and software enabled control for unmanned aerial vehicles, in *Proceedings of the Association of Unmanned Vehicle Systems International*, Baltimore, MD, July 1999.
4. J. Huang, R. Jha, W. Heimerdinger, M. Muhammad, S. Lauzac, B. Kannikeswaran, K. Schwan, Wei Zhao, and R. Bettati, RT-ARM: A real-time adaptive resource management system for distributed mission-critical applications, in *Proceedings of IEEE Workshop on Middleware for Distributed Real-Time Systems and Services*, December 1997.
5. B. Doerr and D. Sharp, Freeing avionics software product lines from execution dependencies, *Software Technology Conference*, May 1999.
6. D. Schmidt The ADAPTIVE Communication Environment: Object-Oriented Network Programming Components for Developing Client/Server Applications, in 12th Annual Sun Users Group Conference, pp. 214–255, San Francisco, CA, 1994.
7. D. Godbole, T. Samad, and V. Gopal, Active multi-model control for dynamic maneuver optimization of unmanned air vehicles. *Proceedings IEEE International Conference on Robotics and Automation*, San Francisco, 2000.
8. T. Samad, D. Gorinevsky, and F. Stoffelen, Dynamic route optimization for uninhabited aerial vehicles, in *Proceedings IEEE International Symposium on Intelligent Control*, Mexico City, 2001.
9. Y. Meyer, *Wavelets: Algorithms and Applications*, SIAM Publications, Philadelphia, 1993.

CHAPTER 7

HETEROGENEOUS MODELING AND DESIGN OF CONTROL SYSTEMS

XIAOJUN LIU, JIE LIU, JOHAN EKER, and EDWARD A. LEE

Editors' Notes

Complex control systems integrate a variety of functions and capabilities, which will in general rely on different computational mechanisms. The plant model may be represented as a set of ordinary differential equations, the mode switching logic may be expressed as a finite state machine, and dataflow models may be used to capture the architecture of a sensor processing subsystem, for example. Design tools are needed that can support these heterogeneous models of computation—and their integration within a single control system.

This chapter describes Ptolemy II, a component-based design environment that allows different models of computation to be hierarchically composed. Individual components are called actors; these can include simple operators such as an AND gate and, through compositionality in the form of arbitrarily nested actor hierarchies, complex functions such as a Kalman filter. Different models of computation can then be realized by imposing a (possibly partial) execution order and a communiation mechanism on the actors that comprise a composite actor.

Ptolemy II can be particularly useful for the design of hybrid dynamical systems—systems that combine discrete event mode logic and continuous time dynamics within each mode. By using Ptolemy II to model a hybrid system, we can choose whichever model of computation is most appropriate for a particular task while ensuring the consistency and integrity of the overall model.

An application of Ptolemy II to the control of the Furuta inverted pendulum is also described in this chapter, with an emphasis on implementation issues relevant to real-time operating systems. As noted in Chapter 4, Ptolemy II has been interfaced with the OCP and used to model a nonlinear actuator as part of an overall vehicle simulation. Chapter 5 also notes an OCP-related application of the Ptolemy project, in this case to support the validation of reconfiguration strategies and other hybrid control aspects.

Software-Enabled Control. Edited by Tariq Samad and Gary Balas
ISBN 0-471-23436-2 © 2003 IEEE

7.1. INTRODUCTION

Computer control is now the standard technique for implementing control systems, mainly for two reasons. First is the exponential reduction in the cost of computing; second is the versatility of implementing control laws in software. Many developments in control systems are only practical with computer control—for example, to implement the nonlinear and time-varying control laws associated with adaptive control. Complicated computations can be incorporated into the control loop—for example, when computer vision is used to guide a robot.

Designing the software for such control systems is hard because the systems are usually *heterogeneous*. They may include subsystems with very different characteristics, such as hydraulic actuators and an inertial navigation system. On the software side the situation is similar. The controller may have several operational modes. The control law in each mode can be specified by difference equations; the mode-switching logic can be specified by a state machine. For vision guidance, complex image processing algorithms need to be programmed.

For each of these subsystems and aspects of the software, formal models that support its modeling, analysis, or programming have been developed. For example, image processing algorithms can be programmed in various dataflow models [1, 2]. Each formal model employs a computational mechanism that dictates what are the components in the model, and how they communicate and execute. Such a mechanism is called a *model of computation.*

Working with heterogeneous systems requires more than one model of computation. This is evident from the trend of adding extensions to existing tools and formal models. For example, both VHDL and Verilog, originally designed for digital circuits and based on the discrete-event model, have been extended to handle analog components [3, 4]. Simulink, a continuous-time environment, has been extended with Stateflow [5] for modeling and designing event-driven systems. Ideal switching elements, controlled by finite-state machines, are added to bond graphs for modeling hybrid systems [6]. However, most of these tools and formal models support just a few models of computation and few choices in the way they can be combined. Further extensions may be awkward or impossible due to the semantic mismatch between the new model of computation and the existing infrastructure.

Ptolemy II [7] is a system-level design environment that supports component-based heterogeneous modeling and design. Its model structure allows a variety of models of computation to be implemented, and to be hierarchically composed in heterogeneous models. This chapter presents Ptolemy II and illustrates its application to control system design. We use several case studies in Section 7.2 to elaborate the challenges in designing complex control software. The Ptolemy II model structure is discussed next. Section 7.4 gives an overview of the models of computation that are useful in control system design. The Ptolemy II modal model structure is presented in Section 7.5. An inverted pendulum controller is used to demonstrate how Ptolemy II can be used in the modeling and design exploration of control systems. In the final section we present conclusions and discuss our ongoing work.

7.2. SOFTWARE COMPLEXITY IN CONTROL SYSTEMS

The use of computers in control systems started in the 1950s. In the 1980s, computer control became the standard technique for implementing new control systems, from simple single-loop controllers to large distributed control systems [8]. The versatility of implementing control laws in software brings many opportunities to control system design. For example, a proportional-integral-derivative (PID) controller can come equipped with automatic tuning and gain scheduling. As another example, to achieve better performance, a controller can be designed to switch among a set of candidate control laws according to the operating region of the controlled process. Such developments bring about increased complexity in control software. As the following cases from the theory and applications of control systems will demonstrate, the capability to build complex and reliable software has become a key enabler of further developments in computer-controlled systems.

Vision-Guided Landing of Unmanned Aerial Vehicles (UAVs) [9]. A UAV equipped with a video camera is to land on a moving landing platform (e.g., the landing pad on a ship). The UAV control system uses computer vision as a sensor in the feedback control loop. The images captured by the camera are processed and relevant features in the field of vision are extracted. The extracted features are further processed by a computer vision algorithm to estimate the motion of the UAV relative to the landing platform. The control software has to perform complex image transformations and analyses in real time.

Model-Based Fault Diagnosis [10]. The goal of fault diagnosis is to detect and isolate faults in physical processes. In the model-based approach to fault diagnosis, a process model is used to predict normal process behavior. Faults are detected when observed process behavior deviates from normal behavior. Based on the deviation, one or more hypothesized faults can be generated for fault isolation. The hypothesized faults are injected into the process model to predict future behavior. The result of fault isolation consists of those faults whose predictions are consistent with the observations. Elaborate process models are needed to achieve greater resolution and coverage in fault diagnosis. When we build a control system that uses model-based fault diagnosis, the software for process modeling and simulation is not only an essential design-time tool, but also a crucial component in the deployed system. A desirable feature in such software is the support of dynamic model modification for fault injection.

Multimodal Control [11]. Many controlled systems have multiple modes of operation. Consider the flight of a helicopter. Each possible maneuver—hover, turn, vertical climb, and so on—corresponds to a mode of operation. To optimize performance, each mode has its own closed-loop feedback controller. The helicopter flight management system can be structured in layers—for example, a trajectory planner layer and a regulation layer. Given a flight task—for example, to fly to a certain location, drop the load, and fly back—the trajectory planner comes up with a sequence of flight modes (maneuvers), the set points for the controller of each mode, and ending conditions. For example, one maneuver in the sequence may be to accelerate horizontally, with the ending condition that the horizontal velocity reaches 150 km/h. The regulation layer switches controllers according to the flight sequence and ending conditions. The software for the regulation layer can be very well structured using the hybrid system formalism [12]. The operation modes and switching among modes are captured by a finite-state machine (FSM). Each mode is represented by a state that contains the controller of that mode.

Using FSMs in a hierarchical model was first made popular by Harel. He proposed *Statecharts* [13], which combine hierarchical FSMs and concurrency. The proposal stimulated many developments in both theory and applications. A recent development, **charts* (pronounced *star charts*) [14], generalizes and unifies Statecharts and hybrid systems.

Embedded Control Systems. Similar to what happened in control engineering, computer technology has been extensively applied to many application domains and opens up many exciting opportunities. For example, the concept of "real-time" enterprises has recently been proposed [15]. In such an enterprise, all the information that is relevant to business decision-making, from inventory to cash flow, is made available at the click of a mouse, not just on a weekly or monthly basis. The enterprise can adapt better to the rapidly changing marketplace. This is enabled by the use of computer and Internet technologies in every aspect of enterprise management. Many computer-control systems will be integrated into a larger context. In this vision, the process control system of a petrochemical plant will become a component of the production management system that also manages the inventory of raw materials and end products and schedules production according to supply and demand. Such integration requires a sound strategy for interface design and abstraction. It is already a hard problem to integrate software systems in the same application domain but from different vendors. Integration across application domains can only be more challenging.

From this brief survey, how to manage heterogeneity emerges as the key question to be answered when designing complex control software. To further illustrate the notion of heterogeneity, let us consider a flight management system of an unmanned helicopter. The system employs multimodal control, fault diagnosis based on a model of the helicopter dynamics, and vision-guided landing. The controllers of some flight modes may be described by difference equations. In the landing mode, the controller needs to perform image transformations that are best programmed using dataflow languages and models. The dynamics model is simulated, possibly by numerically solving ordinary differential equations, to predict the state of the helicopter. An FSM captures the mode-switching logic. Multimodal control and fault diagnosis are assigned to different tasks in a real-time operating system (RTOS). Such a control system is heterogeneous in that its subsystems have very different characteristics. Their components may interact by synchronous rendezvous or asynchronous event notification, may execute sequentially or in parallel, or may communicate via continuous-time signals or streams of data. A disciplined approach is indispensable when composing heterogeneous systems from diverse subsystems.

7.3. THE PTOLEMY II MODEL STRUCTURE

The Ptolemy II modeling and design environment [7] uses a component-based design methodology that is consistent with component-based techniques used in object-oriented design [16]. The components in a Ptolemy II model are called *actors*. A model is a hierarchical composition of actors, as shown in Figure 7.1. The *atomic* actors, such as A1, only appear at the bottom of the hierarchy. Actors that contain other actors, such as A2, are *composite*. A composite actor can be contained by another composite actor, so the hierarchy can be arbitrarily nested.

Atomic actors encapsulate basic computation, from as simple as an AND gate to more complex such as a fast Fourier transform (FFT). Through composition, actors that perform even more complex functions can be built. Actors have *ports*, which are their communication interfaces. For example, in Figure 7.1, A5 receives data from input ports P3 and P4, performs its computation, and sends the result to output port P5. A port can be both an

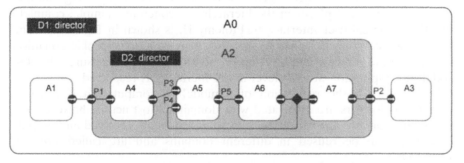

Figure 7.1. The schematic of a hierarchical Ptolemy II model.

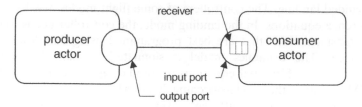

Figure 7.2. A receiver is used to mediate communication between actors.

input and an output. Communication channels among actors are established by connecting ports. A port of a composite actor, such as P1, can have connections both to the inside and to the outside, thus linking inside actors to outside actors.

There is one last part in Figure 7.1 that we have yet to explain: the *directors*. As the names "actor" and "director" suggest, a director controls the execution order of the actors in a composite and mediates their communication. In Figure 7.1, D1 may choose to execute A1, A2, and A3 sequentially. Whenever A2 is executed, D2 takes over and executes A4–A7 accordingly. A director uses *receivers* to mediate actor communication. As shown in Figure 7.2, one receiver is created for each communication channel; it is situated at the input ports, although this makes little difference. When the producer actor sends a piece of data, called a *token* in Ptolemy II, to the output port, the receiver decides how the transaction is completed. It may put the token into a first-in first-out (FIFO) buffer, from which the consumer actor will get data. It may tag the token as an event and put the event in a global event queue. The token will be made available to the consumer when the time comes for the consumer to process the event. Or it may stall the producer to wait for the consumer to be ready.

By choosing an ordering strategy and a communication mechanism, a director implements a model of computation. Within a composite actor, the actors under the immediate control of a director interact homogeneously. Properties of the director's model of computation can be used to reason about the interaction. A heterogeneous system is modeled by using multiple directors in different places in the hierarchy. A concrete example, complete with the graphical user interface to Ptolemy II, is shown in Figure 7.3. The directors are carefully designed so that they provide a polymorphic execution interface to the director one level up in the hierarchy. This ensures that the model of computation at each level in the hierarchy is respected.

In Ptolemy II, the realization of a model of computation is called a *domain*, so directors are associated with domains. Most actors, however, are not. Such actors are agnostic about how their inputs are received and outputs sent. They can be reused in different domains and are called *domain-polymorphic*.

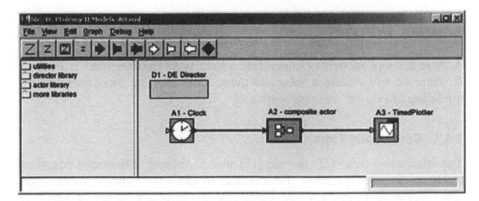

(a)

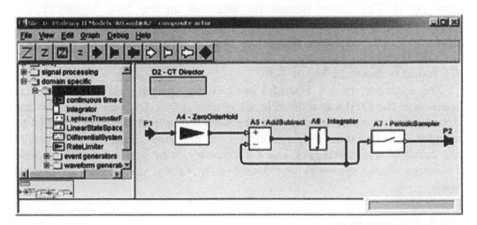

(b)

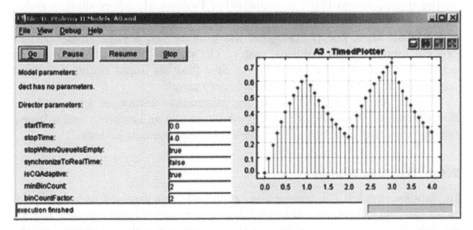

(c)

Figure 7.3. A heterogeneous model realizing the schematic shown in Figure 7.1. (a) Top level, a discrete-event model. (b) A continuous-time model embedded in the above discrete-event model. (c) The window that controls the simulation of the model and displays the result.

7.4. CONCURRENT MODELS OF COMPUTATION FOR CONTROL SYSTEMS

A diverse set of models of computation have been implemented in Ptolemy II. Here we will discuss a subset of them, those that are most useful in the modeling and design of control systems.

7.4.1. Continuous Time

The continuous-time (CT) domain [17] models ordinary differential equations (ODEs), extended to allow the handling of discrete events. Special actors that represent *integrators* are connected in feedback loops in order to represent the ODEs. Each connection in this domain carries a continuous-time signal. The actors denote the relations among these signals. Event generators (e.g., periodic samplers, triggered samplers, and zero-crossing detectors) and waveform generators (e.g., a zero-order hold) are implemented to convert between continuous-time signals and discrete events. A CT model is shown in Figure 7.3b.

The execution of a CT model involves the computation of a numerical solution to the ODEs at a discrete set of time points. In order to support the detection of discrete events and the interaction with discrete models of computation, the time progression and the execution order of a CT model are carefully controlled [17]. The CT domain can be used to model physical processes whose dynamics are described by ODEs, or continuous control laws.

7.4.2. Discrete Event

In the discrete-event (DE) domain [18] of Ptolemy II, actors share a global notion of time and communicate through events that are placed on a (continuous) time line. Each event has a value and a time stamp. Actors process events in chronological order. The output events produced by an actor are required to be no earlier in time than the input events that were consumed. In other words, DE models are *causal*.

Discrete event models, having the continuous notion of time and the discrete notion of events, are widely used in modeling hardware and software timing properties, communication networks, and queuing systems.

7.4.3. Dataflow Models

In dataflow models [2], connections represent data streams and actors are processes that compute their output data streams from input streams. In such models, the order of execution for the processes is only constrained by the data dependency among them. This makes dataflow models amenable to optimized execution (for example, to minimize buffer sizes) or to achieve a higher degree of parallelism. Dataflow models are very useful in designing signal processing algorithms and sampled control laws.

There are many variants of dataflow models, of which synchronous dataflow (SDF) [1] is a particularly restricted special case. In SDF, when an actor executes, it consumes a fixed number of tokens from each input port and produces a fixed number of tokens to each output port. For a consistent SDF model, a static schedule can be computed, such that the actors always have sufficient data before execution. For algorithms with a fixed structure, SDF is very efficient and predictable.

7.4.4. Timed Multitasking

The timed multitasking (TM) domain in Ptolemy II allows designers to explore priority-based scheduling policies such as those found in an RTOS and their effects on real-time software. In this domain, actors are software tasks with priorities. The director of this domain implements a prioritized event dispatching mechanism and invokes tasks according to their feasibility and priority. Both preemptive and nonpreemptive scheduling, as well as static and dynamic priority assignment, can be modeled.

7.4.5. Synchronous / Reactive

In the synchronous/reactive (SR) model of computation [19], the connections represent signals whose values are aligned with global clock ticks. Thus, they are discrete signals, but need not have a value at every clock tick. The actors represent relations between input and output values at each tick, and they are usually partial functions with certain technical restrictions to ensure determinacy. Examples of languages that use the SR model of computation include Esterel [20] and Signal [21]. SR models are excellent for discrete control applications with multiple, tightly-coupled, and concurrent tasks. Because of the tight synchronization, safety-critical real-time applications are a good match.

7.4.6. Finite-State Machines

An FSM has a set of states and transitions among states. An FSM reacts to input by taking a transition from its current state. (The transition may be an implicit transition back to the current state.) Output may be produced by actions associated with the transition. FSMs are very intuitive models of sequential control logic and the discrete evolution of physical processes. FSM models are amenable to in-depth formal analysis and verification. Applications of FSM models include datapath controllers in microprocessors and communication protocols.

In Ptolemy II, the FSM domain provides two modeling mechanisms. One allows designers to create actors whose behaviors are specified by FSMs. We can think of this as a graphical scripting language for writing new actors. The other one applies to *modal models*, which are hierarchical composition, of FSMs with other models of computation.

7.5. MODAL MODELS

Many engineering systems exhibit modes of operation. We call such systems *modal systems*. Defining operational modes is a very useful instrument of abstraction, which helps us to gain a high level understanding of system operation. In control engineering, a modal system operates with continuous dynamics in each mode. When the mode changes, the continuous dynamics change abruptly [22]. Such modal systems are more specifically called *hybrid systems*. One example is a multitank system, a common experimental platform in control engineering. A modal controller, discussed in Section 7.2, is also an example.

A number of approaches to the modeling and simulation of hybrid systems have been proposed [22]. Discrete variables can be introduced into dynamic equations to model the changes in system dynamics. Another possibility is to use discrete components in an otherwise continuous model—for example, the ideal switch in switched bond graphs [26]. These approaches merge the discrete mode changes and continuous dynamics into one form (one set of equations or one monolithic model). The results are usually compact, but capturing all possible system configurations in one form may make it hard to understand. These approaches often do not give explicit representations to operational modes.

In Ptolemy II, the model structure naturally dictates how a hybrid system is to be modeled [12]. The continuous dynamics of each mode is captured by a CT model (a composite actor with a CT director). The discrete mode changes are modeled by an FSM. A modal model actor contains the CT models of all modes and the FSM. Each state of the FSM represents a mode and has as refinement the CT model of that mode. This modal model matches very well the hybrid I/O automata [23]. The schematic of a modal model in Ptolemy II is shown in Figure 7.4. A process is controlled by a modal controller. There are two actors in the top-level CT model. One actor models the process dynamics; the other is a modal model actor. Inside this actor is a two-state FSM. From this FSM, we know that the modal controller employs two alternative control laws, which are modeled by controllers A and B. The conditions for mode changes are annotated on the transitions between states. The conditions are expressed as predicates p and q on the process state y.

The model in Figure 7.4 clearly demonstrates the benefits of orthogonalizing concerns. Modeling discrete mode changes with FSMs yields an easy to understand summary of system operation. In each mode, we can deal with

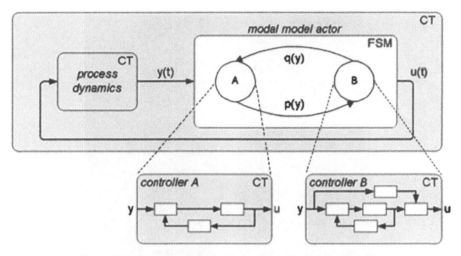

Figure 7.4. The schematic of a modal model in Ptolemy II.

pure continuous dynamics. This is possible in Ptolemy II because of the variety of models of computation supported. When modeling a heterogeneous system, we can choose whichever model of computation best fits with the aspect or subsystem we are working on. Because of the disciplined hierarchical composition of different models of computation, their properties are preserved in a heterogeneous model.

Given the Ptolemy II infrastructure, it is natural to generalize modal models from hybrid system modeling to *charts [14], which allow the heterogeneous hierarchical composition of FSMs and other models of computation. For example, combining FSMs with the synchronous/reactive model of computation yields Statecharts-like models, which are widely used in designing complex discrete control logic. If we combine FSMs with process networks [2], the semantics is similar to that of the Specification and Description Language (SDL) [24], which is widely used in the telecommunications field.

Modal systems from many application domains can be modeled cleanly with *charts. A particularly useful special case of *charts is the composition of FSM and SDF, which we call *heterochronous dataflow* (HDF) [14]. As discussed in Section 7.4.3, SDF models have very desirable properties but only apply to algorithms with a fixed structure. Many components in signal processing systems need to adapt their algorithms to changing environments. For example, in a wireless handset, a simple channel equalization algorithm may suffice when the interference level is low. A more sophisticated algorithm will be used when the interference level becomes high. Such switching of algorithms can be captured by an HDF model. In each mode, the whole signal processing system can still be treated as a hierarchical SDF model.

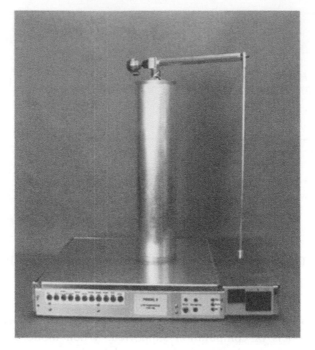

Figure 7.5. The Furuta inverted pendulum.

7.6. APPLICATION: INVERTED PENDULUM CONTROLLER

In this section, we illustrate how the Ptolemy II environment supports the modeling, simulation, and design exploration of control systems, using an inverted pendulum controller as an example. (A more detailed model, a helicopter controller, is described in reference 17.) Controlling the inverted pendulum is a classical problem in control laboratories because the pendulum dynamics is both nonlinear and unstable. The experimental setup is shown in Figure 7.5. The pendulum consists of two moving parts: (1) the arm that rotates in the horizontal plane and (2) the pendulum that moves in the vertical plane. The states of the pendulum process are the angle and angular velocity of the pendulum, and those of the arm. The process is controlled by the acceleration of arm rotation. Complete equations describing the process dynamics can be found in reference 25.

A modal controller can be designed to swing up the pendulum and stabilize it in the upright position. The controller has three modes of operation: a swing-up mode, a catch mode, and a stabilize mode. The swing-up controller moves the pendulum from its initial downward position toward the upright position, using a nonlinear energy-based algorithm. When the pendulum comes close enough to the upright equilibrium, the catch controller takes over. The task of the catch mode is to reduce the speed of the pendulum and

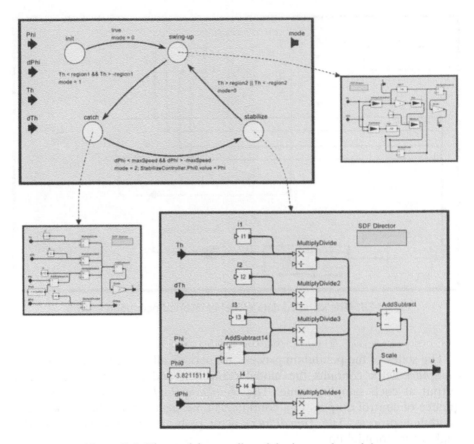

Figure 7.6. The modal controller of the inverted pendulum.

the arm before the stabilize mode is entered. The catch and stabilize controllers use linear state feedback.

The Ptolemy II model of the modal controller is shown in Figure 7.6. The mode switching is modeled by the FSM. (The *init* state of the FSM serves no purpose other than to produce information about the initial mode of the controller.) On the left are the inputs to the FSM, which are the measured states of the pendulum process. For each mode, the computation required by the control law is fixed, and it is modeled with SDF.

When the modal controller is implemented on a computer that controls the pendulum, the whole system becomes a sampled-data system. The Ptolemy II model of such a system is shown in Figure 7.7. The *PendulumDynamics* actor is an instance of the *DifferentialSystem* actor in the continuous-time actor library, which can model dynamic equations of the form

$$\frac{d\bar{x}}{dt} = f(\bar{x}, \bar{u}, t), \qquad \bar{y} = g(\bar{x}, \bar{u}, t), \qquad \bar{x}(0) = \bar{x}_0$$

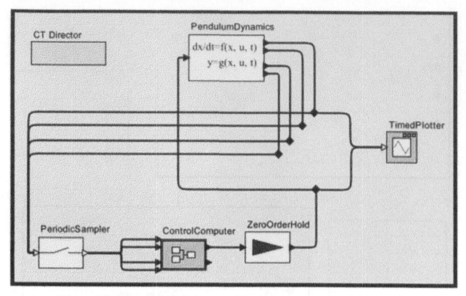

Figure 7.7. The sampled-data model of inverted pendulum control.

The states of the pendulum process are periodically sampled. The *Control-Computer* actor contains the modal controller, and it computes a control output at each sampling time. The *ZeroOrderHold* actor converts the sequence of control outputs to a continuous-time signal that is fed back to the pendulum process. The *TimedPlotter* actor plots the control signal and the angle of the pendulum.

The sampled-data model in Figure 7.7 helps us to verify with simulation that the control laws are effective in a sampled-data system. This is only a first step in the design process from control law specifications to the final implementation, because the model does not capture many issues present in an actual implementation. For example, the model treats the computation in the control computer as taking zero time, and there is no communication delay in the control loop. In Ptolemy II, we can elaborate this model to explore a number of such issues, as shown in Figure 7.8.

Control software today is often implemented on top of an RTOS. In such a realization, the controller runs as a task in the RTOS and competes with other tasks for resources (e.g., CPU time.) This may cause jitter in the input-to-output delay and even changes in the sampling period. We can model these effects with the TM domain in Ptolemy II. As shown in Figure 7.8, the control computer is modeled as a composite actor with a TM director. The modal controller actor is treated as a task by the TM director. Some other tasks, such as one for fault diagnosis and one for I/O, may be added to reflect the dynamics of concurrent tasks running in the same RTOS. With such a model, we can evaluate various scheduling mechanisms in terms of control performance.

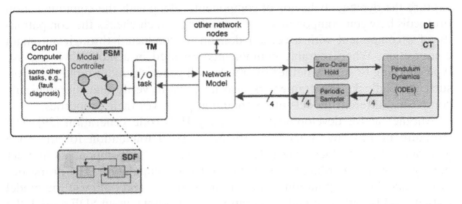

Figure 7.8. An elaborate Ptolemy II model that incorporates many implementation issues.

The computers, sensors, and actuators in a control system are often networked. We can model this by using the discrete event domain at the top level. The pendulum model and the control computer are both connected to a network model actor, along with other network nodes to reflect the contention of network resources. The network model simulates the communication among the nodes in the control system, taking into account contention, transmission delay and loss, and so on. Such a model helps the designers to evaluate different network architectures and protocols.

7.7. CONCLUSION

The modeling and design of complex control systems require tools that support multiple models of computation, along with a software architecture to compose heterogeneous systems. The Ptolemy II environment has a component-based hierarchical structure that meets these requirements. Components in Ptolemy II have a fine-grained interface for execution control, which allows directors to compose the execution of components according to various models of computation. All directors provide the same execution interface to the outside domain, so that when different models of computation are composed in a heterogeneous model, their properties are preserved.

Ptolemy II includes a number of domains that are useful in control system design. Other pertinent models of computation are being implemented, such as Petri nets and port-based objects [26]. With modal models, Ptolemy II provides a clean structure for studying modal systems that are common in control engineering.

A framework for studying the dynamic interaction among actors, receivers, and directors is being developed. The framework [27] extends the concept of type systems in programming languages. Interface automata [28] are used to capture the dynamic behavior of components, along with the communication protocols between components. Type checking, which checks the compatibility of a component with a certain domain, is conducted through automata composition. When reusable components are used in a model, we can use type checking to study whether the model will generate undesirable behavior, such as deadlock. The framework also helps us to verify that hierarchically composing different models of computation preserves their properties.

Another active development in Ptolemy II is code generation [29]. The structure of Ptolemy II models provides a good foundation for the code generator. At each hierarchical level of the model, the components interact homogeneously according to a specific model of computation. The properties of the model of computation can be used to analyze and optimize the model at that level for efficient code generation. For example, in an SDF model, the actors can be statically scheduled. It is possible to merge the functions of a cluster of actors in the schedule, so as to reduce the number of function calls in the generated code. Different models of computation offer different optimizations, which can be applied orthogonally in a heterogeneous model.

The URL of the Ptolemy project homepage is

http://ptolemy.eecs.berkeley.edu/

The Ptolemy II software can be downloaded from there, complete with source code and design documentation. Online demonstrations can be viewed with an applet-enabled Internet browser. Most publications from the Ptolemy group are made available in electronic form.

ACKNOWLEDGMENTS

This work is part of the Ptolemy project, which is supported by DARPA, the State of California MICRO program, and the following companies: Agilent Technologies, Cadence Design Systems, Hitachi, National Semiconductor, and Philips.

REFERENCES

1. E. A. Lee and D. G. Messerschmitt, Synchronous data flow, *Proceedings of the IEEE* **75**(9):1235–1245, September 1987.
2. E. A. Lee and T. M. Parks, Dataflow process networks, *Proceedings of the IEEE* **83**(5):773–801, May 1995.
3. IEEE Computer Society, *IEEE Draft Standard VHDL-AMS Language Reference Manual*, 1997.
4. OVI, *Verilog-AMS Language Reference Manual, Analog & Mixed-Signal Extensions to Verilog HDL*, Version 1.1, March 13, 1998.
5. The MathWorks, Inc., *Stateflow User's Guide*, May 1997.

6. U. Soderman and J.-E. Stromberg, Switched bond graphs: Towards systematic composition of computational models, *International Conference on Bond Graph Modeling and Simulation (ICBGM'95)*, Simulation Series, Vol. 27, SCS Publishing, 1995, pp. 73–79.

7. J. Davis II, C. Hylands, B. Kienhuis, E. A. Lee, J. Liu, X. Liu, L.Muliadi, S. Neuendorffer, J. Tsay, B. Vogel, and Y. Xiong, Heterogeneous Concurrent Modeling and Design in Java, *Technical Memorandum UCB / ERL M01 / 12*, EECS, University of California, Berkeley, March 15, 2001 (http://ptolemy.eecs. berkeley.edu/publications/papers/01/HMAD/).

8. K. J. Astrom and B. Wittenmark, *Computer-Controlled Systems*, 2nd ed., Prentice-Hall, Englewood Cliffs, NJ, 1990.

9. C. S. Sharp, O. Shakernia, and S. Sastry, A vision system for landing an unmanned aerial vehicle, in *International Conference on Robotics and Automation*, Seoul, Korea, May 2001.

10. S. Narasimhan, F. Zhao, G. Biswas, and E. Hung, Fault isolation in hybrid systems combining model-based diagnosis and signal processing, in *Proceedings of IFAC 4th Symposium on Fault Detection, Supervision, and Safety for Technical Processes*, Budapest, Hungary, 2000.

11. T. J. Koo, G. J. Pappas, and S. Sastry, Multi-modal control of systems with constraints, *IEEE Conference on Decision and Control*, Orlando, FL, December 2001.

12. J. Liu, X. Liu, T. J. Koo, B. Sinopoli, S. Sastry, and E. A. Lee, A hierarchical hybrid system model and its simulation, in *38th IEEE Conference on Decision and Control (CDC'99)*, Phoenix, Arizona, December 1999.

13. D. Harel, Statecharts: A visual formalism for complex systems, *Science Computer Programming* **8** 231–274, 1987.

14. A. Girault, B. Lee, and E. A. Lee, Hierarchical finite state machines with multiple concurrency models, *IEEE Transactions on Computer-aided Design of Integrated Circuits and Systems* **18**(6):00–00, June 1999.

15. KnowNow Inc., Vision: Powering the Real-Time Enterprise (http://www.know-now.com/ products/whitepapers/powering.html), 2001.

16. C. Szyperski, *Component Software: Beyond Object-Oriented Programming*, Addison-Wesley, Reading, MA, 1998.

17. J. Liu and Edward A. Lee, Component-based hierarchical modeling of systems with continuous and discrete dynamics, *IEEE Symposium on Computer-Aided Control System Design (CACSD'00)*, Anchorage, Alaska, September 2000.

18. J. Banks, J. S. Carson, B. L. Nelson, and D. M. Nicol, *Discrete Event System Simulation*, Prentice-Hall, Upper Saddle River, NJ, 2000.

19. S. A. Edwards, The Specification and Execution of Heterogeneous Synchronous Reactive Systems, Ph.D. thesis, University of California, Berkeley, May 1997. Available as UCB/ERL M97/31 (http://ptolemy.eecs.berkeley.edu/papers/97/ sedwardsThesis/).

20. G. Berry and G. Gonthier, The Esterel synchronous programming language: Design, semantics, implementation, *Science of Computer Programming* **19**(2):87–152, 1992.

21. A. Benveniste and P. Le Guernic, Hybrid dynamical systems theory and the SIGNAL language, *IEEE Transactions on Automatic Control* **35**(5):525–546, May 1990.

22. P. J. Mosterman, An overview of hybrid simulation phenomena and their support by simulation packages, *Hybrid Systems: Computation and Control '99*, volume 1569 of Lecture Notes in Computer Science, pp. 165–177, 1999.

23. N. Lynch, R. Segala, F. Vaandrager, and H. B. Weinberg, Hybrid I/O automata, *Hybrid Systems III*, number 1066 in LNCS, pp. 496–510, Springer-Verlag, New York, 1996.

24. F. Belina, D. Hogrefe, and A. Sarma, *SDL with Applications from Protocol Specification*, Prentice-Hall, Englewood Cliffs, NJ, 1991.

25. J. Akesson, Safe Manual Control of Unstable Systems, (Master's thesis ISRN LUTFD2/TFRT-5646-SE, Department of Automatic Control, Lund Institute of Technology, Lund, Sweden, September 2000.

26. D. B. Stewart and P. Khosla, Chimera methodology: Designing dynamically reconfigurable real-time software using port-based objects, *Proceedings of the Workshop on Object-Oriented Real-Time Dependable Systems (WORDS)*, pp. 46–53, 1995.

27. E. A. Lee and Y. Xiong, System-level types for component-based design, *First Workshop on Embedded Software, EMSOFT2001*, Springer-Verlag, Lake Tahoe, CA, October 8–10, 2001.

28. L. de Alfaro and T. A. Henzinger, Interface automata, to appear in *Proceedings of the Joint 8th European Software Engineering Conference and 9th ACM SIGSOFT International Symposium on the Foundations of Software Engineering (ESEC/FSE 01)*, Austria, 2001.

29. J. Tsay, C. Hylands, and E. A. Lee, A code generation framework for Java component-based designs, *CASES '00*, San Jose, CA, November 17–19, 2000.

30. T. J. Koo, F. Hoffmann, H. Shim, B. Sinopoli, and S. Sastry, Hybrid control of model helicopter, in *Proceedings of IFAC Workshop on Motion Control*, Grenoble, France, October 1999, pp. 285–290.

CHAPTER 8

EMBEDDED CONTROL SYSTEMS DEVELOPMENT WITH GIOTTO

THOMAS A. HENZINGER, BENJAMIN HOROWITZ,
and CHRISTOPH MEYER KIRSCH

Editors' Notes

Chapter 8 focuses on one particular, and important, model of computation: the time-triggered paradigm. In a time-triggered system, communication between system components is triggered by the tick of a notional global clock. The time-triggered paradigm provides timing predictability. As a result, its hardware and protocol realization—the time-triggered architecture (TTA)—is now being adopted for safety-critical applications (e.g., in the automotive industry) where hard real-time constraints must be met.

Giotto is a software realization of the time-triggered paradigm. Giotto consists of a design methodology and support tools for the software implementation of time-triggered distributed control systems. An important feature of Giotto is that it cleanly separates platform-independent issues from platform-dependent ones. A Giotto program determines the behavior of concurrent periodic tasks and the timing of sensing and actuation—it does not specify platform-dependent aspects such as scheduling. The Giotto compiler then translates the platform-independent program into platform-specific embedded software. Compilation of the same program for platforms with different resource and performance characteristics will result in different task mappings and different task and communication schedules.

Annotations can be included with a Giotto program to help the compiler find a feasible schedule; the scheduling problem is especially difficult for distributed platforms. The annotations fall into three categories, each of which enables the programmer to associate different program constructs with platform-specific information.

Note: This research was supported in part by the DARPA SEC grant F33615-C-98-3614, the MARCO GSRC grant 98-DT-660, the AFOSR MURI grant F49620-00-1-0327, and the NSF ITR grant CCR-0085949.
A preliminary version of this chapter appeared in the *Proceedings of the ACM Workshop on Languages, Compilers, and Tools for Embedded Systems* (LCTES), 2001, pp. 64–72.

Giotto has now been demonstrated for an autonomously flying model helicopter and a multirobot wireless testbed. An appendix to the chapter lists the annotated Giotto code used for a two-robot version of the latter application.

8.1. INTRODUCTION

Embedded software development for control applications consists of two phases: first modeling, then implementation. Modeling control applications is usually done by control engineers with support from tools such as Matlab or MatrixX. While these tools offer limited code-generation facilities, the efficient implementation of control designs remains a challenging subdiscipline of software engineering. Control designs impose hard real-time requirements, which software engineers traditionally meet by tightly coupling model, code, and platform. We advocate a decoupling of these domains.

Throughout this chapter, the term *platform* denotes a hardware configuration, operating system, and communication protocol. Platforms, which may be distributed, consist of sensors, actuators, CPUs, and networks. Platform-independent issues include application functionality and timing. In contrast, platform-dependent issues include scheduling, communication, and physical performance. The key to automating embedded software development is to understand the interface between platform-independent and platform-dependent issues. Such an interface—that is, an abstract programmer's model for embedded systems—enables decoupling software design from implementation, even for distributed platforms and even in the presence of hard real-time requirements.

Giotto provides an abstract programmer's model based on the time-triggered paradigm. In a *time-triggered system*, communication—from sensors, between CPUs, to actuators—is triggered by the tick of a notional global clock. In a distributed system, such a clock can be provided by a clock synchronization service. The time-triggered architecture (TTA) [1] offers a hardware and protocol realization of the time-triggered paradigm, thereby providing a natural platform for a Giotto implementation. The TTA has recently gained momentum in safety-critical automotive applications, where timing predictability is essential. While Giotto offers the predictability of time-triggered systems, it also offers the flexibility of platform independence. In particular, it is possible to use a clock synchronization service on top of any real-time communication protocol to implement Giotto. For example, recently the event-triggered CAN standard has been extended to TTCAN [2], which includes a clock synchronization service and thus offers another attractive platform for implementing Giotto.

The two central ingredients of Giotto are periodic task invocations and time-triggered mode switches. More precisely, a *Giotto program* specifies a set of *modes*. Each mode determines a set of *tasks* and a set of *mode switches*. At every time instant, the program execution is in one specific mode, say, M. Each task of M has a real-time frequency and is invoked at this frequency as long as the mode M remains unchanged. A task invocation typically causes nontrivial computational activity, such as the calculation of desired actuator values. Each mode switch of M has a real-time frequency, a condition that is evaluated at this frequency, and a target mode, say, N: If the condition evaluates to true, then the new mode is N. In the new mode, some tasks may be removed and others added.

Giotto has a formal semantics that specifies the meaning of mode switches, of intertask communication, and of communication with the program environment. The environment consists of sensors and actuators. A Giotto program determines (a) the functionality (input and output behavior) of concurrent periodic tasks and (b) the timing of the program's interaction with its environment. Functionality and timing are the key elements of the interface between control design and implementation. A Giotto program does not specify platform-dependent aspects such as priorities and other scheduling and communication directives. Giotto's strength is its simplicity: Giotto is compatible with any choice of real-time operating system (RTOS), scheduling algorithm, and real-time communication protocol. Moreover, Giotto's simplicity allows us to automate schedule and code generation.

The *Giotto compiler* is an essential part of the methodology. A Giotto program is a platform-independent specification of a control software design, from which the Giotto compiler synthesizes embedded software for a given platform. The Giotto tasks are given, say, as C code. The tasks' worst-case execution times (WCETs) are known by the Giotto compiler.[1] Given a platform, the compiler maps tasks to CPUs. The compiler then computes a task and communication schedule that guarantees the timing requirements of the Giotto program. Compilation of the same program for platforms with different resource and performance characteristics will result in different task mappings and different task and communication schedules. Since the synthesis problem is difficult for distributed platforms, a Giotto compiler may fail to find a feasible schedule, even if such a schedule exists. For this case, we propose *Giotto annotations*, which allow the programmer to give directives that aid the compiler in finding a feasible schedule. A Giotto annotation constrains the compiler to a nonempty subset of the permissible schedules.

Figure 8.1 summarizes the design flow of embedded control systems development with Giotto. First the control and software engineers agree on the functionality and timing of a design, specified as a Giotto program. Then the software engineer uses the Giotto compiler to map the program to a given platform. The Giotto compiler produces an executable, which can then be linked against the Giotto run-time library. The Giotto run-time library provides a layer of scheduling and communication primitives. This layer

---

[1]The difficult problem of estimating WCETs is orthogonal to the problems that Giotto addresses. For example, for complex processor architectures with cache and pipeline mechanisms, abstract interpretation can be used to generate integer linear programs for WCET prediction [3].

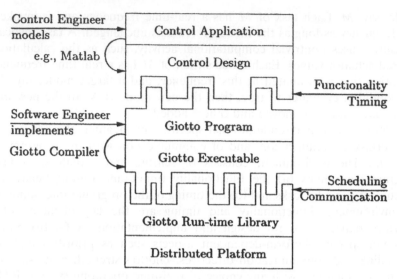

Figure 8.1. Embedded control systems development with Giotto.

defines the interface between the Giotto executable and a platform. We have developed a Giotto run-time library for Wind River's VxWorks RTOS on Intel x86 targets. We are currently in the process of porting the library to other platforms.

Figure 8.2 shows a more detailed picture of the design flow from a Giotto program to a Giotto executable using Giotto annotations. For distributed platforms, if the compiler is unable to find a permissible schedule, the software engineer may use Giotto annotations to give directives to the compiler on how to map tasks to hosts and how to schedule resources. A Giotto program can be gradually refined with more and more specific annotations, until the compiler is able to generate a mapping and a schedule that meet the timing requirements. Giotto annotations fall into three increasingly specific classes of directives: Giotto-H annotations specify the hardware and its performance; Giotto-M annotations map tasks to CPUs and communications to networks; and Giotto-S annotations schedule tasks and communications. It is important to note that the functionality and timing properties of a Giotto program are not affected by Giotto annotations. Rather, the annotations provide a means of incrementally refining a platform-independent Giotto program into an executable for a specific platform.

In the next section we introduce the pure, platform-independent version of the Giotto programming language. In Section 8.3, we use an example to illustrate embedded control systems development using Giotto. The example requires the coordination of a heterogeneous flock of Intel x86 robots and Lego Mindstorms robots. The example is implemented by a Giotto program first discussed in Section 8.4, and then refined in Section 8.5 with Giotto

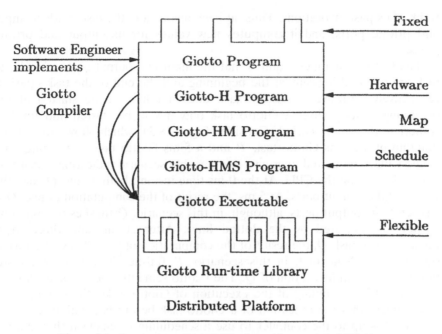

Figure 8.2. Semiautomatic compilation with annotated Giotto.

annotations. We relate Giotto to existing work and conclude in Section 8.6. For a complete and detailed exposition of Giotto, including formal definitions of abstract syntax and semantics, see reference 4.

8.2. THE GIOTTO PROGRAMMING LANGUAGE

Giotto is a programming language that aims at distributed hard real-time applications with periodic behavior, such as control systems. A typical control system periodically reads sensor information, computes control laws, and writes the results to actuators. Moreover, such a control system may react to changes in its environment by switching control laws as well as periodicity. Giotto's language primitives match these requirements of distributed hard real-time control applications.

A *Giotto port* is a physical location in memory. A Giotto port may either be associated with a sensor or actuator or else may be used for intertask communication. A *Giotto task* is a periodic job, say, computing a control law. A Giotto task has input, private, and output ports and has an implementation with known WCET, written in any programming language. Thus Giotto can be seen as a real-time extension of a standard, nonembedded programming language. A Giotto task is invoked with a certain *period*, given in milliseconds (ms). A Giotto task may carry state in its private ports in order to keep track of its past invocations. Thus, at each invocation, the task reads its input and private ports, and it computes new values for its output and private ports.

The Giotto semantics requires that the input and output ports of a Giotto task are updated logically at the beginning and the end of the task's period, respectively. However, a Giotto task does not have to be started at the beginning of its period. A Giotto task only has to be started and finished sometime during its period. Consider Figure 8.3a, which shows the timing diagram of a 20-ms Giotto task P and a 5-ms Giotto task Q running on a single CPU. The dotted lines give one possible scenario indicating which task currently runs on the CPU. At the 0-ms time instant, both P and Q read the values of their input ports. At 5 ms, the result of the computation of task Q is written to its output ports, although, in this scenario, Q finishes its execution earlier (as indicated by the dotted line). After 20 ms and three more invocations of task Q, the result of the computation of task P is written to its output ports. Note that, in this scenario, P finishes its execution already before the fourth invocation of Q. The Giotto semantics does not specify the physical CPU scheduling of the execution of Giotto tasks; there is only the requirement that the execution of a Giotto task be finished within the task's period. It is up to the compiler to use a scheduling mechanism that guarantees the deadlines (or declare the program to be invalid, if no permissible schedule exists; for example, if the WCET exceeds the period of a task). In our scenario, the compiler uses rate-monotonic scheduling and thus assigns a higher priority to the more frequent task Q.

For processing sensor values, sensor ports can be connected to task input ports; for driving actuators, task output ports can be connected to actuator ports; and for communication between Giotto tasks, output ports of one task can be connected to input ports of another task. Connections are established by drivers. A *Giotto driver* computes a function on its source ports, and it passes the result to its destination ports. If the destination ports are actuator ports, we refer to the driver as an *actuator driver*; if the destination ports are the input ports of a task, we call the driver a *task driver*. Every Giotto driver has a *guard*, which is a predicate on its source ports. If the guard evaluates to false, then the driver is not executed; and in the case of a task driver, neither is the associated task. There is an important difference between drivers and tasks: Drivers represent system-level code for preprocessing and transmitting data between local ports; tasks, on the other hand, perform application-level computation. In Giotto's programming model, driver execution is logically instantaneous, while task execution takes time up to the length of the task's period.

Since the result of a task's computation is written at the end of the task's period, task drivers only cause data flow from past task invocations to current invocations, and not between concurrent invocations. Figure 8.3b shows the data flow of a driver for task Q that reads an output port p of task P. The first four invocations of task Q after the 0-ms time instant see the result of

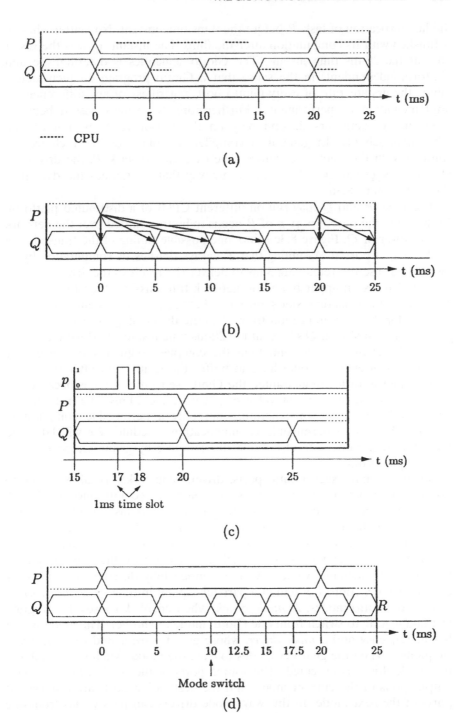

Figure 8.3. Timing diagrams. (a) Two Giotto tasks P and Q. (b) The data flow using a Giotto driver. (c) A transmission from the task output port p. (d) The Giotto mode switch s.

the last invocation of task P before the 0-ms time instant. No matter when P is finished with its computation after the 0-ms time instant, Q sees the result only at the 20-ms time instant. The Giotto semantics is deterministic and platform-independent, in the sense that a Giotto program uniquely determines the update rate of every Giotto port, regardless of any differences in implementation or performance. Furthermore, the values passed between tasks and to actuators depend only on the sensor readings, not on the scheduling scheme. In general, a compiler has many different choices to conform with the Giotto semantics: The execution of task P, the driver for Q, and task Q can be scheduled in any way that guarantees the data flow shown in Figure 8.3b.

The tasks P and Q may run on different CPUs of a distributed platform. In this case, the output port p of P needs to be transmitted over the network to the driver of Q. Figure 8.3c shows one possible timing of the transmission of the output port p. Assuming that P always finishes its computation by 17 ms, we may use the remaining 3 ms to deliver the result to the driver of Q on time. If 1 ms is an upper bound on network transmission, then the compiler may allocate a 1-ms time slot somewhere between the 17-ms and 20-ms time instants for the communication from P to the driver of Q. This ensures that p's value is available at Q's CPU at the 20-ms time instant. If P were to finish before the 15-ms time instant, then the compiler might reserve an earlier time slot. However, it would have to buffer the result until after the 15-ms time instant in order to guarantee the Giotto semantics. This is because the task Q must not see the new value of p before the 20-ms time instant. The compiler is in charge of generating not only a permissible computation schedule but also a permissible communication schedule. We will later see how the programmer can also guide the compiler by giving directives for a specific platform.

So far, we have seen Giotto ports, drivers, and tasks. To allow a Giotto program to react to changes in its environment, we introduce modes. A *Giotto mode* consists of a set of concurrent Giotto tasks with invocation frequencies (which determine the task periods[2]) and task drivers, a set of actuator updates, and a set of mode switches. An *actuator update* has a frequency and an actuator driver, which is executed with the given frequency to update the actuator values. A Giotto mode fully describes the behavior and timing of a control system at a particular point in time. A *Giotto program* is a set of Giotto modes. A *mode switch*, when enabled, causes the program instantaneously to switch from one Giotto mode to another. The mode switch has a frequency and a *mode driver*, whose guard is evaluated with the given frequency. When the guard evaluates to true, the mode switch is enabled and the mode driver is executed. The source ports of the mode driver are task output ports of the current mode, and the destination ports are task output ports of the next mode. In this way, mode drivers can pass values from one

---

[2]For an event with frequency f, the period is $1/f$, and vice versa.

Giotto mode to the next. To guarantee determinism, we require that, for every Giotto mode, the conjunction of any two mode driver guards is unsatisfiable; that is, at most one mode switch can be enabled at any given time.

In Giotto, a task is considered a unit of work, which, once started, must be allowed to complete. A mode switch may cease the periodic invocation of a task if that task's period ends at the time the mode switch guard is evaluated. However, a mode switch must not terminate any task whose period has not ended. These tasks are called *active*. If a task is active when a mode switch occurs, then the Giotto semantics requires that the next mode again contains the task. The least common multiple of all task invocation periods, actuator update periods, and mode switch periods of a Giotto mode determines the *period* of the mode. The execution of a Giotto mode for a single period is called a *round*. When a mode switch occurs in the middle of a round, first the current mode is terminated instantaneously. If t is the time until the periods of all currently active tasks end simultaneously, then the next mode is entered t milliseconds before the start of a new round. This ensures that as little time as necessary elapses before the full functionality of the new mode begins.

Suppose we are given a Giotto mode M containing the Giotto tasks P and Q invoked with 20-ms and 5-ms periods, respectively, and a Giotto mode N containing the Giotto tasks P and R invoked with 20-ms and 2.5-ms periods, respectively. Suppose there is a mode switch s from M to N with a 5-ms period. Then M and N have the same period of 20 ms. Figure 8.3d shows the timing diagram of the mode switch s enabled at the 10-ms time instant in the middle of a round of M. Since both modes M and N contain the task P, it is not terminated but can continue its computation as if nothing happened. However, Q's invocations are replaced by two times as many invocations of the task R. Since N's round has already been completed halfway at the 10-ms time instant, there will be only four invocations of R before a new round of N starts at the 20-ms time instant.

8.3. A DISTRIBUTED HARD REAL-TIME CONTROL PROBLEM

Giotto has been designed specifically for high-performance control applications, which are structured around periodic sensor readings, task invocations, actuator updates, and mode switches, and are implemented on fault-tolerant, distributed platforms. A paradigmatic example of this type is fly-by-wire flight control, with several operational modes (take-off, cruise, etc.) and degraded modes caused by component failures [5]. Giotto formalizes and provides tool support for some of the design principles commonly used in the avionics domain to build this kind of system. In collaboration with Marco Sanvido and Walter Schaufelberger at ETH Zürich, we have implemented in Giotto the on-board control system of an autonomously flying model helicopter.

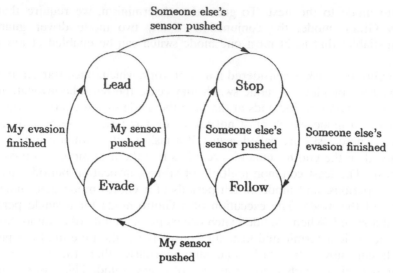

Figure 8.4. The behavior of an n robot system.

For the sake of simplicity, we describe here in detail a simpler control problem, whose solution nonetheless illustrates the essential features offered by Giotto.

As an example of a distributed hard real-time control problem, we consider a set of n robots. Each robot has a CPU, two motors, and a touch sensor. The motors drive wheels and allow the robot to move forward and backward and to rotate. The touch sensor is connected to a bumper. The n robots share a broadcast communication medium. Figure 8.4 shows the behavior of the n robot system, where a circle depicts the state of a robot and an arc is a transition from one state to another. Note that state is here a behavioral concept rather than, say, a Giotto mode. A robot that is in either the lead state or the evade state is called a *leader*. A robot that is in either the follow state or the stop state is called a *follower*. We require that at all times there is only a single robot that is a leader, while the $n - 1$ remaining robots are followers. Upon initialization the leader robot (which is chosen at random) is in the lead state and determines the movements taken by all n robots. For simplicity, the leader tells everyone to move in the same way, resulting in a synchronized "dance." The $n - 1$ followers are in the follow state and listen to the commands of the leader.

Now, there are two possible scenarios. Either the leader's bumper or the bumper of one of the followers is pushed, presumably by an obstacle. Again for simplicity, we assume that no more than a single bumper can be pushed at the same time. Suppose that the leader's bumper is pushed. Then the leader goes into the evade state, while the $n - 1$ followers go into the stop state. A robot in the evade state performs an evasion procedure for a short amount of time, to avoid the obstacle, whereas a robot in the stop state simply stops. When the leader is finished with the evasion procedure, it goes back into the lead state, while the $n - 1$ followers go back into the follow state. Suppose now that the bumper of one of the followers is pushed. Then this robot goes into the evade state while all other robots, including the leader, go into the stop state. Pushing the bumper of a follower makes this robot the new leader. This concludes the description of the n robot system. In the next section, we describe a Giotto implementation.

8.4. A GIOTTO PROGRAM

In order to demonstrate Giotto's applicability for distributed and heterogeneous platforms, we implemented a Giotto program for coordinating five robots in the way described above. Two of the robots feature a credit-card form-factor single-board computer with an Intel 80486 processor and a Lucent WaveLAN wireless Ethernet card. The single-board computers run Wind River's VxWorks RTOS. Both robots use Lego Mindstorms motors and touch sensors. The three other robots are pure Lego Mindstorms robots equipped with Hitachi microcontrollers and infrared transceivers. The microcontrollers run Lego's original firmware. Communication between the different platforms is established through a gateway—a notebook PC—between wireless Ethernet and the infrared link.

For the sake of simplicity, we describe the Giotto program for a two-robot system. The Appendix contains the Giotto program. In this section, any program code in brackets can be disregarded; it belongs to the annotated version of Giotto. We will discuss annotated Giotto in the next section. The Giotto program begins with the global declaration of ports. The sensor port `sensorX` of robot X contains `true` whenever the bumper of robot X is pushed. These ports use the method `c_sensorget`, written in the host programming language, to read the status of the bumper device. We call `c_sensorget` a *device driver*. A device driver like `c_sensorget` implements the association of a sensor port with a device. Every time a sensor port is accessed by a task or mode driver, its associated device driver is called. The implementation of the device driver may, for example, poll the device or read the latest value of an asynchronous update. Similarly to sensor ports, the actuator ports `motorLX` and `motorRX` are associated with the left and right motors of robot X using the device drivers `c_motorLput` and `c_motorRput`, respectively. The remaining ports—`command`, `motor1`, `motor2`, and `finished`—are task output ports, which are not associated with any devices.

The port declarations in the Giotto program are followed by task and driver declarations. Consider, for example, the task declaration `motorCtrl`. This task has a single input port called `com`, which is a formal parameter, and a single output port, which is the globally declared output port `motor1`. In general, the input ports of a task are declared by formal parameters with local scope, whereas the output ports are globally declared. In the body of the declaration, we specify the name c_botCom1 of the procedure that implements the task. In this example, we use ANSI C to implement all tasks and drivers (not shown). Now, consider the driver declaration stop1Drv. This driver has a single source port, which is the globally declared sensor port sensor1, and two destination ports given by the formal parameters who and com. In general, the source ports of a driver are globally declared sensor or task output ports, whereas the destination ports are declared by formal parameters with local scope. In the body of the declaration, we specify the name c_true_port of the procedure that implements the guard of the driver, and we specify the name c_stop1Drv of the procedure that implements the actual driver.

Finally, the Giotto program contains six mode declarations. The Lead1Follow mode is the start mode. Each Giotto mode describes the behavior of the whole system of CPUs and networks. Since one robot is in the lead or evade state and the rest are in the follow or stop state, we use a LeadXFollow mode and an EvadeXStop mode for each leader X. To improve responsiveness of the implementation we also introduce for each robot X a Giotto mode StopX, which allows the robots to stop quickly. In general, for n robots we obtain $3n$ modes.

All modes run with a period of 400 ms. Consider the Lead1Follow mode in which robot 1 is the leader. Actuator updates are indicated by the keyword actfreq; mode switches, by exitfreq; and task invocations, by taskfreq. The botCom1 task runs once per round and computes a command stored in the output port command. There are two more Giotto tasks motorCtr1 and motorCtr2 running with a period of 100 ms four times per round. The two tasks control the motors of both robots according to the command in command. The task driver motorDrv delivers the value in command to the input ports of the tasks motorCtr1 and motorCtr2. The guard motorDrv is always true, so the tasks motorCtr1 and motorCtr2 are always executed. The higher frequency of these tasks allows for smoother control of the motors. For example, the left motor of robot 1 is controlled by the actuator port motorL1, which is updated by the actuator driver motorActL1 at the end of every period of the task motorCtr1. The actuator driver motorActL1 is invoked at the same frequency as motorCtr1; it extracts the necessary information for the left motor from the output port motor1. Figure 8.5a shows the timing diagram for one round in the Lead1Follow mode.

The state of the touch sensors is checked twice every round by the two mode switches. The mode switches employ the mode drivers stop1Drv and stop2Drv, which perform the work of checking the sensors of robot 1 and 2, respectively. We assume that both bumpers cannot be pressed simultaneously. If the bumper of robot 1 is pushed, we switch to the Stop1 mode, in which both robots stop driving. Figure 8.5b shows the timing diagram for one

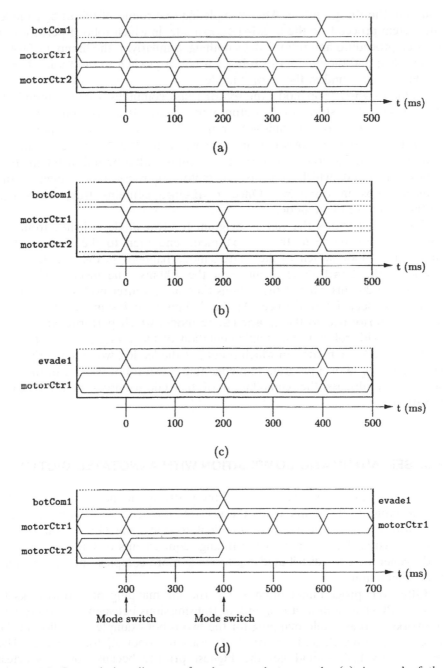

Figure 8.5. Some timing diagrams for the two-robot example. (a) A round of the Lead1Follow mode. (b) A round of the Stop1 mode. (c) A round of the Evade1Stop mode. (d) Mode switches to the Stop1 mode and to the Evade1Stop mode.

round of the Stop1 mode. After completing one round of the Stop1 mode, the system proceeds to the Evade1Stop mode, in which robot 1 performs an evasion procedure and robot 2 does nothing. Similarly, if the bumper of robot 2 is pushed and the bumper of robot 1 is not, we switch to the Evade2Stop mode via one round in the Stop2 mode.

In the Evade1Stop mode, the evade1 task computes once per round the next evasion step (stored in the output port command) and computes whether the evasion maneuver is finished or not (stored in the output port finished). Figure 8.5c shows the timing diagram for the Evade1Stop mode. There is also a Giotto task motorCtr1 running with a period of 100 ms four times per round, which controls the motors of robot 1 according to the evasion steps in command. Once finished contains true, we switch to the Lead1Follow mode.

Figure 8.5d shows the timing diagram for two mode switches from the Lead1Follow mode to the Stop1 mode and then to the Evade1Stop mode. The mode switch to the Stop1 mode happens in the middle of the round of the Lead1Follow mode at the 200-ms time instant. Then the robots stop 200 ms later at the 400-ms time instant after both motor control tasks have been invoked once. At the 400-ms time instant, another mode switch is performed to the Evade1Stop mode, which performs an evasion maneuver with robot 1. The implementation of the Lead2Follow, Stop2, and Evade2Stop modes, in which robot 2 is the leader, works similarly. This completes the Giotto program implementing a two-robot system. In the next section, we discuss the compilation of the Giotto program for a given platform.

8.5. SEMIAUTOMATIC COMPILATION WITH ANNOTATED GIOTTO

In the previous section, we presented the platform-independent aspects of a Giotto program that implements the two-robot system. For a nondistributed platform, this level of detail is sufficient to allow the Giotto compiler to generate code that guarantees the timing requirements of Giotto. However, code generation for distributed platforms is more complex and may require user interaction.

Often the programmer intends a particular mapping of Giotto tasks to hosts (CPUs), as well as a mapping of transmissions between Giotto ports to networks. A reasonable mapping for the two-robot example may allocate the botComX, evadeX, and motorCtrX tasks to robot X, for X = 1, 2. The motorCtrX task should get the highest priority because of its shortest deadline. Also, a reasonable mapping may allocate the mode switches of the modes LeadXFollow, StopX, and EvadeXStop to robot X. Consequently, the values in com, sensorX, and the next mode have to be communicated between the two robots.

Starting with a pure Giotto program, which contains no platform-related information, the programmer may use Giotto annotations to provide the Giotto compiler with details about the target platform. A *Giotto annotation* falls into one of three possible categories: (1) A *Giotto-H annotation* (H for "hardware") specifies a set of hosts, a set of networks, and WCET information; (2) a *Giotto-M annotation* (M for "map") specifies for a Giotto mode the Giotto task-to-host and communication-to-network mappings; and (3) a *Giotto-S annotation* (S for "schedule") specifies scheduling information for each host and network. In our example, the Giotto-S annotations will specify a priority for each task and a time slot for each communication. This style of Giotto-S annotations is suited for static priority RTOS hosts and TDMA networks. For target platforms with different scheduling primitives, different annotations can be developed. For example, for platforms consisting of nonpreemptive RTOS hosts and CAN networks, Giotto-S annotations would specify time slots for tasks and priorities for communications.

The Giotto program in the Appendix contains examples of all three types of Giotto annotations. A Giotto-H annotation at the top of the program provides details on the two-robot platform. There are two hosts called bot1 and bot2, representing the two robot CPUs, which are connected by a network called net12. Specific mapping and scheduling information is given by the Giotto-M and Giotto-S annotations in the example. Note that for flexibility we use symbolic names rather than numbers for task priorities and communication time slots.

For instance, in the Lead1Follow mode, the botCom1 task is assigned to bot1 with the priority p1, which is lower than the priority p0 of the motorCtrl task assigned to bot1 as well. Consider Figure 8.6, which shows the timing diagram for a round of the Lead1Follow mode with scheduling details. The dotted line shows which Giotto task is running on bot1. The lower priority botCom1 task gets the CPU only when the motorCtrl task is

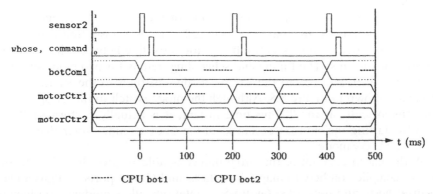

Figure 8.6. The timing diagram for a round of the Lead1Follow mode with scheduling details.

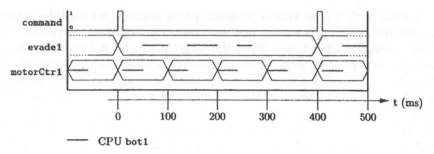

Figure 8.7. The timing diagram for a round of the Evade1Stop mode with scheduling details.

finished. In order to allow bot1 to evaluate the mode switch guards, the value of the sensor2 port has to be transmitted from bot2. This communication from bot2 to bot1 has to occur twice per round because of the mode switch period of 200 ms. Since the sensors are sampled at the 0-ms and 200-ms time instants, the value of sensor2 cannot be transmitted early. The push annotation indicates that the output host bot2 initiates the communication. The dual pull annotation is also available, for example, to support less capable distributed sensors.

The signal at the sensor2 port in Figure 8.6 shows the timing for the transmission of the sensor2 values to bot1. We assume that communication consumes CPU cycles. The 0-ms to 10-ms time slot for the sensor2 value transmission delays the decision of performing a mode switch. Such delays do not affect the Giotto semantics as long as all Giotto tasks still meet their deadlines. The mode switch decision is taken by bot1 between 10 ms and 20 ms. Then, between 20 ms and 30 ms, bot1 sends its decision to bot2. The communication between 20 ms and 30 ms from bot1 to bot2 also transmits the value of the output port command computed in the previous round and the value of the port whose, which indicates whose bumper was pressed.

Figure 8.7 shows the timing for a round of the Evade1Stop mode with scheduling details. In this mode, command has to be transmitted from bot1 to bot2 once per round, as shown by the top-most line between the 0-ms and 10-ms time instants. The timing and scheduling details for two mode switches from the Lead1Follow to the Stop1 mode and then to the Evade1Stop mode are shown in Figure 8.8. Although the mode switch to the Stop1 mode happens logically at the 200-ms time instant, it is actually performed 30 ms later by starting the motorCtrX tasks of the Stop1 mode rather than of the Lead1Follow mode.

With annotated Giotto, the programmer is able to give directives to the Giotto compiler on how to map Giotto tasks and Giotto ports to a given platform of hosts and networks. Giotto-S annotations allow guidance on how to schedule computation and communication resources. Most importantly, Giotto

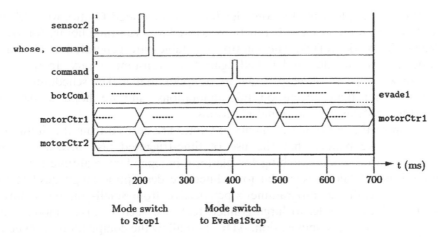

Figure 8.8. The timing diagram for mode switches from the Lead1Follow mode to the Stop1 mode and to the Evade1Stop mode with scheduling details.

annotations refine a given nonannotated Giotto program without affecting the functionality and timing specification. In this way, Giotto separates program behavior from the use of a particular scheduling or communication scheme: If incomplete directives are given, then different compilers may use different scheduling schemes. Indeed, one compiler, using one particular scheduling scheme, may fail, whereas another, "smarter" compiler may succeed in compiling a given Giotto program on a given distributed platform.

8.6. SUMMARY AND RELATED WORK

We have presented a tool-supported development methodology for embedded control software which is based on the programming language Giotto. In Giotto, the programmer specifies the functionality and timing of a control design, leaving the specification of scheduling schemes to the Giotto compiler. The Giotto compiler automates the implementation of embedded control systems, by taking over the tedious and error-prone task of producing computation and communication schedules. Given a Giotto program and a particular platform, the Giotto compiler may (or may not) be able to generate Giotto executables that obey the timing requirements of the program. When targeting complex distributed platforms, the programmer may also give explicit scheduling directives to the compiler using Giotto annotations.

Giotto has a time-triggered semantics. Task invocations as well as observations of the environment in a Giotto system are triggered by the tick of a notional global clock. Consequently, the timing behavior of a Giotto system is highly predictable, which makes Giotto particularly well-suited for safety-critical applications with hard real-time constraints.

We have implemented a compiler for fully annotated Giotto, with drivers and tasks that are given as C functions, as well as a run-time library for Wind River's VxWorks RTOS. The Giotto executables are generated as C source code that is compiled and linked against the run-time library. In the near future we hope to develop a Giotto compiler that makes scheduling decisions, rather than relying on full annotations, and we hope to develop run-time systems for additional platforms.

Many of the individual elements of Giotto are derived from the literature. However, we believe that the use of time-triggered task invocation plus time-triggered mode switching for platform-independent real-time programming is novel. Giotto is similar to architecture description languages (ADLs) [6]. ADLs shift the programmer's perspective from small-grained features, such as lines of code, to large-grained features, such as tasks, modes, and intercomponent communication. ADLs also allow the compilation of scheduling code to connect tasks written in conventional programming languages. The design methodology for the MARS system, a precursor of the TTA, similarly distinguishes *programming-in-the-large* from *programming-in-the-small* [7]. Giotto's intertask communication semantics is particularly similar to the MetaH language [8], which is designed for real-time, distributed avionics applications. MetaH supports periodic tasks, multimodal control, and distributed implementations. Giotto can be viewed as capturing a time-triggered fragment of MetaH in an abstract and formal way. Unlike MetaH, the Giotto abstraction does not constrain the implementation to a particular scheduling paradigm as long as the timing requirements of a Giotto program are guaranteed. Since the semantics of Giotto is defined formally, the behavioral properties of a Giotto program may be subject to formal verification [9].

The goal of Giotto—to provide a platform-independent programming abstraction for real-time systems—is shared by the family of synchronous reactive programming languages [10], such as Esterel [11], Lustre [12], and Signal [13]. While the synchronous reactive languages are designed around zero-delay computation, Giotto is based on the formally weaker notion of unit-delay computation, because the execution of a Giotto task has a positive duration. This avoids the complications involved with fixed-point semantics and shifts the emphasis to code generation under WCET constraints. Giotto can be seen as identifying a class of synchronous reactive programs that support both typical real-time control applications and distributed code generation.

ACKNOWLEDGMENTS

We thank Rupak Majumdar for implementing a prototype Giotto compiler for Lego Mindstorms robots. We thank Dmitry Derevyanko and Winthrop Williams for building our Intel x86 robots. We thank Edward Lee and Xiaojun Liu for help with a Ptolemy II [14] implementation of Giotto.

APPENDIX: A GIOTTO PROGRAM WITH ANNOTATIONS

```
[host bot1 address 192.168.0.1 priorities p0 > p1;
 host bot2 address 192.168.0.2 priorities q0 > q1;
 network n12 address 192.168.0.0 connects bot1, bot2]

// Sensor ports

sensor

// True means pushed
c_bool sensor1 uses c_sensorget [host bot1];  // bot1 touch sensor
c_bool sensor2 uses c_sensorget [host bot2];  // bot2 touch sensor

// Actuator ports

actuator

c_int motorL1 uses c_motorLput [host bot1];  // bot1 left motor
c_int motorR1 uses c_motorRput [host bot1];  // bot1 right motor
c_int motorL2 uses c_motorLput [host bot2];  // bot2 left motor
c_int motorR2 uses c_motorRput [host bot2];  // bot2 right motor

// Output ports

output

// command port
c_int command := c_zero;

// bot1 motor control port
c_int motor1 := c_zero;

// bot2 motor control port
c_int motor2 := c_zero;

// Evade maneuver finished
c_bool finished := false;

// Task declarations

task botCom1() output (command)
{ schedule c_botCom1(command) }

task botCom2() output (command)
{ schedule c_botCom2(command) }

task motorCtr1(c_int com) output (motor1)
{ schedule c_motorCtr(com, motor1) }

task motorCtr2(c_int com) output (motor2)
{ schedule c_motorCtr(com, motor2) }
```

```
task evade1() output (command, finished)
{ schedule c_evade1(command, finished) }

task evade2() output (command, finished)
{ schedule c_evade2(command, finished) }

// Driver declarations

driver motorDrv(command) output (c_int com)
{ if c_true() then c_motorDrv(command, com) }

driver motorActL1(motor1) output (c_int mot)
{ if c_true() then c_motorActL(motor1, mot) }

driver motorActR1(motor1) output (c_int mot)
{ if c_true() then c_motorActR(motor1, mot) }

driver motorActL2(motor2) output (c_int mot)
{ if c_true() then c_motorActL(motor2, mot) }

driver motorActR2(motor2) output (c_int mot)
{ if c_true() then c_motorActR(motor2, mot) }

driver stop1Drv(sensor1) output (c_int who, c_int com)
{ if c_true_port(sensor1) then c_stop1Drv(who, com) }

driver stop2Drv(sensor2) output (c_int who, c_int com)
{ if c_true_port(sensor2) then c_stop2Drv(who, com) }

driver motorStopDrv() output (c_int com)
{ if c_true() then c_motorStopDrv(com) }

driver evade1StopDrv(whose) output (c_int com)
{ if c_evade1StopGrd(whose) then c_evadeStopDrv(com) }

driver evade2StopDrv(whose) output (c_int com)
{ if c_evade2StopGrd(whose) then c_evadeStopDrv(com) }

driver leadFollowDrv(finished) output (c_int com)
{ if c_true_port(finished) then c_leadFollowDrv(com) }

start Lead1Follow(c_zero) {
 mode Lead1Follow(command) period 400
  [network n12 slots s0 (0,10), s1 (20,30),
                s2 (200,210), s3 (220,230)] {

  actfreq 4 do motorL1(motorActL1);
  actfreq 4 do motorR1(motorActR1);
  actfreq 4 do motorL2(motorActL2);
  actfreq 4 do motorR2(motorActR2);

  [push (sensor2) from bot2 to (bot1) in n12 slots s0,s2]
```

```
  exitfreq 2 do Stop1(stop1Drv)
  [host bot1; push (whose,command) to (bot2) in n12 slots s1,s3];
  exitfreq 2 do Stop2(stop2Drv)
  [host bot1; push (whose,command) to (bot2) in n12 slots s1,s3];

  taskfreq 1 do botCom1()              [host bot1 priority p1];
  taskfreq 4 do motorCtr1(motorDrv)    [host bot1 priority p0];
  taskfreq 4 do motorCtr2(motorDrv)    [host bot1 priority q0];
}

mode Stop1(whose, command) period 400
  [network n12 slots s0 (0,10)] {

  actfreq 2 do motorL1(motorActL1);
  actfreq 2 do motorR1(motorActR1);
  actfreq 2 do motorL2(motorActL2);
  actfreq 2 do motorR2(motorActR2);

  exitfreq 1 do Evade1Stop(evade1StopDrv)
  [host bot1; push (command) to (bot2) in n12 slots s0];
  exitfreq 1 do Evade2Stop(evade2StopDrv)
  [host bot1; push (command) to (bot2) in n12 slots s0];

  taskfreq 1 do botCom1()                 [host bot1 priority p1];
  taskfreq 2 do motorCtr1(motorStopDrv)   [host bot1 priority p0];
  taskfreq 2 do motorCtr2(motorStopDrv)   [host bot2 priority q0];
}

mode Evade1Stop(command) period 400
  [network n12 slots s0 (0,10)] {

  actfreq 4 do motorL1(motorActL1);
  actfreq 4 do motorR1(motorActR1);

  exitfreq 1 do Lead1Follow(leadFollowDrv)
  [host bot1; push (command) to (bot2) in n12 slots s0];

  taskfreq 1 do evade1()               [host bot1 priority p1];
  taskfreq 4 do motorCtr1(motorDrv)    [host bot1 priority p0];
}

mode Lead2Follow(command) period 400
  [network n12 slots s0 (0,10), s1 (20,30),
                s2 (200,210), s3 (220,230)] {

  actfreq 4 do motorL1(motorActL1);
  actfreq 4 do motorR1(motorActR1);
  actfreq 4 do motorL2(motorActL2);
  actfreq 4 do motorR2(motorActR2);

  [push (sensor1) from bot1 to (bot2) in n12 slots s0,s2]

  exitfreq 2 do Stop1(stop1Drv)
  [host bot2; push (whose,command) to (bot1) in n12 slots s1,s3];
```

```
exitfreq 2 do Stop2(stop2Drv)
[host bot2; push (whose,command) to (bot1) in n12 slots s1,s3];

taskfreq 1 do botCom2()             [host bot2 priority q1];
taskfreq 4 do motorCtr1(motorDrv)   [host bot1 priority p0];
taskfreq 4 do motorCtr2(motorDrv)   [host bot2 priority q0];
}

mode Stop2(whose, command) period 400
[network n12 slots s0 (0,10)] {

actfreq 2 do motorL1(motorActL1);
actfreq 2 do motorR1(motorActR1);
actfreq 2 do motorL2(motorActL2);
actfreq 2 do motorR2(motorActR2);

exitfreq 1 do Evade1Stop(evade1StopDrv)
[host bot2; push (command) to (bot1) in n12 slots s0];
exitfreq 1 do Evade2Stop(evade2StopDrv)
[host bot2; push (command) to (bot1) in n12 slots s0];

taskfreq 1 do botCom2()               [host bot2 priority q1];
taskfreq 2 do motorCtr1(motorStopDrv) [host bot1 priority p0];
taskfreq 2 do motorCtr2(motorStopDrv) [host bot2 priority q0];
}

mode Evade2Stop(command) period 400
[network n12 slots s0 (0,10)] {

actfreq 4 do motorL2(motorActL2);
actfreq 4 do motorR2(motorActR2);

exitfreq 1 do Lead2Follow(leadFollowDrv)
[host bot2; push (command) to (bot1) in n12 slots s0];

taskfreq 1 do evade2()                [host bot2 priority q1];
taskfreq 4 do motorCtr2(motorDrv)     [host bot2 priority q0];
}
}
```

REFERENCES

1. H. Kopetz, *Real-Time Systems: Design Principles for Distributed Embedded Applications*. Kluwer Academic Publishers, Dordrecht, 1997.
2. T. Führer, B. Müller, W. Dieterle, F. Hartwich, and R. Hugel, Time-triggered communication on CAN (TTCAN), in *Proceedings of the 7th International CAN Conference*, 2000.
3. H. Theiling, C. Ferdinand, and R. Wilhelm, Fast and precise WCET prediction by separated cache and path analyses, *Real-Time Systems* 18(2/3):157–179, 2000.
4. T. A. Henzinger, B. Horowitz, and C. M. Kirsch. Giotto: A time-triggered language for embedded programming, in *Proceedings of the First International Workshop on Embedded Software, Lecture Notes in Computer Science*, Vol. 2211, Springer-Verlag, New York, 2001, pp. 175–194.

5. R. P. G. Collinson, Fly-by-wire flight control. *Computing and Control Engineering*, **10**:141–152, 1999.

6. P. Clements, A survey of architecture description languages, in *Proceedings of the 8th International Workshop on Software Specification and Design*, IEEE Computer Society Press, New York, 1996, pp. 16–25.

7. H. Kopetz, R. Zainlinger, G. Fohler, H. Kantz, P. Puschner, and W. Schütz, The design of real-time systems: From specification to implementation and verification, *IEE/BCS Software Engineering Journal* **6**(3):72–82, 1991.

8. S. Vestal, MetaH support for real-time multiprocessor avionics, in *Proceedings of the 5th International Workshop on Parallel and Distributed Real-Time Systems*, IEEE Computer Society Press, New York, 1997, pp. 11–21.

9. T. A. Henzinger, Masaccio: A formal model for embedded components, in *Proceedings of the First IFIP International Conference on Theoretical Computer Science, Lecture Notes in Computer Science*, Vol. 1872, Springer-Verlag, New York, 2000, pp. 549–563.

10. N. Halbwachs, *Synchronous Programming of Reactive Systems*, Kluwer Academic Publishers, Dordrecht, 1993.

11. G. Berry, The foundations of Esterel, in G. Plotkin, C. Stirling, and M. Tofte, editors, *Proof, Language and Interaction: Essays in Honour of Robin Milner*, MIT Press, Cambridge, MA, 2000, pp. 425–454.

12. N. Halbwachs, P. Caspi, P. Raymond, and D. Pilaud, The synchronous dataflow programming language Lustre, *Proceedings of the IEEE* **79**(9):1305–1320, 1991.

13. A. Benveniste, P. Le Guernic, and C. Jacquemot, Synchronous programming with events and relations: The Signal language and its semantics, *Science of Computer Programming* **16**(2):103–149, 1991.

14. J. Davis, M. Goel, C. Hylands, B. Kienhuis, E. A. Lee, J. Liu, X. Liu, L. Muliadi, S. Neuendorffer, J. Reekie, N. Smyth, J. Tsay, and Y. Xiong, *Ptolemy II: Heterogeneous Concurrent Modeling and Design in Java*, Technical Report UCB/ERL-M99/44, University of California, Berkeley, 1999.

PART III

ONLINE MODELING AND CONTROL

CHAPTER 9

ONLINE CONTROL CUSTOMIZATION VIA OPTIMIZATION-BASED CONTROL

RICHARD M. MURRAY, JOHN HAUSER, ALI JADBABAIE, MARK B. MILAM, NICOLAS PETIT, WILLIAM B. DUNBAR, and RYAN FRANZ

Editors' Notes

An advanced control technology that has had tremendous practical impact over the last two decades or so is model predictive control (MPC). By embedding an optimization solution in each sampling instant within the control calculation, MPC applications have demonstrated dramatic improvements in control performance. To date, this impact has largely been limited to the process industries. The reasons for the domain-specific benefit have to do with the relatively slow time constants of most industrial processes and their relatively benign dynamics (e.g., their open-loop stability). For aerospace systems to avail of the promise of MPC, research is needed in extending the technology so that it can be applied to systems with nonlinear, unstable, and fast dynamics.

This chapter presents a new framework for MPC and optimization-based control for flight control applications. The MPC formulation replaces the traditional terminal constraint with a terminal cost based on a control Lyapunov function. This reduces computational requirements and allows proofs of stability under a variety of realistic assumptions on computation.

The authors also show how differential flatness in a system can be used to computational advantage. (A system is differentially flat if, roughly, it can be modeled as a dynamical equation in one variable and its derivatives.) In this case the optimization can be done over a space of parametrized basis functions and a constrained nonlinear program can be solved using collocation points. A software package has been developed to implement these theoretical developments.

There is an experimental component to this research as well. A tethered ducted fan testbed has been developed at Caltech that mimics the longitudinal dynamics of an aircraft. High-performance maneuvers can safely be flown with the ducted fan and an interface to high-end workstations allows complex control schemes to be solved in real-time. The chapter presents results from several experiments, detailing among

Research for this chapter was supported in part by DARPA contract F33615-98-C-3613.

Software-Enabled Control. Edited by Tariq Samad and Gary Balas
ISBN 0-471-23436-2 © 2003 IEEE

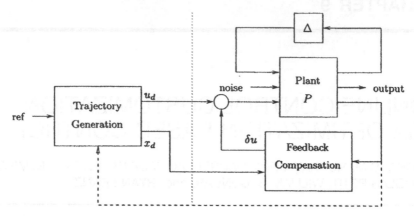

Figure 9.1. Two-degree-of-freedom controller design for a plant P with uncertainty Δ. See text for a detailed explanation.

other things the effects of different MPC optimization horizons on computing time and dynamic performance. Another model predictive control approach is discussed in Chapter 10.

9.1. INTRODUCTION

A large class of industrial and military control problems consist of planning and following a trajectory in the presence of noise and uncertainty. Examples include unmanned airplanes and submarines for surveillance and combat, mobile robots in factories and on the surface of Mars, and medical robots performing inspection and manipulation tasks inside the human body under the control of a surgeon. All of these systems are highly nonlinear and demand accurate performance.

To control such systems, we make use of the notion of *two-degree-of freedom* controller design. This is a standard technique in linear control theory that separates a controller into a feedforward compensator and a feedback compensator. The feedforward compensator generates the nominal input required to track a given reference trajectory. The feedback compensator corrects for errors between the desired and actual trajectories. This is shown schematically in Figure 9.1.

In a nonlinear setting, two-degree-of-freedom controller design decouples the trajectory generation and asymptotic tracking problems. Given a desired output trajectory, we first construct a state space trajectory x_d and a nominal input u_d that satisfy the equations of motion. The error system can then be written as a time-varying control system in terms of the error, $e = x - x_d$. Under the assumption that that tracking error remains small, we can linearize this time-varying system about $e = 0$ and stabilize the $e = 0$ state. A more detailed description of this approach, including references to some of the related literature, is given in reference 1.

In optimization-based control, we use the two-degree-of-freedom paradigm with an optimal control computation for generating the feasible trajectory. In addition, we take the extra step of updating the generated trajectory based on the current state of the system. This additional feedback path is denoted by a dashed line in Figure 9.1 and allows the use of so-called *receding horizon* control techniques: a (optimal) feasible trajectory is computed from the current position to the desired position over a finite-time T horizon, used for a short period of time $\delta < T$, and then recomputed based on the new position.

Many variations on this approach are possible, blurring the line between the trajectory generation block and the feedback compensation. For example, if $\delta \ll T$, one can eliminate all or part of the "inner loop" feedback compensator, relying on the receding horizon optimization to stabilize the system. A local feedback compensator may still be employed to correct for errors due to noise and uncertainty on the fastest time scales. In this chapter, we will explore both (a) the case where we have a relatively large δ, in which case we consider the problem to be primarily one of trajectory generation, and (b) the case where we have a relatively small δ, where optimization is used for stabilizing the system.

A key advantage of optimization-based approaches is that they allow the potential for customization of the controller based on changes in *mission*, *condition*, and *environment*. Because the controller is solving the optimization problem online, updates can be made to the cost function, to change the desired operation of the system; to the model, to reflect changes in parameter values or damage to sensors and actuators; and to the constraints, to reflect new regions of the state space that must be avoided due to external influences. Thus, many of the challenges of designing controllers that are robust to a large set of possible uncertainties become embedded in the online optimization.

Development and application of receding horizon control (also called model predictive control, or MPC) originated in process control industries where plants being controlled are sufficiently slow to permit its implementation. An overview of the evolution of commercially available MPC technology is given in reference 2, and a survey of the current state of stability theory of MPC is given in reference 3. Closely related to the work in this chapter, Singh and Fuller [4] have used MPC to stabilize a linearized simplified UAV helicopter model around an open-loop trajectory, while respecting state and input constraints.

In the remainder of this chapter, we give a survey of the tools required to implement online control customization via optimization-based control. Section 9.2 introduces some of the mathematical results and notation required for the remainder of the chapter. Section 9.3 gives the main theoretical results of the chapter, where the problem of receding horizon control using a control Lyapunov function (CLF) as a terminal cost is described. In Section 9.4, we provide a computational framework for computing optimal trajectories in real-time, a necessary step toward implementation of optimization-based control in many applications. Finally, in Section 9.5, we present an experimental implementation of both real-time trajectory generation and model-predictive control on a flight control experiment.

The results in this chapter are based in part on work presented elsewhere. The work on receding horizon control using a CLF terminal cost was developed by Jadbabaie et al. [5]. The real-time trajectory generation framework, and the corresponding software, was developed by Milam and coworkers [6–8]. The implementation of model predictive control given in this chapter is based on the work of Dunbar, Milam, and Franz [9, 10].

9.2. MATHEMATICAL PRELIMINARIES

In this section we provide some mathematical preliminaries and establish the notation used through the chapter. We consider a nonlinear control system of the form

$$\dot{x} = f(x, u) \qquad (9.1)$$

where the vector field $f \colon \mathbb{R}^n \times \mathbb{R}^m \to \mathbb{R}^n$ is at least C^2 and possesses a linearly controllable equilibrium point at the origin; for example, $f(0, 0) = 0$ and $(A, B) := (D_1 f(0, 0), D_2 f(0, 0))$ is controllable.

9.2.1. Differential Flatness and Trajectory Generation

For optimization-based control in applications such as flight control, a critical need is the ability to compute optimal trajectories *very* quickly, so that they can be used in a real-time setting. For general problems this can be very difficult, but there are classes of systems for which simplifications can be made that vastly reduce the computational requirements for generating trajectories. We describe one such class of systems here, the so-called differentially flat systems.

Roughly speaking, a system is said to be differentially flat if all of the feasible trajectories for the system can be written as functions of a flat output $z(\cdot)$ and its derivatives. More precisely, given a nonlinear control system (9.1) we say the system is *differentially flat* if there exists a function $z(x, u, \dot{u}, \ldots, u^{(p)})$ such that all feasible solutions of the differential equation (9.1) can be written as

$$
\begin{aligned}
x &= \alpha(z, \dot{z}, \ldots, z^{(q)}) \\
u &= \beta(z, \dot{z}, \ldots, z^{(q)})
\end{aligned}
\qquad (9.2)
$$

Differentially flat systems were originally studied by Fliess et al. [11]. See reference 1 for a description of the role of flatness in control of mechanical systems, and see reference 12 for more information on flatness applied to flight control systems.

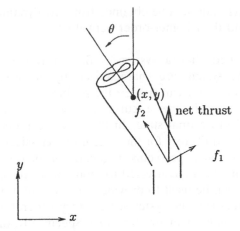

Figure 9.2. Planar ducted fan engine. Thrust is vectored by moving the flaps at the end of the duct.

Example 1 (Planar Ducted Fan). Consider the dynamics of a planar, vectored thrust flight control system as shown in Figure 9.2. This system consists of a rigid body with body fixed forces and is a simplified model for the Caltech ducted fan described in Section 9.5. Let (x, y, θ) denote the position and orientation of the center of mass of the fan. We assume that the forces acting on the fan consist of (a) a force f_1 perpendicular to the axis of the fan acting at a distance r from the center of mass and (b) a force f_2 parallel to the axis of the fan. Let m be the mass of the fan, J the moment of inertia, and g the gravitational constant. We ignore aerodynamic forces for the purpose of this example.

The dynamics for the system are

$$m\ddot{x} = f_1 \cos \theta - f_2 \sin \theta$$
$$m\ddot{y} = f_1 \sin \theta + f_2 \cos \theta - mg \qquad (9.3)$$
$$J\ddot{\theta} = r f_1$$

Martin et al. [13] showed that this system is differentially flat and that one set of flat outputs is given by

$$z_1 = x - (J/mr)\sin \theta$$
$$z_2 = y + (J/mr)\cos \theta \qquad (9.4)$$

Using the system dynamics, it can be shown that

$$\ddot{z}_1 \cos \theta + (\ddot{z}_2 + g)\sin \theta = 0 \qquad (9.5)$$

and thus given $z_1(t)$ and $z_2(t)$ we can find $\theta(t)$ except for an ambiguity of π and away from the singularity $\ddot{z}_1 = \ddot{z}_2 + g = 0$. The remaining states and the forces $f_1(t)$ and $f_2(t)$ can then be obtained from the dynamic equations, all in terms of z_1, z_2, and their higher-order derivatives.

Having determined that a system is flat, it follows that all feasible trajectories for the system are characterized by the evolution of the flat outputs. Using this fact, we can convert the problem of point to point motion generation to one of finding a curve $z(\cdot)$ which joins an initial state $z(0), \dot{z}(0), \ldots, z^{(q)}(0)$ to a final state. In this way, we reduce the problem of generating a feasible trajectory for the system to a classical algebraic problem in interpolation. Similarly, problems in generation of trajectories to track a reference signal can also be converted to problems involving curves $z(\cdot)$, and algebraic methods can be used to provide real-time solutions [12, 14].

Thus, for differentially flat systems, trajectory generation can be reduced from a dynamic problem to an algebraic one. Specifically, one can parameterize the flat outputs using basis functions $\phi_i(t)$,

$$z = \sum a_i \phi_i(t) \tag{9.6}$$

and then write the feasible trajectories as functions of the coefficients a:

$$
\begin{aligned}
x_d &= \alpha(z, \dot{z}, \ldots, z^{(q)}) = x_d(a) \\
u_d &= \beta(z, \dot{z}, \ldots, z^{(q)}) = u_d(a)
\end{aligned}
\tag{9.7}
$$

Note that no ODEs need to be integrated in order to compute the feasible trajectories (unlike optimal control methods, which involve parameterizing the *input* and then solving the ODEs). This is the defining feature of differentially flat systems. The practical implication is that nominal trajectories and inputs which satisfy the equations of motion for a differentially flat system can be computed in a computationally efficient way (solution of algebraic equations).

9.2.2. Control Lyapunov Functions

For the optimal control problems that we introduce in the next section, we will make use of a terminal cost that is also a control Lyapunov function for the system. Control Lyapunov functions are an extension of standard Lyapunov functions and were originally introduced by Sontag [15]. They allow constructive design of nonlinear controllers and the Lyapunov function that proves their stability. A more complete treatment is given in reference 16.

Definition 1. Control Lyapunov Function. A locally positive function $V: \mathbb{R}^n \to \mathbb{R}_+$ is called a *control Lyapunov function* (CLF) for a control system (9.1) if

$$\inf_{u \in \mathbb{R}^m} \left(\frac{\partial V}{\partial x} f(x, u) \right) < 0 \qquad \text{for all } x \neq 0$$

In general, it is difficult to find a CLF for a given system. However, for many classes of systems, there are specialized methods that can be used. One of the simplest is to use the Jacobian linearization of the system around the desired equilibrium point and generate a CLF by solving an LQR problem.

It is a well-known result that the problem of minimizing the quadratic performance index,

$$J = \int_0^\infty \left(x^T(t)Qx(t) + u^T Ru(t) \right) dt \qquad \text{subject to} \quad \dot{x} = Ax + Bu, \quad x(0) = x_0$$

(9.8)

results in finding the positive definite solution of the following Riccati equation:

$$A^T P + PA - PBR^{-1}B^T P + Q = 0 \qquad (9.9)$$

The optimal control action is given by

$$u = -R^{-1}B^T Px$$

and $V = x^T Px$ is a CLF for the system.

In the case of the nonlinear system $\dot{x} = f(x, u)$, A and B are taken as

$$A = \frac{\partial f(x, u)}{\partial x}\bigg|_{(0,0)}, \qquad B = \frac{\partial f(x, u)}{\partial u}\bigg|_{(0,0)}$$

where the pairs (A, B) and $(Q^{\frac{1}{2}}, A)$ are assumed to be stabilizable and detectable respectively. Obviously the obtained CLF $V(x) = x^T Px$ will be valid only in a region around the equilibrium $(0, 0)$.

More complicated methods for finding control Lyapunov functions are often required and many techniques have been developed. An overview of some of these methods can be found in reference 17.

9.3. OPTIMIZATION-BASED CONTROL

In receding horizon control, a finite horizon optimal control problem is solved, generating an open-loop state and control trajectories. The resulting control trajectory is then applied to the system for a fraction of the horizon length. This process is then repeated, resulting in a sampled data feedback law. Although receding horizon control has been successfully used in the process control industry, its application to fast, stability critical nonlinear systems has been more difficult. This is mainly due to two issues. The first is that the finite horizon optimizations must be solved in a relatively short period of time. Second, it can be demonstrated using linear examples that a naive application of the receding horizon strategy can have disastrous effects, often rendering a system unstable. Various approaches have been proposed to tackle this second problem; see reference 3 for a comprehensive review of this literature. The theoretical framework presented here also addresses the stability issue directly, but is motivated by the need to relax the computational demands of existing stabilizing MPC formulations.

A number of approaches in receding horizon control employ the use of terminal state equality or inequality constraints, often together with a terminal cost, to ensure closed loop stability. In Primbs et al. [18], aspects of a stability-guaranteeing, global control Lyapunov function were used, via state and control constraints, to develop a stabilizing receding horizon scheme. Many of the nice characteristics of the CLF controller together with better cost performance were realized. Unfortunately, a global control Lyapunov function is rarely available and often not possible.

Motivated by the difficulties in solving constrained optimal control problems, we have developed an alternative receding horizon control strategy for the stabilization of nonlinear systems [5]. In this approach, closed-loop stability is ensured through the use of a terminal cost consisting of a control Lyapunov function that is an incremental upper bound on the optimal cost to go. This terminal cost eliminates the need for terminal constraints in the optimization and gives a dramatic speed-up in computation. Also, questions of existence and regularity of optimal solutions (very important for online optimization) can be dealt with in a rather straightforward manner. In the remainder of this section, we review the results presented in reference 5.

9.3.1. Finite Horizon Optimal Control

We first consider the problem of optimal control over a finite-time horizon. Given an initial state x_0 and a control trajectory $u(\cdot)$ for a nonlinear control system $\dot{x} = f(x, u)$, the state trajectory $x^u(\cdot; x_0)$ is the (absolutely continuous) curve in $\mathbb{R}^n$ satisfying

$$x^u(t; x_0) = x_0 + \int_0^t f(x^u(\tau; x_0), u(\tau)) \, d\tau$$

for $t \geq 0$.

The performance of the system will be measured by a given incremental cost $q: \mathbb{R}^n \times \mathbb{R}^m \to \mathbb{R}$ that is C^2 and fully penalizes both state and control according to

$$q(x, u) \geq c_q(\|x\|^2 + \|u\|^2), \qquad x \in \mathbb{R}^n, u \in \mathbb{R}^m$$

for some $c_q > 0$ and $q(0,0) = 0$. It follows that the quadratic approximation of q at the origin is positive definite, $D^2 q(0,0) \geq c_q I > 0$.

To ensure that the solutions of the optimization problems of interest are well-behaved, we impose some convexity conditions. We require the set $f(x, \mathbb{R}^m) \subset \mathbb{R}^n$ to be convex for each $x \in \mathbb{R}^n$. Letting $p \in \mathbb{R}^n$ represent the co-state, we also require that the pre-Hamiltonian function $u \mapsto p^T f(x, u) + q(x, u) := K(x, u, p)$ be strictly convex for each $(x, p) \in \mathbb{R}^n \times \mathbb{R}^n$ and that there be a C^2 function $\bar{u}^*\colon \mathbb{R}^n \times \mathbb{R}^n \to \mathbb{R}^m\colon (x, p) \mapsto \bar{u}^*(x, p)$ providing the global minimum of $K(x, u, p)$. The Hamiltonian $H(x, p) := K(x, \bar{u}^*(x, p), p)$ is then C^2, ensuring that extremal state, co-state, and control trajectories will all be sufficiently smooth (C^1 or better). Note that these conditions are trivially satisfied for control affine f and quadratic q.

The cost of applying a control $u(\cdot)$ from an initial state x over the infinite time interval $[0, \infty)$ is given by

$$J_\infty(x, u(\cdot)) = \int_0^\infty q(x^u(\tau; x), u(\tau)) \, d\tau$$

The optimal cost (from x) is given by

$$J_\infty^*(x) = \inf_{u(\cdot)} J_\infty(x, u(\cdot))$$

where the control functions $u(\cdot)$ belong to some reasonable class of admissible controls (e.g., piecewise continuous or measurable). The function $x \mapsto J_\infty^*(x)$ is often called the *optimal value function* for the infinite horizon optimal control problem.

For the class of f and q considered, we know that $J_\infty^*(\cdot)$ is a positive definite C^2 function on a neighborhood of the origin. This follows from the *geometry* of the corresponding Hamiltonian system (see reference 19 and the references therein). In particular, since $(x, p) = (0, 0)$ is a hyperbolic critical point of the C^1 Hamiltonian vector field $X_H(x, p) := (D_2 H(x, p), -D_1 H(x, p))^T$, the local properties of $J_\infty^*(\cdot)$ are determined by the linear-quadratic approximation to the problem; moreover, $D^2 J_\infty^*(0) = P > 0$, where P is the stabilizing solution of the appropriate algebraic Riccati equation.

For practical purposes, we are interested in finite horizon approximations of the infinite horizon optimization problem. In particular, let $V(\cdot)$ be a nonnegative C^2 function with $V(0) = 0$ and define the finite horizon cost (from x using $u(\cdot)$) to be

$$J_T(x, u(\cdot)) = \int_0^T q(x^u(\tau; x), u(\tau)) \, d\tau + V(x^u(T; x)) \tag{9.10}$$

and denote the optimal cost (from x) as

$$J_T^*(x) = \inf_{u(\cdot)} J_T(x, u(\cdot))$$

As in the infinite horizon case, one can show, by geometric means, that $J_T^*(\cdot)$ is locally smooth (C^2). Other properties will depend on the choice of V and T.

Let Γ^∞ denote the domain of $J_\infty^*(\cdot)$ (the subset of $\mathbb{R}^n$ on which J_∞^* is finite). It is not too difficult to show that the cost functions $J_\infty^*(\cdot)$ and $J_T^*(\cdot)$, $T \geq 0$, are continuous functions on Γ_∞ [16]. For simplicity, we will allow $J_\infty^*(\cdot)$

to take values in the extended real line so that, for instance, $J_\infty^*(x) = +\infty$ means that there is no control taking x to the origin.

We will assume that f and q are such that the minimum value of the cost functions $J_\infty^*(x)$, $J_T^*(x)$, $T \geq 0$, is attained for each (suitable) x. That is, given x and $T > 0$ (including $T = \infty$ when $x \in \Gamma^\infty$), there is a (C^1 in t) optimal trajectory $(x_T^*(t; x), u_T^*(t; x))$, $t \in [0, T]$, such that $J_T(x, u_T^*(\cdot; x)) = J_T^*(x)$. For instance, if f is such that its trajectories can be bounded on finite intervals as a function of its input size—for example, there is a continuous function β such that $\|x^u(t; x_0)\| \leq \beta(\|x_0\|, \|u(\cdot)\|_{L_1[0, t]})$,—then (together with the conditions above) there will be a minimizing control (cf. reference 20). Many such conditions may be used to good effect; see reference 17 for a more complete discussion.

It is easy to see that $J_\infty^*(\cdot)$ is proper on its domain so that the sublevel sets

$$\Gamma_r^\infty := \left\{ x \in \Gamma^\infty : J_\infty^*(x) \leq r^2 \right\}$$

are compact and path connected and moreover $\Gamma^\infty = \bigcup_{r \geq 0} \Gamma_r^\infty$. Note also that Γ^∞ may be a proper subset of $\mathbb{R}^n$ since there may be states that cannot be driven to the origin. We use r^2 (rather than r) here to reflect the fact that our incremental cost is quadratically bounded from below. We refer to sublevel sets of $J_T^*(\cdot)$ and $V(\cdot)$ using

$$\Gamma_r^T := \text{path connected component of } \left\{ x \in \Gamma^\infty : J_T^*(x) \leq r^2 \right\} \text{ containing } 0$$

and

$$\Omega_r := \text{path connected component of } \left\{ x \in \mathbb{R}^n : V(x) \leq r^2 \right\} \text{ containing } 0$$

These results provide the technical framework needed for receding horizon control.

9.3.2. Receding Horizon Control with CLF Terminal Cost

Receding horizon control provides a practical strategy for the use of model information through online optimization. Every δ seconds, an optimal control problem is solved over a T second horizon, starting from the current state. The first δ seconds of the optimal control $u_T^*(\cdot; x(t))$ is then applied to the system, driving the system from $x(t)$ at current time t to $x_T^*(\delta, x(t))$ at the next sample time $t + \delta$ (assuming no model uncertainty). We denote this receding horizon scheme as $\mathcal{RH}(T, \delta)$.

In defining (unconstrained) finite horizon approximations to the infinite horizon problem, the key design parameters are the terminal cost function $V(\cdot)$ and the horizon length T (and, perhaps also, the increment δ). What choices will result in success?

It is well known (and easily demonstrated with linear examples) that simple truncation of the integral (i.e., $V(x) \equiv 0$) may have disastrous effects if $T > 0$ is too small. Indeed, although the resulting value function may be nicely behaved, the "optimal" receding horizon closed-loop system can be unstable.

A more sophisticated approach is to make good use of a suitable terminal cost $V(\cdot)$. Evidently, the best choice for the terminal cost is $V(x) = J_x^*(x)$ since then the optimal finite and infinite horizon costs are the same. Of course, if the optimal value function were available, there would be no need to solve a trajectory optimization problem. What properties of the optimal value function should be retained in the terminal cost? To be effective, the terminal cost should account for the discarded tail by ensuring that the origin can be reached from the terminal state $x^u(T; x)$ in an efficient manner (as measured by q). One way to do this is to use an appropriate control Lyapunov function which is also an upper bound on the cost-to-go.

The following theorem shows that the use of a particular type of CLF is in fact effective, providing rather strong and specific guarantees.

Theorem   9.1 [5]. *Suppose that the terminal cost $V(\cdot)$ is a control Lyapunov function such that*

$$\min_{u \in \mathbb{R}^m} (\dot{V} + q)(x, u) \leq 0 \qquad (9.11)$$

for each $x \in \Omega_{r_v}$ for some $r_v > 0$. Then, for every $T > 0$ and $\delta \in (0, T]$, the resulting receding horizon trajectories go to zero exponentially fast. For each $T > 0$, there is an $\bar{r}(T) \geq r_v$ such that $\Gamma_{\bar{r}(T)}^T$ is contained in the region of attraction of $\mathcal{RH}(T, \delta)$. Moreover, given any compact subset Λ of Γ^∞, there is a T^* such that $\Lambda \subset \Gamma_{\bar{r}(T)}^T$ for all $T \geq T^*$.

Theorem 9.1 shows that for *any* horizon length $T > 0$ and *any* sampling time $\delta \in (0, T]$, the receding horizon scheme is exponentially stabilizing over the set $\Gamma_{r_v}^T$. For a given T, the region of attraction estimate is enlarged by increasing r beyond r_v to $\bar{r}(T)$ according to the requirement that $V(x_T^*(T; x)) \leq r_v^2$ on that set. An important feature of the above result is that, for operations with the set $\Gamma_{\bar{r}(T)}^T$, there is no need to impose stability ensuring constraints that would likely make the online optimizations more difficult and time consuming to solve.

An important benefit of receding horizon control is its ability to handle state and control constraints. While the above theorem provides stability guarantees when there are no constraints present, it can be modified to include constraints on states and controls as well. In order to ensure stability when state and control constraints are present, the terminal cost $V(\cdot)$ should be a local CLF satisfying $\min_{u \in \mathcal{U}} \dot{V} + q(x, u) \leq 0$ where $\mathcal{U}$ is the set of controls where the control constraints are satisfied. Moreover, one should also require that the resulting state trajectory $x^{CLF}(\cdot) \in \mathcal{X}$, where $\mathcal{X}$ is the set of states where the constraints are satisfied. (Both $\mathcal{X}$ and $\mathcal{U}$ are assumed to be compact with origin in their interior). Of course, the set Ω_{r_v} will end up being smaller than before, resulting in a decrease in the size of the guaranteed region of operation (see reference 3 for more details).

9.4. REAL-TIME TRAJECTORY GENERATION AND DIFFERENTIAL FLATNESS

In this section we demonstrate how to use differential flatness to find fast numerical algorithms for solving the optimal control problems required for the receding horizon control results of the previous section. We consider the affine nonlinear control system

$$\dot{x} = f(x) + g(x)u \tag{9.12}$$

where all vector fields and functions are smooth. For simplicity, we focus on the single input case, $u \in \mathbb{R}$. We wish to find a trajectory of Eq. (9.12) that minimizes the performance index (9.10), subject to a vector of initial, final, and trajectory constraints

$$lb_0 \le \psi_0\big(x(t_0), u(t_0)\big) \le ub_0$$

$$lb_f \le \psi_f\big(x(t_f), u(t_f)\big) \le ub_f \tag{9.13}$$

$$lb_t \le S(x, u) \le ub_t$$

respectively. For conciseness, we will refer to this optimal control problem as

$$\min_{(x,u)} J(x, u) \quad \text{subject to} \quad \begin{cases} \dot{x} = f(x) + g(x)u \\ lb \le c(x, u) \le ub \end{cases} \tag{9.14}$$

9.4.1. Numerical Solution Using Collocation

A numerical approach to solving this optimal control problem is to use the direct collocation method outlined in Hargraves and Paris [21]. The idea behind this approach is to transform the optimal control problem into a nonlinear programming problem. This is accomplished by discretizing time into a grid of $N - 1$ intervals

$$t_0 = t_1 < t_2 < \cdots < t_N = t_f \tag{9.15}$$

and approximating the state x and the control input u as piecewise polynomials $\hat{x}$ and $\hat{u}$, respectively. Typically, a cubic polynomial is chosen for the states and a linear polynomial for the control on each interval. Collocation is then used at the midpoint of each interval to satisfy Eq. (9.12). Let $\hat{x}(x(t_1), \ldots, x(t_N))$ and $\hat{u}(u(t_1), \ldots, u(t_N))$ denote the approximations to x and u, respectively, depending on $(x(t_1), \ldots, x(t_N)) \in \mathbb{R}^{nN}$ and $(u(t_1), \ldots, u(t_N)) \in \mathbb{R}^N$ corresponding to the value of x and u at the grid points. Then one solves the following finite-dimension approximation of the original control problem [9.14]:

$$\min_{y \in \mathbb{R}^M} F(y) = J(\hat{x}(y), \hat{u}(y)) \quad \text{subject to} \quad \begin{cases} \dot{\hat{x}} - f(\hat{x}(y)) + g(\hat{x}(y))\hat{u}(y) = 0 \\ lb \le c(\hat{x}(y), \hat{u}(y)) \le ub \\ \forall t = \dfrac{t_j + t_{j+1}}{2} \quad j = 1, \dots, N-1 \end{cases}$$

$$(9.16)$$

where $y = (x(t_1), u(t_1), \dots, x(t_N), u(t_N))$, and $M = \dim y = (n+1)N$.

Seywald [22] suggested an improvement to the previous method (see also reference 23). Following this work, one first solves a subset of system dynamics in (9.14) for the the control in terms of combinations of the state and its time derivative. Then one substitutes for the control in the remaining system dynamics and constraints. Next all the time derivatives $\dot{x}_i$ are approximated by the finite-difference approximations

$$\dot{\bar{x}}(t_i) = \frac{x(t_{i+1}) - x(t_i)}{t_{i+1} - t_i}$$

to get

$$\left. \begin{array}{l} p(\bar{x}(t_i), x(t_i)) = 0 \\ q(\bar{x}(t_i), x(t_i)) \le 0 \end{array} \right\} \quad i = 0, \dots, N-1$$

The optimal control problem is turned into

$$\min_{y \in \mathbb{R}^M} F(y) \quad \text{subject to} \quad \begin{cases} p(\dot{\bar{x}}(t_i), x(t_i)) = 0 \\ q(\dot{\bar{x}}(t_i), x(t_i)) \le 0 \end{cases} \qquad (9.17)$$

where $y = (x(t_1), \dots, x(t_N))$, and $M = \dim y = nN$. As with the Hargraves and Paris method, this parameterization of the optimal control problem (9.14) can be solved using nonlinear programming.

The dimensionality of this discretized problem is lower than the dimensionality of the Hargraves and Paris method, where both the states and the input are the unknowns. This induces substantial improvement in numerical implementation.

9.4.2. Differential-Flatness-Based Approach

The results of Seywald give a constrained optimization problem in which we wish to minimize a cost functional subject to $n - 1$ equality constraints, corresponding to the system dynamics, at each time instant. In fact, it is usually possible to reduce the dimension of the problem further. Given an output, it is generally possible to parameterize the control and a part of the state in terms of this output and its time derivatives. In contrast to the previous approach, one must use more than one derivative of this output for this purpose.

When the whole state and the input can be parameterized with one output, the system is differentially flat, as described in Section 9.2. When the parameterization is only partial, the dimension of the subspace spanned by the output and its derivatives is given by r the *relative degree* of this output [24]. In this case, it is possible to write the system dynamics as

$$x = \alpha(z, \dot{z}, \ldots, z^{(q)})$$

$$u = \beta(z, \dot{z}, \ldots, z^{(q)}) \qquad (9.18)$$

$$\Phi(z, z, \ldots, z^{n-r}) = 0$$

where $z \in \mathbb{R}^p$, $p > m$ represents a set of outputs that parameterize the trajectory and $\Phi: \mathbb{R}^n \times \mathbb{R}^m$ represents $n - r$ remaining differential constraints on the output. In the case that the system is flat, $r = n$ and we eliminate these differential constraints.

Unlike the approach of Seywald, it is not realistic to use finite-difference approximations as soon as $r > 2$. In this context, it is convenient to represent z using B-splines. B-splines are chosen as basis functions because of their ease of enforcing continuity across knot points and ease of computing their derivatives. A pictorial representation of such an approximation is given in Figure 9.3. Doing so, we get

$$z_j = \sum_{i=1}^{p_j} B_{i,k_j}(t) C_i^j, \qquad p_j = l_j(k_j - m_j) + m_j$$

where $B_{i,k_j}(t)$ is the B-spline basis function defined in reference 25 for the output z_j with order k_j, C_i^j are the coefficients of the B-spline, l_j is the number of knot intervals, and m_j is number of smoothness conditions at the

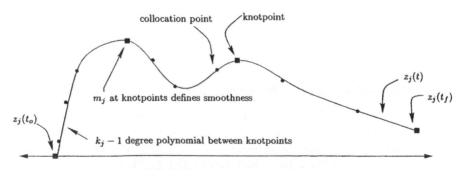

Figure 9.3. Spline representation of a variable.

knots. The set $(z_1, z_2, \ldots, z_{n-r})$ is thus represented by $M = \Sigma_{j \in \{1, r+1, \ldots, n\}} P_j$ coefficients.

In general, w collocation points are chosen uniformly over the time interval $[t_o, t_f]$ (though optimal knots placements or Gaussian points may also be considered). Both dynamics and constraints will be enforced at the collocation points. The problem can be stated as the following nonlinear programming form:

$$\min_{y \in \mathbb{R} R^M} F(y) \quad \text{subject to} \quad \begin{cases} \Phi\left(z(y), \dot{z}(y), \ldots, z^{(n-r)}(y)\right) = 0 \\ lb \leq c(y) \leq ub \end{cases} \quad (9.19)$$

where

$$y = \left(C_1^1, \ldots, C_{p_1}^1, C_1^{r+1}, \ldots, C_{p_{r+1}}^{r+1}, \ldots, C_1^n, \ldots, C_{p_n}^n\right)$$

The coefficients of the B-spline basis functions can be found using nonlinear programming.

A software package called Nonlinear Trajectory Generation (NTG) has been written to solve optimal control problems in the manner described above (see reference 7 for details). The sequential quadratic programming package NPSOL by reference 26 is used as the nonlinear programming solver in NTG. When specifying a problem to NTG, the user is required to state the problem in terms of some choice of outputs and its derivatives. The user is also required to specify the regularity of the variables, the placement of the knot points, the order and regularity of the B-splines, and the collocation points for each output.

9.5. IMPLEMENTATION ON THE CALTECH DUCTED FAN

To demonstrate the use of the techniques described in the previous section, we present an implementation of optimization-based control on the Caltech Ducted Fan, a real-time, flight control experiment that mimics the longitudinal dynamics of an aircraft. The experiment is shown in Figure 9.4.

9.5.1. Description of the Caltech Ducted Fan Experiment

The Caltech ducted fan is an experimental testbed designed for research and development of nonlinear flight guidance and control techniques for Uninhabited Combat Aerial Vehicles (UCAVs). The fan is a scaled model of the longitudinal axis of a flight vehicle, and flight test results validate that the dynamics replicate qualities of actual flight vehicles [27].

The ducted fan has three degrees of freedom: The boom holding the ducted fan is allowed to operate on a cylinder, 2 m high and 4.7 m in diameter, permitting horizontal and vertical displacements. Also, the wing/fan

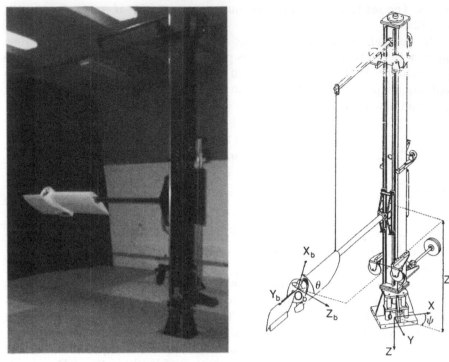

Figure 9.4. Caltech ducted fan.

assembly at the end of the boom is allowed to rotate about its center of mass. Optical encoders mounted on the ducted fan, gearing wheel, and the base of the stand measure the three degrees of freedom. The fan is controlled by commanding a current to the electric motor for fan thrust and by commanding RC servos to control the thrust vectoring mechanism.

The sensors are read and the commands sent by a dSPACE multiprocessor system, comprised of a D/A card, a digital I/O card, two Texas Instruments C40 signal processors, two Compaq Alpha processors, and an ISA bus to interface with a PC. The dSPACE system provides a real-time interface to the four processors and I/O card to the hardware. The NTG software resides on both of the Alpha processors, each capable of running real-time optimization.

The ducted fan is modeled in terms of the position and orientation of the fan, along with their velocities. Letting x represent the horizontal translation, z the vertical translation, and θ the rotation about the boom axis, the equations of motion are given by

$$m\ddot{x} + F_{X_a} - F_{X_b}\cos\theta - F_{Z_b}\sin\theta = 0$$

$$m\ddot{z} + F_{Z_a} + F_{X_b}\sin\theta - F_{Z_b}\cos\theta = mg_{\text{eff}} \qquad (9.20)$$

$$J\ddot{\theta} - M_a + \frac{1}{r_s}I_p\Omega\dot{x}\cos\theta - F_{Z_b}r_f = 0$$

where $F_{X_a} = D \cos \gamma + L \sin \gamma$ and $F_{Z_a} = -D \sin \gamma + L \cos \gamma$ are the aerodynamic forces and F_{X_b} and F_{Z_b} are thrust vectoring body forces in terms of the lift (L), drag (D), and flight path angle (γ). I_p and Ω are the moment of inertia and angular velocity of the ducted fan propeller, respectively. J is the moment of ducted fan and r_f is the distance from center of mass along the X_b axis to the effective application point of the thrust vectoring force. The angle of attack α can be derived from the pitch angle θ and the flight path angle γ by

$$\alpha = \theta - \gamma$$

The flight path angle can be derived from the spatial velocities by

$$\gamma = \arctan \frac{-\dot{z}}{\dot{x}}$$

The lift (L), drag (D), and moment (M) are given by

$$L = qSC_L(\alpha), \qquad D = qSC_D(\alpha), \qquad M = \bar{c}SC_M(\alpha)$$

respectively. The dynamic pressure is given by $q = \frac{1}{2}\rho V^2$. The norm of the velocity is denoted by V, S is the surface area of the wings, and ρ is the atmospheric density. The coefficients of lift $(C_L(\alpha))$, drag $(C_D(\alpha))$ and the moment coefficient $(C_M(\alpha))$ are determined from a combination of wind tunnel and flight testing and are described in more detail in reference 27, along with the values of the other parameters.

9.5.2. Real-Time Trajectory Generation

In this section we demonstrate the trajectory generation results by using NTG to generate minimum time trajectories in real time. An LQR-based regulator is used to stabilize the system, allowing us to focus on the trajectory generation properties. We focus in this section on aggressive, forward flight trajectories. The next section extends the controller to use a receding horizon controller, but on a simpler class of trajectories.

Stabilization Around Reference Trajectory. The results in this section rely on the traditional two-degree-of-freedom design paradigm described in Section 9.1. In this approach, a local control law (inner loop) is used to stabilize the system around the trajectory computed based on a nominal model. This compensates for uncertainties in the model, which are predominantly due to aerodynamics and friction. Elements such as the ducted fan flying through its own wake, ground effects, and thrust not modeled as a function of velocity and angle of attack contribute to the aerodynamic uncertainty. The friction in the vertical direction is also not considered in the model. The prismatic joint has an unbalanced load creating an effective moment on the bearings. The vertical frictional force of the ducted fan stand varies with the vertical acceleration of the ducted fan as well as the forward velocity. Actuation models are not used when generating the reference trajectory, resulting in another source of uncertainty.

Since only the position of the fan is measured, we must estimate the velocities. We use an extended Kalman filter in which the optimal gain matrix is gain scheduled on the (estimated) forward velocity. The Kalman filter outperformed other methods that computed the derivative using only the position data and a filter.

The stabilizing LQR controllers were gain scheduled on pitch angle, θ, and the forward velocity, $\dot{x}$. The pitch angle was allowed to vary from $-\pi/2$ to $\pi/2$ and the velocity ranged from 0 to 6 m/s. The weights were chosen differently for the hover-to-hover and forward flight modes. For the forward flight mode, a smaller weight was placed on the horizontal (x) position of the fan compared to the hover-to-hover mode. Furthermore, the z weight was scheduled as a function of forward velocity in the forward flight mode. There was no scheduling on the weights for hover-to-hover. The elements of the gain matrices for each of the controller and observer are linearly interpolated over 51 operating points.

Nonlinear Trajectory Generation Parameters. We solve a minimum time optimal control problem to generate a feasible trajectory for the system. The system is modeled using the nonlinear equations described above, and it computed the open-loop forces and state trajectories for the nominal system. This system is not known to be differentially flat (due to the aerodynamic forces), and hence we cannot completely eliminate the differential constraints.

We choose three outputs, $z_1 = x$, $z_2 = z$, and $z_3 = \theta$, which results in a system with one remaining differential constraint. Each output is parameterized with four sixth-order, C^4 (multiplicity), piecewise polynomials over the time interval scaled by the minimum time. A fourth output, $z_4 = T$, is used to represent the time horizon to be minimized and is parameterized by a scalar. By choosing the outputs to be parameterized in this way, we are in effect controlling the frequency content of inputs. Since we are not including the actuators in the model, it would be undesirable to have inputs with a bandwidth higher than the actuators. There are a total of 37 variables in this optimization problem. The trajectory constraints are enforced at 21 equidistant breakpoints over the scaled time interval.

There are many considerations in the choice of the parameterization of the outputs. Clearly there is a trade between the parameters (variables, initial values of the variables, and breakpoints) and measures of performance (convergence, run time, and conservative constraints). Extensive simulations were run to determine the right combination of parameters to meet the performance goals of our system.

Forward Flight. To obtain the forward flight test data, the operator commanded a desired forward velocity and vertical position with the joysticks.

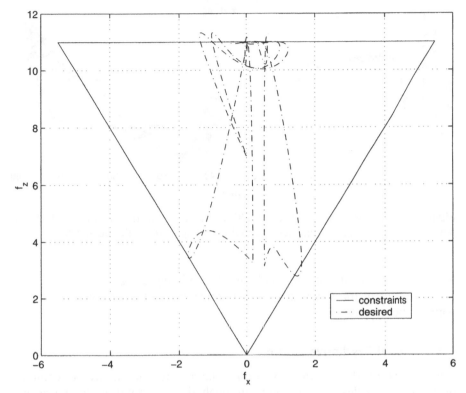

Figure 9.5. Forward flight-test case: (a) θ and $\dot{x}$ desired and actual, (b) desired F_{X_b} and F_{Z_b} with bounds.

We set the trajectory update time, δ, to 2 s. By rapidly changing the joysticks, NTG produces high-angle-of-attack maneuvers. Figure 9.5a depicts the reference trajectories and the actual θ and $\dot{x}$ over 60 s. Figure 9.5b shows the commanded forces for the same time interval. The sequence of maneuvers corresponds to the ducted fan transitioning from near hover to forward flight, then following a command from a large forward velocity to a large negative velocity, and finally returning to hover.

Figure 9.6 is an illustration of the ducted fan altitude and x position for these maneuvers. The airfoil in the figure depicts the pitch angle (θ). It is apparent from this figure that the stabilizing controller is not tracking well in the z direction. This is due to the fact that unmodeled frictional effects are significant in the vertical direction. This could be corrected with an integrator in the stabilizing controller.

An analysis of the run times was performed for 30 trajectories; the average computation time was less than 1 s. Each of the 30 trajectories converged to an optimal solution and was approximately between 4 and 12 s in length. A random initial guess was used for the first NTG trajectory computation. Subsequent NTG computations used the previous solution as an initial guess.

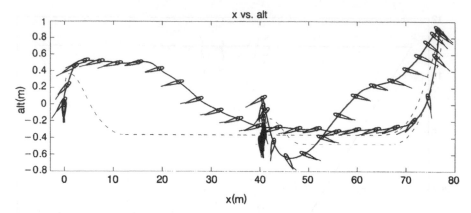

Figure 9.6. Forward flight-test case: altitude and x position (actual (solid) and desired (dashed)). Airfoil represents actual pitch angle (θ) of the ducted fan.

Much improvement can be made in determining a "good" initial guess. Improvement in the initial guess will improve not only convergence but also computation times.

9.5.3. Model Predictive Control

The results of the previous section demonstrate the ability to compute optimal trajectories in real time, although the computation time was not sufficiently fast for closing the loop around the optimization. In this section, we make use of a shorter update time δ, a fixed horizon time T with a quadratic integral cost, and a CLF terminal cost to implement the receding horizon controller described in Section 9.3. We also limit the operation of the system to near hover, so that we can use the local linearization to find the terminal CLF.

We have implemented the receding horizon controller on the ducted fan experiment where the control objective is to stabilize the hover equilibrium point. The quadratic cost is given by

$$q(x,u) = \tfrac{1}{2}\hat{x}^T Q\hat{x} + \tfrac{1}{2}\hat{x}^T R\hat{u}$$
$$V(x) = \gamma\hat{x}^T P\hat{x} \tag{9.21}$$

where

$$\hat{x} = x - x_{eq} = \left(x, z, \theta - \pi/2, \dot{x}, \dot{z}, \dot{\theta}\right)$$
$$\hat{u} = u - u_{eq} = \left(F_{X_b} - mg, F_{Z_b}\right)$$
$$Q = \text{diag}\{4, 15, 4, 1, 3, 0.3\}$$
$$R = \text{diag}\{0.5, 0.5\}$$

$\gamma = 0.075$ and P is the unique stable solution to the algebraic Riccati equation corresponding to the linearized dynamics of Eq. (9.3) at hover and the weights Q and R. Note that if $\gamma = 1/2$, then $V(\cdot)$ is the CLF for the system corresponding to the LQR problem. Instead V is a relaxed (in magnitude) CLF, which achieved better performance in the experiment. In either case, V is valid as a CLF only in a neighborhood around hover because it is based on the linearized dynamics. We do not try to compute offline a region of attraction for this CLF. Experimental tests omitting the terminal cost and/or the input constraints leads to instability. The results in this section show the success of this choice for V for stabilization. An inner-loop PD controller on θ, $\dot{\theta}$ is implemented to stabilize to the receding horizon states θ_T^*, $\dot{\theta}_T^*$. The θ dynamics are the fastest for this system and although most receding horizon controllers were found to be nominally stable without this inner-loop controller, small disturbances could lead to instability.

The optimal control problem is set-up in NTG code by parameterizing the three position states (x, z, θ), each with eight B-spline coefficients. Over the receding horizon time intervals, 11 and 16 breakpoints were used with horizon lengths of 1, 1.5, 2, 3, 4 and 6 s. Breakpoints specify the locations in time where the differential equations and any constraints must be satisfied, up to some tolerance. The value of $F_{X_b}^{\max}$ for the input constraints is made conservative to avoid prolonged input saturation on the real hardware. The logic for this is that if the inputs are saturated on the real hardware, no actuation is left for the inner-loop θ controller and the system can go unstable. The value used in the optimization is $F_{X_b}^{\max} = 9$ N.

Computation time is non-negligible and must be considered when implementing the optimal trajectories. The computation time varies with each optimization as the current state of the ducted fan changes. The following notational definitions will facilitate the description of how the timing is set up:

i	Integer counter of MPC computations
t_i	Value of current time when MPC computation i started
$\delta_c(i)$	Computation time for computation i
$u_T^*(i)(t)$	Optimal output trajectory corresponding to computation i, with time interval $t \in [t_i, t_i + T]$

A natural choice for updating the optimal trajectories for stabilization is to do so as fast as possible. This is achieved here by constantly resolving the optimization. When computation i is done, computation $i + 1$ is immediately started, so $t_{i+1} = t_i + \delta_c(i)$. Figure 9.7 gives a graphical picture of the timing setup as the optimal input trajectories $u_T^*(\cdot)$ are updated. As shown in the figure, any computation i for $u_T^*(i)(\cdot)$ occurs for $t \in [t_i, t_{i+1}]$ and the resulting trajectory is applied for $t \in [t_{i+1}, t_{i+2}]$. At $t = t_{i+1}$, computation, $i + 1$ is started for trajectory $u_T^*(i + 1)(\cdot)$, which is applied as soon as it is available

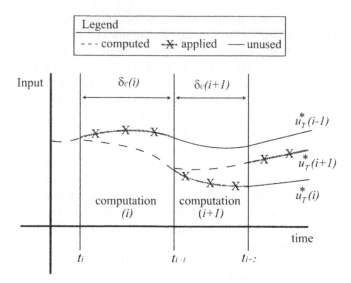

Figure 9.7. Receding horizon input trajectories.

$(t = t_{i+2})$. For the experimental runs detailed in the results, $\delta_c(i)$ is typically in the range of $[0.05, 0.25]$ seconds, meaning 4 to 20 optimal control computations per second. Each optimization i requires the current measured state of the ducted fan and the value of the previous optimal input trajectories $u_T^*(i-1)$ at time $t = t_i$. This corresponds to, respectively, six initial conditions for state vector x and two initial constraints on the input vector u. Figure 9.7 shows that the optimal trajectories are advanced by their computation time prior to application to the system. A dashed line corresponds to the initial portion of an optimal trajectory and is not applied since it is not available until that computation is complete. The figure also reveals the possible discontinuity between successive applied optimal input trajectories, with a larger discontinuity more likely for longer computation times. The initial input constraint is an effort to reduce such discontinuities, although some discontinuity is unavoidable by this method. Also note that the same discontinuity is present for the six open-loop optimal state trajectories generated, again with a likelihood for greater discontinuity for longer computation times. In this description, initialization is not an issue because we assume that the receding horizon computations are already running prior to any test runs. This is true of the experimental runs detailed in the results.

The experimental results show the response of the fan with each controller to a 6-m horizontal offset, which is effectively engaging a step response to a change in the initial condition for x. The following details the effects of different receding horizon control parameterizations, namely as the horizon changes, and the responses with the different controllers to the induced offset.

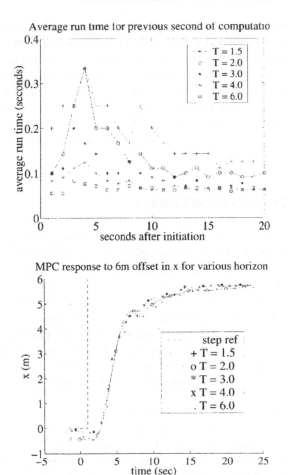

Figure 9.8. Receding horizon control: (a) moving one second average of computation time for MPC implementation with varying horizon time, (b) response of MPC controllers to 6-m offset in x for different horizon lengths.

The first comparison is between different receding horizon controllers, where time horizon is varied to be 1.5, 2.0, 3.0, 4.0 or 6.0 s. Each controller uses 16 breakpoints. Figure 9.8a shows a comparison of the average computation time as time proceeds. For each second after the offset was initiated, the data corresponds to the average run time over the previous second of computation. Note that these computation times are substantially smaller than those reported for real-time trajectory generation, due to the use of the CLF terminal cost versus the terminal constraints in the minimum-time, real-time trajectory generation experiments.

There is a clear trend toward shorter average computation times as the time horizon is made longer. There is also an initial transient increase in average computation time that is greater for shorter horizon times. In fact, the 6-s horizon controller exhibits a relatively constant average computation time. One explanation for this trend is that, for this particular test, a 6-s horizon is closer to what the system can actually do. After 1.5 s, the fan is still far from the desired hover position and the terminal cost CLF is large, likely far from its region of attraction. Figure 9.8b shows the measured x response for these different controllers, exhibiting a rise time of 8–9 s independent of the controller. So a horizon time closer to the rise time results in a more feasible optimization in this case.

9.6. SUMMARY AND CONCLUSION

This chapter has given a survey of some basic concepts required to analyze and implement online control customization via optimization-based control. By making use of real-time trajectory generation algorithms that exploit geometric structure and implementing receding horizon control using control Lyapunov functions as terminal costs, we have been able to demonstrate closed-loop control on a flight control experiment. These results build on the rapid advances in computational capability over the past decade, combined with careful use of control theory, system structure, and numerical optimization. A key property of this approach is that it explicitly handles constraints in the input and state vectors, allowing complex nonlinear behavior over large operating regions.

The framework presented here is a first step toward a fundamental shift in the way that control laws are designed and implemented. By moving the control design into the system itself, it becomes possible to implement much more versatile controllers that respond to changes in the system dynamics, mission intent, and environmental constraints. Experimental results have validated this approach in the case of manually varied end points, a particularly simple version of change in mission.

Future control systems will continue to be more complex and more interconnected. An important element will be the networked nature of future control systems, where many individual agents are combined to allow cooperative control in dynamic, uncertain, and adversarial environments. While many traditional control paradigms will not operate well for these classes of systems, the optimization-based controllers presented here can be transitioned to systems with strongly nonlinear behavior, communications delays, and mixed continuous and discrete states. Thus, while traditional frequency domain techniques are likely to remain useful for isolated systems, design of controllers for large-scale, complex, networked systems will increasingly rely on techniques based on Lyapunov theory and (closed-loop) optimal control.

However, there are still many gaps in the theory and practice of optimization-based control. Guaranteed robustness, the hallmark of modern control theory, is largely absent from our present formulation and will require substantial work in extending the theory. Existing approaches such as differential games are not likely to work in an online environment, due to the extreme computational cost required to solve such problems. Furthermore, while the extension to hybrid systems with mixed continuous and discrete variables (in both states and time) is conceivable at the theoretical level, effective computational tools for mixed integer programs must be developed that exploit the system structure to achieve fast computation.

Finally, we note that existing optimization-based techniques are still primarily aimed at the lowest levels of control, despite their potential to apply more broadly. Higher-level protocols for control and decision making must be developed that build on the strength of optimization-based control, but they are likely to require substantially new paradigms and approaches. At the same time, new methods in designing software systems that take into account external dynamics and environmental factors are also required.

ACKNOWLEDGMENTS

The authors would like to thank Mario Sznaier and John Doyle for many helpful discussions on the results presented here. The support of the Software Enabled Control (SEC) program and the SEC team members is also gratefully acknowledged.

REFERENCES

1. R. M. Murray, Nonlinear control of mechanical systems: A Lagrangian perspective, *Annual Reviews in Control* **21**:31–45, 1997.
2. S. J. Qin and T. A. Badgwell, An overview of industrial model predictive control technology, In J. C. Kantor, C. E. Garcia, and B. Carnahan, editors, *Fifth International Conference on Chemical Process Control*, pp. 232–256, 1997.
3. D. Q. Mayne, J. B. Rawlings, C. V. Rao, and P. O. M. Scokaert, Constrained model predictive control: Stability and optimality, *Automatica*, **36**(6):789–814, 2000.
4. L. Singh and J. Fuller, Trajectory generation for a uav in urban terrain, using nonlinear MPC, in *Proceedings of the American Control Conference*, 2001.
5. A. Jadbabaie, J. Yu, and J. Hauser, Unconstrained receding horizon control of nonlinear systems, *IEEE Transactions on Automatic Control*, **46**:776–783, May 2001.
6. M. B. Milam, R. Franz, and R. M. Murray, Real-time constrained trajectory generation applied to a flight control experiment, in *Proceedings of the IFAC World Congress*, 2000.
7. M. B. Milam, K. Mushambi, and R. M. Murray, A computational approach to real-time trajectory generation for constrained mechanical systems, in *Proceedings of the IEEE Control and Decision Conference*, 2000.

8. N. Petit, M. B. Milam, and R. M. Murray, Inversion based trajectory optimization, in *IFAC Symposium on Nonlinear Control Systems Design* (*NOLCOS*), 2001.

9. W. B. Dunbar, M. B. Milam, R. Franz, and R. M. Murray, Model predictive control of a thrust-vectored flight control experiment, in *Proceedings of the IFAC World Congress*, 2002.

10. R. Franz, M. B. Milam, and J. E. Hauser, Applied receding horizon control of the Caltech ducted fan, in *Proceedings of the American Control Conference*, 2002.

11. M. Fliess, J. Levine, P. Martin, and P. Rouchon, On differentially flat nonlinear systems, *C. R. Acad. Sci. Paris Ser. I*, **315**:619–624, 1992.

12. M. J. van Nieuwstadt and R. M. Murray, Rapid hover to forward flight transitions for a thrust vectored aircraft, *Journal of Guidance, Control, and Dynamics*, **21**(1):93–100, 1998.

13. P. Martin, S. Devasia, and B. Paden, A different look at output tracking—Control of a VTOL aircraft, *Automatica* **32**(1):101–107, 1994.

14. M. J. van Nieuwstadt and R. M. Murray, Real time trajectory generation for differentially flat systems, *International Journal of Robust and Nonlinear Control*, **8**(11):995–1020, 1998.

15. E. D. Sontag, A {Lyapunov-like characterization of asymptotic controllability, *SIAM Journal of Control and Optimization*, **21**:462–471, 1983.

16. M. Krstić, I. Kanellakopoulos, and P. Kokotović, *Nonlinear and Adaptive Control Design*, Wiley, New York, 1995.

17. A. Jadbabaie, *Nonlinear Receding Horizon Control: A Control Lyapunov Function Approach*, Ph.D. thesis, California Institute of Technology, Control and Dynamical Systems, 2001.

18. J. A. Primbs, V. Nevistić, and J. C. Doyle, A receding horizon generalization of pointwise min-norm controllers, *IEEE Transactions on Automatic Control*, **45**:898–909, June 2000.

19. J. Hauser and H. Osinga, On the geometry of optimal control: The inverted pendulum example, in *Proceedings of the American Control Conference*, 2001.

20. E. B. Lee and L. Markus, *Foundations of Optimal Control Theory*, Wiley, New York, 1967.

21. C. Hargraves and S. Paris, Direct trajectory optimization using nonlinear programming and collocation, *AIAA Journal of Guidance and Control*, **10**:338–342, 1987.

22. H. Seywald, Trajectory optimization based on differential inclusion, *J. Guidance, Control and Dynamics*, **17**(3):480–487, 1994.

23. A. E. Bryson, *Dynamic Optimization*, Addison-Wesley, Reading, MA, 1999.

24. A. Isidori, *Nonlinear Control Systems*, 2nd edition, Springer-Verlag, Berlin, 1989.

25. C. de Boor, *A Practical Guide to Splines*, Springer-Verlag, Berlin, 1978.

26. P. E. Gill, W. Murray, M. A. Saunders, and M. Wright, *User's Guide for NPSOL 5.0: A Fortran Package for Nonlinear Programming*, Systems Optimization Laboratory, Stanford University, Stanford, CA 94305.

27. M. B. Milam and R. M. Murray, A testbed for nonlinear flight control techniques: The Caltech ducted fan, In *Proceedings of the IEEE International Conference on Control and Applications*, 1999.

CHAPTER 10

MODEL PREDICTIVE NEURAL CONTROL FOR AGGRESSIVE HELICOPTER MANEUVERS

ERIC A. WAN, ALEXANDER A. BOGDANOV, RICHARD KIEBURTZ, ANTONIO BAPTISTA, MAGNUS CARLSSON, YINGLONG ZHANG, and MIKE ZULAUF

Editors' Notes

This chapter shares with Chapter 9 the adoption of a model predictive control (MPC) framework for flight control applications, but the details differ substantially. In particular, the control feedback in this case is a superposition of a neural-network-based nonlinear mapping and a nonlinear state-dependent Riccati equation (SDRE) controller. The neural network is optimized (trained) online for high performance using a high-fidelity dynamic simulation model of the vehicle. The SDRE controller design, repeated at every sample time, provides initial local asymptotic stability. The relative contributions of each controller vary depending on the training error of the neural network.

The application considered is maneuver control of autonomous helicopters. The controller is multivariable with five actuator command outputs. Simulation results for a variety of maneuvers are presented, including rapid take-off and landing and a difficult "elliptic" maneuver. The authors also incorporate wind effects in the optimization. Instead of the conventional tactic of treating environmental conditions as a random disturbance, this allows control commands to be optimized for localized wind flows. This approach relies on wind predictions over the optimization horizon; sensors and models to support such predictions at the appropriate scale and resolution are the object of intense research today.

Both the SDRE design and the neural network training are computationally demanding. The authors consider possible tradeoffs between computational effort and controller performance in order to approach real-time feasibility. Various approximations to some of the especially complex calculations are considered. Computing time and control performance data are presented, comparing different neural net-

Software-Enabled Control. Edited by Tariq Samad and Gary Balas
ISBN 0-471-23436-2 © 2003 IEEE

work complexities, optimization horizons, and update intervals. Substantial CPU time speedup can be achieved with minor loss of performance.

The autonomous helicopter control theme is continued in Chapter 12 where also neural networks feature in the solution approach.

10.1. INTRODUCTION

Advances in technology and modeling in commercially available software make possible highly accurate simulations of aircraft and their environmental interactions. With increased on-board computational resources, this allows for the design of sophisticated nonlinear controllers that exploit these simulations online, in order to achieve high-performance autonomous control of vehicles capable of rapid adaptation and aggressive maneuvering. In this chapter, we describe a method for helicopter control through the use of the *FlightLab* simulator [1] coupled with nonlinear control techniques. FlightLab is a commercial software product developed by Advanced Rotorcraft Technologies. Details on the modeling capabilities of FlightLab are given in Section 10.2.2. A challenge in using FlightLab and similar flight simulators to design controllers is that the governing dynamic equations are not readily available (i.e., the aircraft represents a "black-box" model). This precludes the use of most traditional nonlinear control approaches that require an analytic model. Our methodology is based on the model predictive control (MPC) approach [2, 3]. MPC is an optimization-based framework for learning a stabilizing control sequence that minimizes a specified cost function. For general nonlinear systems, this requires a numerical optimization procedure involving iterative forward (and backward) simulations of the model dynamics. The resulting control sequence represents an ''open-loop" control law, which can then be reoptimized online at periodic *update intervals* to improve robustness.

Our approach to MPC, referred to as *model predictive neural control* (MPNC), utilizes a combination of a state-dependent Riccati equation (SDRE) controller and an optimal neural controller. In contrast to traditional MPC, the architecture implements an explicit *feedback* controller. The SDRE technique [4, 5] is an improvement over traditional linearization-based linear quadratic (LQ) controllers (SDRE control will be elaborated on in a later section). SDRE design, however, requires an *analytic* representation of an aircraft model. To provide an analytic representation, the numeric simulator model is approximated by a six-degree-of-freedom (6DOF) rigid-body dynamic model, providing a set of governing equations at each time instant necessary to design the SDRE. In our framework, the SDRE controller provides an initial stabilizing controller, and it is then augmented by a neural network (NN) controller. The NN controller is optimized online using a calculus of variations approach to minimize the MPC cost function. Note that this differs from either (a) the use of a NN for *system identification* of the nonlinear plant as part of a traditional MPC approach (see reference 6) or (b) the use of a NN for error feedback to account for model uncertainty (see references 7 and 8). We also make use of the SDRE solution to provide a control Lyapunov function (CLF) for use in a receding horizon approach to MPC. In addition, we explore a number of numeric approximations in order to improve the computational performance of the approach. The basic framework of our approach has been described in references 9 and 10.

10.2. MPC CONTROL

The general MPC optimization problem involves minimizing a cost function

$$J_t = \sum_{k=t}^{t_{\text{final}}} L_k(\mathbf{x}_k, \mathbf{u}_k)$$

which represents the accumulated cost of the sequence of states $\mathbf{x}_k$ and controls $\mathbf{u}_k$ from the current discrete time t to the final time, t_{final}. For *regulation* problems, $t_{\text{final}} = \infty$. Optimization is done with respect to the control sequence subject to constraints of the system dynamics,

$$\mathbf{x}_{k+1} = \mathbf{f}(\mathbf{x}_k, \mathbf{u}_k) \tag{10.1}$$

As an example, choosing $L_k(\mathbf{x}_k, \mathbf{u}_k) = \mathbf{x}_k^T \mathbf{Q} \mathbf{x}_k + \mathbf{u}_k^T \mathbf{R} \mathbf{u}_k$ corresponds to the standard *linear quadratic* cost. For *linear* systems, this leads to linear state-feedback control, which is found by solving a Riccati equation [11]. More general costs allow for inequality constraints on state and control, minimum time, minimum control (fuel), and so on. In this chapter we consider general multi-input–multi-output (MIMO) nonlinear systems with *tracking* error costs of the form

$$L_k(\mathbf{x}_k, \mathbf{u}_k) = \mathbf{e}_k^T \mathbf{Q} \mathbf{e}_k + \mathbf{u}_k^T \mathbf{R} \mathbf{u}_k + (\mathbf{u}_k^{\text{over}})^T \mathbf{R}_{\text{sat}} \mathbf{u}_k^{\text{over}} \tag{10.2}$$

where $\mathbf{e}_k = \mathbf{x}_k - \mathbf{x}_k^{\text{des}}$, with $\mathbf{x}_k^{\text{des}}$ corresponding to a desired reference state trajectory. The last term assesses a penalty for exceeding control saturation level $\mathbf{u}^{\text{sat}}$, where each element $(j = 1, \ldots, m)$ of the vector $\mathbf{u}_k^{\text{over}}$ is defined as

$$u_{k_j}^{\text{over}} \begin{cases} 0 & \text{if } |u_{k_j}| \le u_j^{\text{sat}} \\ u_{k_j} - u_j^{\text{sat}} \text{sign}(u_{k_j}), & \text{otherwise} \end{cases} \tag{10.3}$$

In general, a numerical optimization approach is used to solve for the sequence of controls, $\{\mathbf{u}_k\}_t^{t_{\text{final}}} = \arg \min J_t$, corresponding to an *open-loop* control law, which can then be reoptimized online at periodic update intervals. The complexity of the approach is a function of the final time t_{final}, which determines the length of the optimal control sequence. In practice, we can reduce the number of computations by taking a *receding horizon* (RH) approach, in which optimization is performed over a shorter fixed-length time interval. This is accomplished by rewriting the cost function as

$$J_t = \sum_{k=t}^{t+N-1} L_k(\mathbf{x}_k, \mathbf{u}_k) + V(\mathbf{x}_{t+N}) \tag{10.4}$$

where the last term $V(\mathbf{x}_{t+N})$ denotes the *cost-to-go* from time $t + N$ to time t_{final}. Note that for trajectory following, the cost-to-go is also implicitly a function of the given desired trajectory $\mathbf{x}_{t+N}^{\text{des}}$ through $\mathbf{x}_{t_{\text{final}}}^{\text{des}}$ (and resulting error). The advantage is that this yields an optimization problem of fixed length N. In practice, the true value of $V(\mathbf{x}_{t+N})$ is unknown and must be approximated. Most common is to simply set $V(\mathbf{x}_{t+N}) = 0$; however, this may lead to reduced stability and poor performance for short horizon lengths [12]. Alternatively, we may include a control Lyapunov function (CLF), which guarantees stability if the CLF is an upper bound on the cost-to-go and also results in a region of attraction for the MPC of at least that of the CLF [12].

10.2.1. Helicopter Modeling and Design Considerations

10.2.1.1. FlightLab Helicopter Model. The FlightLab package provides a simulation tool for multidisciplinary concurrent engineering. Consisting of a number of modular and reconfigurable modeling primitives (called "components"), it uses a high-level interpretive language (Scope) for rapid prototyping. With these reconfigurable components, the user is able to construct a model of selective fidelity, as well as create or link custom components. Numerical integration of dynamics is accomplished through iterative techniques (e.g., Newton–Raphson) to solve systems of equations, using a selective compartmental approach for integrating subsystem components. FlightLab also has a number of GUI tools, such as editors for developing models and control systems, and a dynamic system analyzer. It allows real-time interfacing with external hardware or software components. One of the outstanding features of the FlightLab is its ability to accurately model the main rotor and the interaction between it and other components, such as the tail rotor (through the use of empirical formulas), external wind (e.g., through an external wind server), ground effects (including both in- and out-of-ground effects with an image method, a free-wake model, and a horse-shoe ground vortex model), and so on. It also includes the modeling of dynamic stall, transonic flow, aeroelastic response, vortex wake, and blade element aerodynamics, and it can also provide finite element structure analysis.

For our research purposes, we generated a high-fidelity helicopter model having, a rigid fuselage with empirical airloads, elastic blades with quasi-unsteady airloads,[1] 3-state inflow,[2] and direct control of the swashplate angles. The model is presented as a numerical discrete-time nonlinear system with 92 internal state variables.

10.2.1.2. Control Specifications.

For helicopter control, we define the state vector $\mathbf{x}_k$ to correspond to the standard states of a 6DOF rigid body model. This 12-dimensional state vector consists of Cartesian coordinates in the inertial frame x, y, z, Euler angles ψ, ϕ, θ (yaw, roll, pitch), linear velocities u, v, w, and angular velocities p, q, r in the body coordinate frame. Technically, this represents a reduced state-space, as our FlightLab model utilizes a total of 92 internal state variables (e.g., main and tail rotor states). However, we treat these as both unobservable and uncontrollable for purposes of deriving the controller. There are four control inputs, $\Theta_0, \Theta_{1C}, \Theta_{1S}, \Theta_{T0}$, corresponding to the main collective, lateral cyclic, longitudinal cyclic, and tail collective controls (incident angles of rotor blades).[3]

The tracking error for the helicopter, $\mathbf{e}_k = \mathbf{x}_k - \mathbf{x}_k^{des}$, is determined by the trajectory of a reference target (provided by a *mission planner* for higher-level control coordination). The target specifies desired coordinates, velocities, and attitude in the inertial frame. The reference state is then projected into the body frame to produce the desired state,

$$\mathbf{x}_k^{des} = \mathbf{T}(\mathbf{x}_k)\mathbf{x}_k^{tar} \tag{10.5}$$

where $\mathbf{T}(\mathbf{x}_k)$ is an appropriate projection matrix consisting of necessary rotation operators:

$$\mathbf{T}(\mathbf{x}) = \begin{pmatrix} c\psi c\theta & s\psi s\phi + c\psi c\phi s\theta & 0 & 0 & -s\psi c\phi + c\psi s\theta s\phi & \\ -s\theta & c\theta c\phi & 0 & 0 & c\theta s\phi & \\ 0 & 0 & 1 & 0 & 0 & \mathbf{O}_{5,7} \\ 0 & 0 & 0 & 1 & 0 & \\ s\psi c\theta & -c\psi s\phi + s\theta s\psi c\phi & 0 & 0 & c\psi c\phi + s\phi s\theta s\psi & \\ & \mathbf{O}_{7,5} & & & & \mathbf{I}_{7,7} \end{pmatrix}$$

[1] Quasi-unsteady airloads incorporates both theory and table look-up for the calculation of airloads for the effective angle of attack, side-slip angle and Mach number; certain dynamics (e.g., dynamic stall) are simplified or neglected.
[2] The general finite-state inflow model allows for the variation of the rotor induced flow with arbitrary harmonics azimuthally and an arbitrary order of polynomial radially. The three-state model is truncated from the higher-order finite-state inflow model. Its first state is for uniform inflow (both azimuthally and radially), the second state is the first cosine harmonics of azimuthal variation with linear radial variation, and the third state is for the first sine harmonics of azimuthal variation with the similar linear radial variation [13].
[3] We set control constraints as follows: max $\Theta_0 = \pm 20°$, max $\Theta_{1C} = \pm 30°$, max $\Theta_{1S} = \pm 32°$, max $\Theta_{T0} = \pm 39°$.

with $s(\cdot)$, $c(\cdot)$ denoting sin and cos, $\mathbf{x}_k = (u, w, q, \theta, v, p, \phi, r, \psi, x, y, z)^T$. Minimization of the tracking error causes the helicopter to move in the direction of the target motion.

10.3. MPNC

Difficulties with application of traditional MPC include (a) the need for a "good" initial sequence of controls that is capable of stabilizing the model and (b) the need to reoptimize at short intervals to avoid problems associated with open-loop control (e.g., lack of robustness to model uncertainty or disturbances). We address these issues by directly implementing a *feedback* controller as a combination of an SDRE stabilizing controller and a neural controller:

$$\mathbf{u}_k = \alpha \cdot Nnet(\mathbf{x}_k, \mathbf{e}_k, \mathbf{w}) + (1 - \alpha) \cdot \mathbf{K}(\mathbf{x}_k)\mathbf{e}_k \qquad (10.6)$$

where $0 < \alpha < 1$ is a relative weighting constant. A standard multilayer feedforward neural network is used for its universal mapping capabilities (not for its biological motivation). The SDRE controller $\mathbf{K}(\mathbf{x}_k)\mathbf{e}_k$ provides a robust stabilizing control, while the weights of the neural network, $\mathbf{w}$, are optimized to minimize the overall receding horizon MPC cost. Note that this control structure is a function of the state $\mathbf{x}_k$ (and tracking error $\mathbf{e}_k$), providing explicit feedback. As stated earlier, the SDRE controller requires a set of governing equations, which are not available in FlightLab. Thus we derive a 6DOF rigid body model as an *analytical approximation*. The NN controller, however, is designed using the full-flight simulator model. The NN is trained online, and it is updated for each horizon interval. For this, the SDRE solution and the 6DOF model are again utilized in a number of different ways in order to implement terms used for minimizing the MPC cost. Design of the MPNC controller, which includes the SDRE and neural controller, is detailed in the following subsections.

10.3.1. SDRE Controller

Referring to the system state-space equation (10.1), an SDRE controller [4] is designed by reformulating $\mathbf{f}(\mathbf{x}_k, \mathbf{u}_k)$ as

$$\mathbf{f}(\mathbf{x}_k, \mathbf{u}_k) = \mathbf{\Phi}(\mathbf{x}_k)\mathbf{x}_k + \mathbf{\Gamma}(\mathbf{x}_k)\mathbf{u}_k$$

This representation is *not* a linearization. To illustrate the principle, consider a simple scalar example, $x_{k+1} = \sin x_k + x_k \cos x_k u_k$. A valid state-space representation is then $\Phi(x_k) = \sin x_k / x_k$ and $\Gamma(x_k) = x_k \cos x_k$.

Based on the new state-space representation, we design an optimal LQ controller to track the desired state $\mathbf{x}_k^{\text{des}}$. This leads to the nonlinear controller,[4]

$$\mathbf{u}_k^{sd} = -\mathbf{R}^{-1}\boldsymbol{\Gamma}^T(\mathbf{x}_k)\mathbf{P}(\mathbf{x}_k)\big(\mathbf{x}_k - \mathbf{x}_k^{\text{des}}\big) \equiv \mathbf{K}(\mathbf{x}_k)\mathbf{e}_k$$

where $\mathbf{P}(\mathbf{x}_k)$ is a solution of the standard Riccati Equations using state-dependent matrices $\boldsymbol{\Phi}(\mathbf{x}_k)$ and $\boldsymbol{\Gamma}(\mathbf{x}_k)$, which are treated as being constant. The procedure is repeated at every time step at the current state $\mathbf{x}_k$ and provides local asymptotic stability of the plant [4]. In practice, the approach has been found to be far more robust than LQ controllers based on standard linearization techniques. See reference 4 for a discussion on the class of dynamic equations that can be presented in a state-dependent form.

Dynamic equations for the FlightLab helicopter model are not available. Thus we use the simplified dynamics given by a 6DOF rigid-body model,

$$\dot{u} = -(wq - vr) - g\sin\theta + F_x/M_a,$$

$$\dot{v} = -(ur - wp) + g\cos\theta\sin\phi + F_y/M_a,$$

$$\dot{w} = -(vp - uq) + g\cos\theta\cos\phi + F_z/M_a,$$

$$I_{xx}\dot{p} = (I_{yy} - I_{zz})qr + I_{xz}(\dot{r} + pq) + L,$$

$$I_{yy}\dot{q} = (I_{zz} - I_{xx})rp + I_{xz}(r^2 - p^2) + M,$$

$$I_{zz}\dot{r} = (I_{xx} - I_{yy})pq + I_{xz}(\dot{p} - qr) + N$$

$$\dot{\phi} = p + q\sin\phi\tan\theta + r\cos\phi\tan\theta$$

$$\dot{\theta} = q\cos\phi - r\sin\phi$$

$$\dot{\psi} = q\sin\phi\sec\theta + r\cos\phi\sec\theta$$

$$\big(\dot{x} \quad \dot{y} \quad \dot{z}\big)^T = \mathbf{Rot}_1(\psi, \phi, \theta)\big(u \quad v \quad w\big)^T$$

where $\mathbf{Rot}_1(\psi, \phi, \theta)$ is a rotation matrix (coordinate transformation) from body frame to inertial frame, M_a is the aircraft mass, I_{xx}, I_{yy}, I_{zz} are moments of inertia; I_{xz} is the product inertia, and F_x, F_y, F_z, L, M, N are rotor-induced forces and moments. The forces and moments are nonlinear functions of helicopter states and control inputs. We then rewrite this into a state-dependent continuous canonical representation $\dot{\mathbf{x}} = \mathbf{A}(\mathbf{x})\mathbf{x} + \mathbf{B}(\mathbf{x})\mathbf{u}$. The

---

[4] In the case where the SDRE is used as a stand-alone controller, we formulate $\mathbf{u}_k^{sd} = \mathbf{K}(\mathbf{x}_k)\mathbf{e}_k + u_k^{tr}$, where $\mathbf{u}_k^{tr}$ is scheduled to compensate for steady-state errors related to various trim conditions. Alternatively, way may also include an integral control in the SDRE framework. Note that when the full MPNC control is used, the trim control is accounted for automatically through the optimization of the NN.

matrix $\mathbf{A}(\mathbf{x})$ is given explicitly as

$\mathbf{A}(\mathbf{x}) =$

$$
\begin{pmatrix}
0 & -q/2 & -w/2 & -g\frac{s\theta}{\theta} & r/2 & 0 & 0 & v/2 & 0 & 0 & 0 & 0 \\
q/2 & 0 & u/2 & g\frac{c\theta-1}{\theta} & -p/2 & -v/2 & gc\theta\frac{c\phi-1}{\phi} & 0 & 0 & 0 & 0 & g/z \\
0 & 0 & 0 & 0 & 0 & a_0 r - a_1 p & 0 & a_0 p + a_1 r & 0 & 0 & 0 & 0 \\
0 & 0 & c\phi & 0 & 0 & 0 & 0 & -s\phi & 0 & 0 & 0 & 0 \\
-r/2 & p/2 & 0 & 0 & 0 & w/2 & gc\theta\frac{s\phi}{\phi} & -u/2 & 0 & 0 & 0 & 0 \\
0 & 0 & a_2 p + a_3 r & 0 & 0 & a_2 q & 0 & a_3 q & 0 & 0 & 0 & 0 \\
0 & 0 & s\phi\tan\theta & 0 & 0 & 1 & 0 & c\phi\tan\theta & 0 & 0 & 0 & 0 \\
0 & 0 & a_4 p + a_5 r & 0 & 0 & a_4 q & 0 & a_5 q & 0 & 0 & 0 & 0 \\
0 & 0 & s\phi/c\theta & 0 & 0 & 0 & 0 & c\phi/c\theta & 0 & 0 & 0 & 0 \\
c\psi c\theta & \begin{matrix}s\psi s\phi +\\ +c\psi c\phi s\theta\end{matrix} & 0 & 0 & \begin{matrix}-s\psi c\phi +\\ +c\psi s\theta s\phi\end{matrix} & 0 & 0 & 0 & 0 & 0 & 0 & 0 \\
s\psi c\theta & \begin{matrix}-c\psi s\phi +\\ +s\theta s\psi c\phi\end{matrix} & 0 & 0 & \begin{matrix}c\psi c\phi +\\ +s\phi s\theta s\psi\end{matrix} & 0 & 0 & 0 & 0 & 0 & 0 & 0 \\
-s\theta & c\theta c\phi & 0 & 0 & c\theta s\phi & 0 & 0 & 0 & 0 & 0 & 0 & 0
\end{pmatrix}
$$

where $s(\cdot)$, $c(\cdot)$ denote sin and cos,

$$
a_0 = \frac{I_{zz} - I_{xx}}{2I_{yy}}, \qquad a_1 = \frac{I_{xz}}{I_{yy}}, \qquad a_2 = \frac{I_{xz}(I_{xx} - I_{yy} + I_{zz})}{2(I_{xx}I_{zz} - I_{xz}^2)},
$$

$$
a_3 = \frac{I_{yy}I_{zz} - I_{zz}^2 - I_{xz}^2}{2(I_{xx}I_{zz} - I_{xz}^2)}, \qquad a_4 = \frac{I_{xx}^2 - I_{yy}I_{xx} + I_{xz}^2}{2(I_{xx}I_{zz} - I_{xz}^2)},
$$

$$
a_5 = \frac{I_{xz}(-I_{xx} + I_{yy} - I_{zz})}{2(I_{xx}I_{zz} - I_{xz}^2)}, \qquad \mathbf{x} = (u, w, q, \theta, v, p, \phi, r, \psi, x, y, z)^T
$$

$\mathbf{\Phi}(\mathbf{x}_k)$ is then obtained from $\mathbf{A}(\mathbf{x})$ by discretization at each time step (i.e., $\mathbf{\Phi}(\mathbf{x}_k) = e^{A(\mathbf{x}_k)\Delta t})$.[5]

Explicit analytic equations are unavailable for the rotor-induced forces and moments, which would relate the 6DOF model to the FlightLab model. These terms are highly nonlinear and include vehicle-specific aerodynamic look-up tables. Thus $\mathbf{B}(\mathbf{x})$ cannot be specified in closed-form. Therefore, we approximate $\mathbf{\Gamma}(\mathbf{x}_k)$ with a constant $\mathbf{\Gamma}$ by linearizing[6] the full FlightLab model with respect to the control inputs $\mathbf{u}_k$ around the hover trim state.[7]

Finally given $\mathbf{\Phi}(\mathbf{x}_k)$ and $\mathbf{\Gamma}$, we can design the SDRE control gain $\mathbf{K}(\mathbf{x}_k)$ at each time step. Note that while $\mathbf{\Phi}(\mathbf{x}_k)$ is based on the 6DOF model, the state

[5] Parameter settings to match FlightLab are $M_a = 16{,}308$ lb, and $I_{xx} = 9969$ lb·ft^2, $I_{yy} = 44{,}493$ lb·ft^2, $I_{zz} = 44{,}265$ lb·ft^2, $I_{xz} = -1478$ lb·ft^2.
[6] Linearization is performed over multiple rotor revolutions with averaging.
[7] Alternatively, we may explicitly schedule $\mathbf{\Gamma}(\mathbf{x}_k)$ using combinations of different trim states, as is often done with *linear parameter varying* (LPV) approaches [14–16].

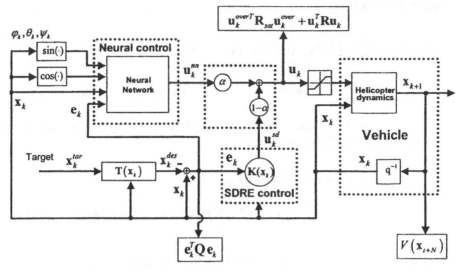

Figure 10.1. MPNC signal flow diagram.

argument $\mathbf{x}_k$ comes directly from the FlightLab model. We have found that this mixed approach using the approximate nonlinear model plus the linearized control matrix, Γ, is far more robust than simply using a standard LQ approach based on linearization of all system matrices (see Section 10.4).

10.3.2. Neural Network Controller

The neural controller is designed using the full FlightLab helicopter model and augments the SDRE controller. The overall flowgraph of the system is shown in Figure 10.1. The neural controller is specified as follows:

$$\mathbf{u}^{nn} = Nnet(u, v, w, p, q, r, z, s\psi, s\theta, s\phi, c\psi, c\theta, c\phi, \mathbf{e}, \mathbf{w}) \quad (10.7)$$

where we have included sines and cosines of the yaw, pitch, and roll angles. This is motivated since the simplified helicopter dynamics depend on such trigonometric functions of the Euler angles. The coordinates x and y of the aircraft in the inertial frame do not influence dynamics, and they are excluded as inputs (the altitude, z, is included due to modeling of ground effects and air density).

The NN represents an optimal feedback controller. Optimization is performed by learning the weights, $\mathbf{w}$, of the NN in order to minimize the receding horizon MPC cost function [Eq. (10.4)] subject to the system dynamics and the composite form of the overall feedback controller [Eq. (10.6)]. The problem is solved by taking a standard calculus of variations approach, where λ_k and μ_k are vectors of Lagrange multipliers in the augmented cost function

$$J_t = \sum_{k=t}^{t+N-1} \left\{ L_k(\mathbf{x}_k, \mathbf{u}_k) + \lambda_k^T \left(\mathbf{x}_{k+1} - \mathbf{f}(\mathbf{x}_k, \mathbf{u}_k) \right) \right.$$

$$\left. + \mu_k^T \left(\mathbf{u}_k - \alpha \mathbf{u}_k^{nn} - (1-\alpha) \mathbf{K}(\mathbf{x}_k) \mathbf{e}_k \right) \right\} + V(\mathbf{x}_{t+N}) \quad (10.8)$$

where $\mathbf{u}_k^{nn} = Nnet(\mathbf{x}_k, \mathbf{e}_k, \mathbf{w})$. The cost-to-go $V(\mathbf{x}_{t+N})$ is approximated using the solution of the SDRE at time $t+N$,

$$V(\mathbf{x}_{t+N}) \approx \mathbf{e}_{t+N}^T \mathbf{P}(\mathbf{x}_{t+N}) \mathbf{e}_{t+N}$$

$$\equiv \sum_{k=N}^{\infty} \left\{ \left(\mathbf{x}_k - \mathbf{x}_{t+N}^{des} \right)^T \mathbf{Q} \left(\mathbf{x}_k - \mathbf{x}_{t+N}^{des} \right) + \mathbf{u}_k^T \mathbf{R} \mathbf{u}_k \right\} \quad (10.9)$$

This CLF provides the exact cost-to-go for *regulation* assuming a linear system at the horizon time. A similar formulation was used for nonlinear regulation control in reference 17.

We can now derive the recurrent Euler–Lagrange equations

$$\lambda_k = \left(\frac{\partial \mathbf{f}(\mathbf{x}_k, \mathbf{u}_k)}{\partial \mathbf{x}_k} \right)^T \lambda_{k+1} + \left(\frac{\partial L_k(\mathbf{x}_k, \mathbf{u}_k)}{\partial \mathbf{x}_k} \right)^T$$

$$+ \left(\alpha \frac{\partial Nnet(\mathbf{x}_k, \mathbf{e}_k, \mathbf{w})}{\partial \mathbf{x}_k} + \alpha \frac{\partial Nnet(\mathbf{x}_k, \mathbf{e}_k, \mathbf{w})}{\partial \mathbf{e}_k} \frac{\partial \mathbf{e}_k}{\partial \mathbf{x}_k} \right.$$

$$\left. + (1-\alpha) \frac{\partial \left(\mathbf{K}(\mathbf{x}_k) \mathbf{e}_k \right)}{\partial \mathbf{x}_k} \right)^T \mu_k \quad (10.10)$$

$$\mu_{k_i} = \begin{cases} \left[\left(\frac{\partial \mathbf{f}(\mathbf{x}_k, \mathbf{u}_k)}{\partial \mathbf{u}_k} \right)^T \lambda_{k+1} + \left(\frac{\partial L_k(\mathbf{x}_k, \mathbf{u}_k)}{\partial \mathbf{u}_k} \right)^T \right]_i & \text{if } |u_{k_i}| \le u_i^{sat} \\[4mm] \left(\frac{\partial L_k(\mathbf{x}_k, \mathbf{u}_k)}{\partial \mathbf{u}_k} \right)_i^T & \text{if } |u_{k_i}| > u_i^{sat} \end{cases}$$

with

$$\lambda_{t+N} = \left(\frac{\partial V(\mathbf{x}_{t+N})}{\partial \mathbf{x}_{t+N}} \right)^T \equiv \left(\frac{\partial V(\mathbf{x}_{t+N})}{\partial \mathbf{x}_{t+N}} + \frac{\partial V(\mathbf{x}_{t+N})}{\partial \mathbf{e}_{t+N}} \frac{\partial \mathbf{e}_{t+N}}{\partial \mathbf{x}_{t+N}} \right)^T,$$

$$k = (t+N) - 1, (t+N) - 2, \ldots, t,$$

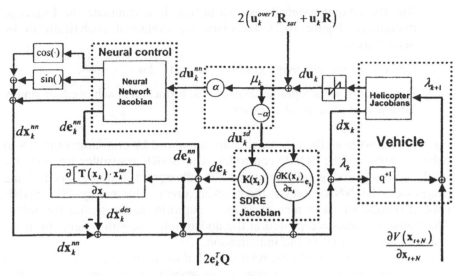

Figure 10.2. Adjoint system.

and $i = 1, \ldots, m$ (m is the dimension of control vector $\mathbf{u}$). From Eqs. (10.2) and (10.5),

$$\frac{L_k(\mathbf{x}_k, \mathbf{u}_k)}{\partial \mathbf{x}_k} = \frac{\partial L_k(\mathbf{x}_k, \mathbf{u}_k)}{\partial \mathbf{e}_k} \frac{\partial \mathbf{e}_k}{\partial \mathbf{x}_k} = 2\mathbf{e}_k^T \mathbf{Q} \left(\mathbf{I} - \frac{\partial \left[\mathbf{T}(\mathbf{x}_k) \mathbf{x}_k^{tar} \right]}{\partial \mathbf{x}_k} \right) \quad (10.11)$$

where the partial is evaluated analytically by specifying $\mathbf{T}(\mathbf{x}_k)$. Each element i of the gradient vector

$$\left(\frac{\partial L_k(\mathbf{e}_k, \mathbf{u}_k)}{\partial \mathbf{u}_k} \right)_i = 2 \left(\mathbf{u}_k^T \mathbf{R} \right)_i \quad \text{if } |u_{k_i}| \leq u_i^{\text{sat}}.$$

These equations correspond to an adjoint system (shown graphically in Figure 10.2), with optimality condition

$$\frac{\partial J_t}{\partial \mathbf{w}} = \sum_{k=t}^{t+N-1} \alpha \mu_k^T \frac{\partial Nnet(\mathbf{x}_k, \mathbf{e}_k, \mathbf{w})}{\partial \mathbf{w}} = 0$$

The overall training procedure for the NN can now be summarized as follows:

1. Simulate the system forward in time for N time steps (Figure 10.1). Note that the SDRE controller is updated at each time step.

2. Run the adjoint system backward in time to accumulate the Lagrange multipliers (Figure 10.2). Jacobians are evaluated analytically or by perturbation.

3. Update the weights using gradient descent,[8] $\Delta \mathbf{w} = -\gamma \frac{\partial J_k}{\partial \mathbf{w}}$.

4. Repeat until an acceptable level of cost reduction is achieved, or simply for a preset number of iterations.

For the first horizon, the NN weights are initialized by pretraining the NN to behave similar to the SDRE controller. The SDRE controller provides for stable tracking and good conditions for subsequent training. As training progresses, the NN decreases the tracking error. This reduces the SDRE control output, which in turn gives more authority to the neural controller. The training process is repeated at the update interval, with weights from the previous horizon used as the initialization.

Stability of MPNC is closely related to that of traditional MPC. Ideally, in the case of unconstrained optimization, stability is guaranteed provided $V(\mathbf{x}_{t+N})$ is a CLF and is an (incremental) upper bound on the cost-to-go [12]. In this case, the minimum region of attraction of the receding horizon optimal control is determined by the CLF used and horizon length. The guaranteed region of operation contains that of the CLF controller and may be made as large as desired by increasing the optimization horizon (restricted to the infinite horizon domain) [19]. In our case, the minimum region of attraction of the receding horizon MPNC is determined by the SDRE solution used as the CLF to approximate the terminal cost. In addition, we also restrict the controls to be of the form given by Eq. (10.6), and the optimization is performed with respect to the NN weights $\mathbf{w}$. In theory, the *universal mapping* capability of neural networks implies that the stability guarantees are equivalent to that of the traditional MPC framework. However, in practice, stability is affected by the chosen size of the NN (which affects the actual mapping capabilities), as well as the horizon length and update interval length (how often the NN is reoptimized). This can be clearly traced in Table 10.2 (Section 10.4), where additional hidden neurons result in a higher overall performance. When the horizon is short, performance is more affected by the chosen CLF. On the other hand, when the horizon is long, performance is limited by the NN properties. An additional factor affecting stability is the specific algorithm used for numeric optimization. Gradient descent, which we use to minimize the cost function [Eq. (10.8)], is guaranteed to converge to only a local minimum (the cost function is not guaranteed convex with respect to the NN weights), and thus it depends on the initial conditions. In addition, convergence is assured only if the learning rate is kept sufficiently small. To summarize these points, stability of the MPNC is guaranteed under certain restricted ideal conditions. In practice,

---

[8] In practice we use an adaptive learning rate for each weight in the network using a procedure similar to delta-bar-delta [18].

the designer must select appropriate settings and perform sufficient flight experiments to assure stability and performance over a desired flight envelope.

10.3.3. Engine Speed Control

Typically, engine speed is maintained by a separate throttle regulator. If this is not implemented correctly, a stall situation may arise due to an improper coupling with the main flight controller. During aggressive maneuvers, increased engine load may result in a reduced rotor speed and a loss of lift. To overcome the altitude loss, the controller reacts by increasing the collective. However, this results in even higher loads and further slowing down of the rotor, and thus further loss of lift and altitude. To prevent this, one must decrease the engine load and gain rotor speed back, while momentarily sacrificing tracking performance.

In our implementation, we chose to bypass a separate throttle regulator and build this directly into the MPNC framework. We accomplish this by adding an additional state variable corresponding to the main rotor speed as well as a direct throttle (rotor speed) control command. For the SDRE controller, we simply augment $\Phi(\mathbf{x}_k)$ with an additional row and column found by linearization with respect to the rotor speed at hover. The neural network is provided with extra inputs corresponding to the main rotor speed and its error, as well as an additional throttle (rotor speed) command output. Optimization within the MPC framework remains the same.

10.3.4. Incorporating Wind Flows

With the increased sophistication in atmospheric modeling and sensing, it is becoming more feasible to directly optimize for wind effects. This is in contrast to traditional approaches that often treat wind as a disturbance to be countered through feedback error correction. Modeling predictions of atmospheric conditions are well established at global scales and coarse resolution, and they are the subject of intense research at regional scales and fine resolution.[9] Global and regional predictions, however, are most supportive of mission-planning level control of autonomous vehicles. For aggressive ma-

---

[9] The medium-range forecast model, MRF [20, 21], maintained by the National Oceanic and Atmospheric Administration, is an example of a global model with a long track record. It covers the entire earth surface and extends in the vertical to a layer at the pressure of 2 hPa. The horizontal grid has 512×256 cells, each roughly equivalent to 0.7×0.7 degree latitude/longitude. In the vertical, the grid has 42 unequally spaced sigma levels. Recent efforts in regional prediction are exemplified by the Advanced Regional Prediction System (ARPS), based on a multiscale nonhydrostatic atmospheric model [22]. ARPS targets storm-scale and mountain-scale predictions, with horizontal and vertical resolutions as fine as 1 km and 100 m, respectively. Applications at this level of resolution have the potential for simulation of boundary layer eddies and wind gusts. The impetus for predictions at these scales is credited to the deployment of Doppler radars in the United States, as well as to techniques for retrieving unobserved quantities from Doppler radar data to yield mass and wind fields appropriate for initialization of storm-scale prediction models [23].

neuvers, data from on-board or land-based radar appear to offer the most promising path to realistically incorporating atmospheric conditions in online control strategies.

FlightLab has the built-in capability of incorporating some forms of winds (e.g., sinusoidal gusts, stochastic atmospheric turbulent wind, etc.) into the calculation of total airloads on the rotors as well as fuselage. For our purposes, we replaced the existing atmospheric turbulent wind component (ATM-TUR) with one we developed to allow inputs from an external wind server. The new component was compiled separately and linked to other components that require wind information (e.g., the aerodynamic components for main rotor, etc.). This enables us to study the responses of a helicopter to controlled wind patterns (e.g., wind shear and large-eddy simulations (LES)—Section 10.4.3), as well as future integration with measurements from on-board or land-based radar.

Given the helicopter and wind interaction model, no additional modifications are necessary for incorporation into the MPNC framework. In Section 10.4.3 we report a number of simulation experiments illustrating performance tradeoffs associated with how the wind is approximated.

10.3.5. Computational Simplifications

Generally, MPC design implies multiple simulations of the system forward in time and the adjoint system backward in time. Computations scale with the number of training epochs and the horizon length. The most computationally demanding operations correspond to solving the Riccati equation for the SDRE controller in the forward simulation and solving the numeric (i.e., by perturbation) computation of FlightLab Jacobians in the backward simulation. In order to approach real-time feasibility, we consider possible tradeoffs between computational effort and controller performance through a number of successive simplifications:

1. The SDRE Jacobian is approximated as $\dfrac{\partial (\mathbf{K}(\mathbf{x}_k)\mathbf{e}_k)}{\partial \mathbf{x}_k} \approx \mathbf{K}(\mathbf{x}_k)\dfrac{\partial \mathbf{e}_k}{\partial \mathbf{x}_k}$, where $\dfrac{\partial \mathbf{e}_k}{\partial \mathbf{x}_k}$, can then be calculated analytically as in Eq. (10.11).

2. Same as 1, with the additional simplification that the matrices $\mathbf{K}(\mathbf{x}_k)$ are *memorized* (i.e., stored) at each time step k during the first epoch, and again used for the all subsequent epochs within the current horizon. This simplification allows us to avoid resolving the SDRE in the multiple forward simulations.

3. Same as 2, with the addition that the Jacobians $\dfrac{\partial \mathbf{f}(\mathbf{x}_k, \mathbf{u}_k)}{\partial \mathbf{x}_k}$ and $\dfrac{\partial \mathbf{f}(\mathbf{x}_k, \mathbf{u}_k)}{\partial \mathbf{u}_k}$ (which must be calculated by perturbation of FlightLab) are memorized during the first epoch and again used for all subsequent epochs within the current horizon. This assumes that the aircraft trajectory and thus the Jacobians are not changing significantly during training within the same horizon from epoch to epoch.

In the following section, we will provide experimental results to illustrate the performance of the MPNC approach, including the computational versus performance tradeoffs associated with these simplifications.

10.4. EXPERIMENTAL RESULTS

Figure 10.3 shows a test trajectory for the helicopter (vertical rise, forward flight, left turn, u-turn, forward flight to hover). The figure compares tracking

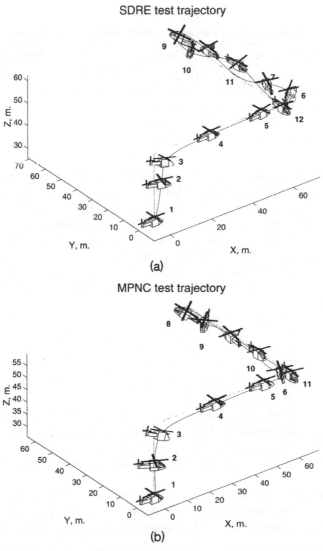

Figure 10.3. Test trajectory: (a) SDRE trajectory, cost = 100.00; (b) MPNC trajectory, cost = 49.25.

Table 10.1. Simplification Levels Versus CPU Time and Performance Costs (Pentium-3 750 MHz, Linux)[a]

Simplification Level	Cost	CPU Time (seconds)
Accurate	49.25	28,882
1	51.66	3,048
2	51.49	1,793
3	52.83	694

[a]Note that the maneuver corresponds to 40 s in actual time.

Table 10.2. Performance Cost Comparisons for Various MPNC Options

Neurons in hidden layer:	50 Neurons				200 Neurons			
Horizon/update interval:	5/1	10/2	25/5	50/10	5/1	10/2	25/5	50/10
MPNC cost with $V(\mathbf{x}_{t+N})$:	69.27	74.55	64.84	70.74	75.84	52.83	54.75	68.15
MPNC cost w/o $V(\mathbf{x}_{t+N})$:	—	126.34	69.39	73.54	—	—	62.72	69.60
SDRE cost:	100.00							

performance at a velocity of 6.0 m/s for the MPNC system (SDRE + NN) versus a pure SDRE controller (MPNC settings are: *horizon* = 10, *update interval* = 2, *training epochs* = 10, *sampling time* = 0.097 s, α = 0.7). (Note that a standard LQ controller based on linearization exhibits loss of tracking in executing this maneuver and crashes for velocities above 3.0 m/s.) The smaller tracking error for the MPNC controller is apparent (note the reduced overshooting and oscillations at mode transitions). The total *normalized* accumulated cost[10] for the MPNC is 49.25 in comparison to 100.00 for the SDRE controller.

Table 10.1 illustrates the tradeoffs between computational effort and control performance (accumulated cost) for the simplifications discussed in the previous section. Clearly, substantial speed-up in CPU time can be achieved with only a minor loss of performance. The actual CPU times should only be viewed as indicative of relative requirements. The experiments were performed in MATLAB (with a *C* module for the vehicle model generated by FlightLab) and were not optimized for efficient implementation or hardware consideration. Note that with all simplifications the MPNC control still achieves a substantial performance improvement over the standard SDRE controller. For all subsequent simulations we use simplification level 3.

Table 10.2 summarizes comparisons of the accumulated cost with respect to variation of the horizon length and MPNC update interval, number of neurons in the hidden layer, and omitting the use of the cost-to-go, $V(\mathbf{x}_{t+N})$.

[10]For comparisons, we specify the normalized accumulated cost: $1/(t_{\text{final}} - t_0)\Sigma_{k=t_o}^{t_{\text{final}}} \mathbf{e}_k^T \mathbf{Q} \mathbf{e}_k + \mathbf{u}_k^T \mathbf{R} \mathbf{u}_k + (\mathbf{u}_k^{\text{over}})^T \mathbf{R}_{\text{sat}} \mathbf{u}_k^{\text{over}}$.

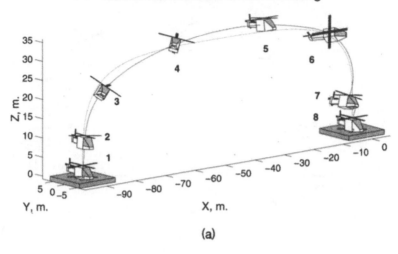

(a)

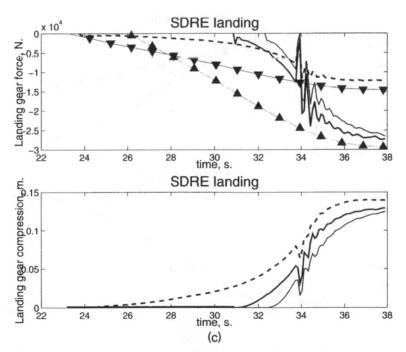

(c)

Figure 10.4. Take-off and landing: (a) SDRE, cost = 2254.90; (desired trajectory is plotted for comparison). Landing trajectories: (c) SDRE. Desired forces at landing are plotted as downward triangles for the tail LG and upward triangles for the left and right LG. Dashed line is the actual force readings at the tail LG, solid lines are for the left and right LG.

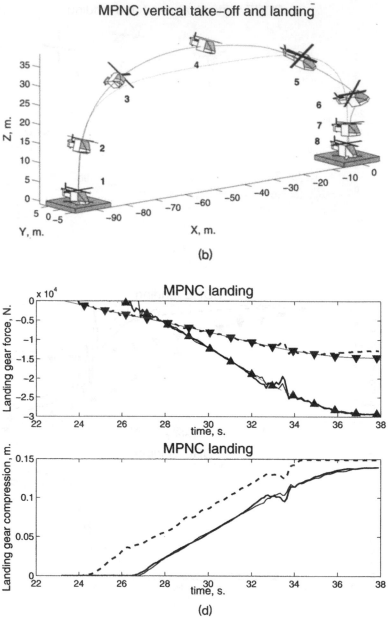

(b)

(d)

Figure 10.4. Take-off and landing: (b) MPNC, cost = 158.88 (desired trajectory is plotted for comparison). Landing trajectories: (d) MPNC. Desired forces at landing are plotted as downward triangles for the tail LG and upward triangles for the left and right LG. Dashed line is the actual force readings at the tail LG, solid lines are for the left and right LG.

The weights of the NN are trained for 10 epochs for each horizon (number of simulated trajectories). Missing data in the table corresponds to a case where states and control inputs exceeded the envelope of the FlightLab model consistency. Results indicate that a sufficient horizon length is between 10 and 25 time steps (1–2.5 s). The importance of the cost-to-go function is apparent for short horizon lengths. On the other hand, inclusion of the cost-to-go does not appear to help for longer horizons. Overall, significant performance improvement is clearly achieved with the MPNC controller relative to the pure SDRE controller.

10.4.1. Rapid Take-off and Landing

The next example illustrates performance of a rapid take-off, side slip, and landing. The tracking ability for this aggressive maneuver is illustrated in Figures 10.4a and 10.4b. While in the static figure the SDRE trajectory appears very accurate, the actual trajectory of the controlled aircraft lags the desired trajectory in time. In contrast, the MPNC-controlled helicopter maintains the desired velocity of approximately 15 m/s throughout the flight. Note that this maneuver involves a number of "mode transitions" (take-off, vertical rise to side slip, landing, etc.) that would have required a combination of different controllers and appropriate gain scheduling using more traditional approaches.

During the landing, the FlightLab model includes aerodynamic ground effects and provides information on landing gear (LG) forces and compression. To achieve a smooth landing, we simply add a quadratic cost associated with deviations from a desired smooth force curve. Referring to Figure 10.1, the NN is provided with two additional inputs corresponding to the LG forces (as output from the helicopter model) and the LG force deviation vector. Then in Figure 10.2, an additional Lagrange multiplier vector is propagated backward through the vehicle and NN Jacobians. Note that such additional constraints are not possible with the SDRE controller using the analytic 6DOF model (which does not include ground effects, force dynamics, or the loss of degrees of freedom at touch down and hence cannot be directly optimized for landing). Figures 10.4c and 10.4d show the desired and actual LG forces (3 points of contact), indicating a much smoother landing for the MPNC controller.

10.4.2. Elliptic Maneuver

In this example, we execute an extremely difficult "elliptic" maneuver,[11] consisting of (hover to) straight flight at 22.8 m/s while performing a constant yaw rotation of 120 deg/sec. Trajectories are illustrated for the MPNC and SDRE controllers in Figure 10.5. Note the tight execution performance of the MPNC which has a total cost of 14.43 versus 267.28 for

---

[11] The mass of the helicopter was reduced to $M_a = 4999$ lb.

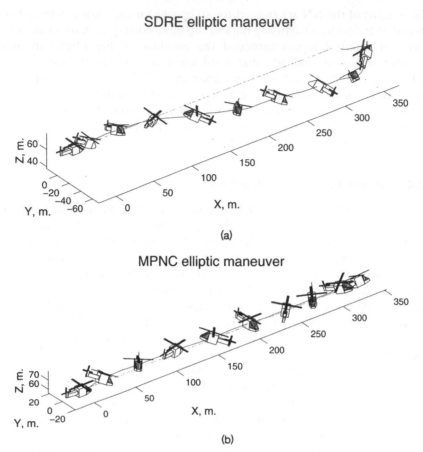

(a)

(b)

Figure 10.5. Elliptic maneuver: 22.8-m/s straight flight, 120-deg/s yaw rate. (a) SDRE, (b) MPNC.

the SDRE. The SDRE had a maximum *yaw* lag error of up to 84 degrees. In contrast, the MPNC had a maximum *yaw* error of only 12 degrees.

10.4.3. Wind Disturbance

Figure 10.6 shows the effect of an artificial wind shear (± 15 m/s) on the helicopter trajectory (3-m/s vertical rise). The displacement with the MPNC (< 1.9 m) is noticeably improved over the SDRE (< 7.5 m). In addition to the simple wind shear experiment, output from a large-eddy simulation (LES) model was used as forcing for the external wind server. This model [24] was specifically designed to examine small-scale atmospheric flows, especially those involving cumulus convection, entrainment, and turbulence. In this instance, the LES was used to simulate a microburst downdraft of the type associated with strong convective storms. The modeled domain enclosed an

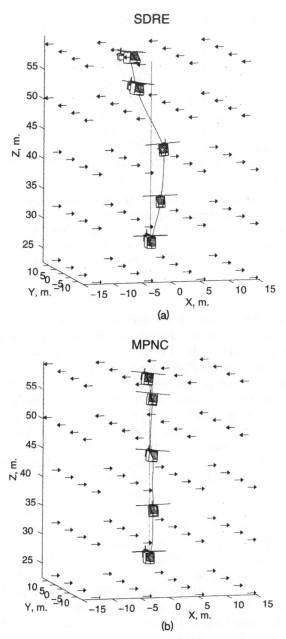

Figure 10.6. Vertical lift through wind shear. (a) SDRE, cost = 33.24, (b) MPNC, cost = 16.40.

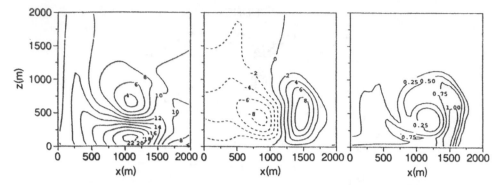

Figure 10.7. $x - z$ cross sections focusing on the gust front formed in conjunction with a microburst downdraft. The panels display the resolved x and z velocity components in m/s (left and center, respectively) and show the subgrid-scale turbulent kinetic energy in m^2/s^2 (right). The subgrid-scale turbulent kinetic energy is decomposed into a high-resolution turbulent velocity field that is summed with the resolved velocities.

area of approximately 8 km × 8 km horizontally and 4.2 km vertically. The grid spacing was 60 m in all directions. In the core downdraft, vertical velocities reached a magnitude of over 16 m/s. Upon impacting the surface, the downdraft spread out radially, forming a gust front—an area of strong turbulence which also includes substantial vertical and horizontal wind shear. In addition to the resolved wind fields, the LES also predicts subgrid-scale turbulent kinetic energy, which is decomposed into a rapidly evolving small-scale turbulence field. This small-scale turbulence was defined using continuous functions over a series of length scales ranging from the resolved grid scale down to viscous scales (i.e., 1 cm), and it varied at time scales on the order of 10 s (largest scale subgrid turbulence) down to less than a second (smallest scale subgrid turbulence). Due to the nature of turbulence within the inertial subrange, the majority of the turbulent kinetic energy resides in the larger scales. Features of the wind field are displayed in Figure 10.7.

The helicopter flight path (straight trajectory at 24 m/s) took it through the gust front (with winds exceeding 18 m/s). The relative performance and improvement with the MPNC controller is illustrated in Figure 10.8. Note that for this simulation, the MPNC uses the resolved wind field for prediction and optimization, while the smaller-scale turbulence is assumed unknown.

In general, we consider four possible scenarios for how to approximate the wind flow with the MPNC design:

1. MPNC trained with no information available about wind.
2. MPNC trained using constant wind fields as measured from the start of each horizon.
3. MPNC trained using resolved wind field (turbulence is assumed unknown and neglected for training purposes).

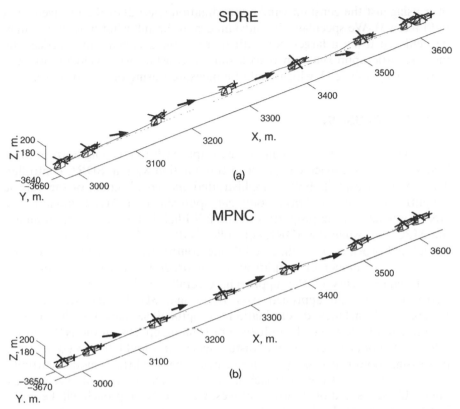

Figure 10.8. Trajectory through the gust front. (a) SDRE, (b) MPNC.

4. MPNC trained using knowledge of both resolved and actual turbulent wind flow (ideal case).

Table 10.3 summarizes the different performance using the experimental flight path as before. It is clearly seen that with even partial wind flow information the performance is greatly increased. However, perfect knowledge of the wind flow (ideal case 4) does not provide significant improvement

Table 10.3. Cost Comparisons for Wind Flow Incorporation

MPNC	Aircraft Speed	
Training Scenario	18 m/s	24 m/s
1	49.46	178.02
2	28.32	57.16
3	26.51	38.61
4	24.23	48.75
SDRE	111.49	162.69

over using just the constant wind approximation (case 2) or the resolved wind field (case 3). We speculate that advantages of the full wind prediction might be achieved using a larger NN with more training epochs. In practice, the most realistic approximation is to assume a constant wind velocity over the horizon length (1–2 s), which could be measured using on-board sensors.

10.5. CONCLUSION

In this chapter, we have presented a new approach to receding horizon MPC based on a NN feedback controller in combination with an SDRE controller. The approach exploits both a sophisticated *numerical* model of the vehicle (FlightLab) and its *analytical* nonlinear approximation (6DOF model). The NN is optimized using properties of the full FlightLab simulator to minimize the MPC cost, while the SDRE controller is designed using the approximate model and provides a baseline stabilizing control trajectory. In addition, we considered a number of simplifications in order to improve the computational requirements of the approach. Overall, results verify the superior performance of the approach over traditional SDRE (and LQ) control. Future work includes determination of optimal settings for the horizon length and update intervals, influence of training epochs on the NN, and the effects of modeling errors and disturbances. It should be noted that the processing power necessary to implement the FlightLab model currently precludes the use of this approach in real-time. However, with ever increasing CPU speeds and on-board resources, we feel this approach will be viable in the near future. In the interim, our plans are to investigate an alternative model designed specifically for small helicopters that can be run in real-time [25] (see also Chapter 15 of this text). Using this model, we are working toward a field demonstration using a model RC helicopter, as well as software integration and algorithm optimization on the open control platform (OCP) described in this book.

ACKNOWLEDGMENTS

This work was sponsored by DARPA under SEC grant F33615-98-C-3516. The authors would also like to thank Andy Moran for his programming assistance in the early phases of this research.

REFERENCES

1. Flightlab release note—version 2.8.4. Advanced Rotorcraft Technology, Inc., 1999.
2. S. J. Qin and T. A. Badgwell, An overview of industrial model predictive control technology. *Chemical Process Control—AIChE Symposium Series*, pp. 232–256, 1997.

3. E. S. Meadow and J. B. Rawlings, *Nonlinear Process Control*, chapter Model predictive control, Prentice-Hall, 1997.

4. J. R. Cloutier, C. N. D'Souza, and C. P. Mracek, Nonlinear regulation and nonlinear H-infinity control via the state-dependent Riccati equation technique: Part 1, Theory, in *Proceedings of the International Conference on Nonlinear Problems in Aviation and Aerospace*, FL, May 1996.

5. J. R. Cloutier, C. N. D'Souza, and C. P. Mracek, Nonlinear regulation and nonlinear H-infinity control via the state-dependent Riccati equation technique: Part 2, Examples, in *Proceedings of the International Conference on Nonlinear Problems in Aviation and Aerospace*, Daytona Beach, FL, May 1996.

6. S. W. Piche, B. Sayyar-Rodsari, D. Johnson, and M. Gerules, Nonlinear model predictive control using neural networks, *IEEE Control Systems Magazine* **20**(3):53–62, 2000.

7. A. Calise and Rysdyk, Nonlinear adaptive flight control using neural networks, *IEEE Control System Magazine* **18**(6):14–25, December 1998.

8. E. Johnson, A. Calise, R. Rysdyk, and H. El-Shirbiny, Feedback linearization with neural network augmentation applied to X-33 attitude control, in *Proceedings of the AIAA Guidance, Navigation, and Control Conference*, 2000.

9. E. A. Wan and A. A. Bogdanov, Model predictive neural control with applications to a 6 DoF helicopter model, in *Proceedings of IEEE American Control Conference*, Arlington, VA, June 2001.

10. A. A. Bogdanov, E. A. Wan, M. Carlsson, Y. Zhang, R. Kieburtz, and A. Baptista, Model predictive neural control of a high fidelity helicopter model, in *AIAA Guidance Navigation and Control Conference*, Montreal, Quebec, Canada, August 2001.

11. G. F. Franklin, J. D. Powell, and M. L. Workman, *Digital Control of Dynamic Systems*, 2nd edition, Addison-Wesley, Reading, MA, 1990.

12. A. Jadbabaie, J. Yu, and J. Hauser, Stabilizing receding horizon control of nonlinear systems: A control Lyapunov function approach, in *Proceedings of American Control Conference*, 1999.

13. D. Peters and C. He, Finite state induced flow models part II, three dimensional rotor disk, *Journal of Aircraft*, **32**(2):323–333, March–April 1995.

14. F. Wu, A. Packard, and G. Balas, LPV control design for pitch-axis missile autopilots, in *Proceedings of the 34th IEEE Conference on Decision and Control*, pp. 53–56, New Orleans, LA, 1995.

15. J. Shamma and J. Cloutier, Gain-scheduled missile autopilot design using linear parameter varying transformations, in *AIAA J. on Guidance, Control an Dynamics* **16**(2):256–263, 1993.

16. G. Balas, I. Fialho, A. Packard, J. Renfrow, and C. Mullaney, On the design of LPV controllers for the F-14 aircraft lateral-directional axis during powered approach, in *American Control Conference*, 1997.

17. M. Sznaizer, J. Cloutier, R. Hull, D. Jacques, and C. Mracek, Receding horizon control Lyapunov function approach to suboptimal regulation of nonlinear systems, *The Journal of Guidance, Control, and Dynamics*, **23**(3):399–405, May–June 2000.

18. R. A. Jacobs, Increasing rates of convergence through learning rate adaptation, *Neural Networks* **1**(4):295–307, 1988.

19. A. Jadbabaie, J. Yu, and J. Hauser, Unconstrained receding horizon control of nonlinear systems, in *Proceedings of IEEE Conference on Decision and Control*, 1999.

20. M. K. Kalnay and W. Baker, Global numerical weather prediction at the national meteorological center. In *Bull. Am. Meteor. Soc.* **71**:1410–1428, 1990.

21. M. Kanamitsu, J. Alpert, K. Campana, P. Caplan, D. Deaven, M. Iredell, B. Katz, H.-L. Pan, J. Sela, and G. White, Recent changes implemented into the global forecast system at NMC, *Weather and Forecasting* **6**:425–435, 1991.

22. M. Xue, K. K. Droegemeier, and V. Wong, The advanced regional prediction system (ARPS)—a multiscale nonhydrostatic atmospheric simulation and prediction tool. part i: Model dynamics and verification, *Meteorology and Atmospheric Physics* **75**:161–193, 2000.

23. A. Shapiro, L. Zhao, S. Weygandt, K. Brewster, S. Lazarus, and K. Droegemeir, Initial forecast fields from single-doppler wind retrieval, thermodynamic retrieval and ADAS, in *11th Conference on Numerical Weather Prediction, American Meteorological Society*, pages 119–121, Norfolk, VA, 1996.

24. M. A. Zulauf, Modeling the Effects of Boundary Layer Circulations Generated by Cumulus Convection and Leads on Large-Scale Surface Fluxes, Ph.D. thesis University of Utah, Salt Lake City, UT 84112, 2001.

25. V. Gavrilets, B. Mettler, and E. Feron, Nonlinear model for a small-size acrobatic helicopter, in *AIAA Guidance Navigation and Control Conference*, Montreal, Quebec, Canada, August 2001.

CHAPTER 11

ACTIVE MODEL ESTIMATION FOR COMPLEX AUTONOMOUS SYSTEMS

MARK E. CAMPBELL, EELCO SCHOLTE, and SHELBY BRUNKE

Editors' Notes

The other chapters in Part III are concerned with modeling, control, and optimization. Here, the focus is on state estimation, another topic central to control engineering. The distinctions between these subjects are not always clear cut and in fact the state estimation methods discussed here are also used for model estimation but they suggest different emphases. Work in estimation often focuses on noise, uncertainty, and stochastic models.

The methods presented in this chapter are useful not only for estimating the true values of states whose measurements are noise-corrupted, but also uncertainty bounds on the estimates. Failure effects are considered. A rapid change in measurement arising from a sudden fault will result in increased uncertainty bounds that will gradually converge as more data is gathered.

The classical state estimator for nonlinear systems is the extended Kalman filter (EKF). Although widely used, EKF's have some well-known deficiencies, including a susceptibility to bias and divergence in the state estimates. In this chapter an alternative nonlinear estimator is presented, the unscented Kalman filter (UKF). Instead of linearizing the nonlinear dynamics as with the EKF, the UKF uses the nonlinear model directly—the distribution of the propagated state is approximated using a finite set of points. The UKF is also much simpler to implement and therefore well suited for online application in systems with fast dynamics. Simulation results for the UKF estimator are presented for an F-15-like model, for estimation of altitude, lift force, and aerodynamic coefficients. State estimation under fault conditions is also demonstrated.

The chapter also describes an alternative nonlinear estimation filter that overcomes a drawback of the UKF-based method: the heuristic (nonrigorous) nature of the uncertainty bounds. This filter, which bears some similarities to the EKF, is called the extended set-membership filter (ESMF) and estimates hard bound, ellipsoidal uncertainties for nonlinear systems. A simplified multi-UAV position estimation problem with intermittent sensing serves to illustrate the use of an ESMF.

Software-Enabled Control. Edited by Tariq Samad and Gary Balas
ISBN 0-471-23436-2 © 2003 IEEE

11.1. INTRODUCTION

Autonomy implies a degree of self-regulation inherent in a system's operation, and it is a key technology for future high-performance and remote operation applications. The "promise" of autonomy includes (1) a reduced need for human intervention, (2) increased performance range and capabilities, (3) extended operation life, and (4) decreased costs. Examples of important future applications that will rely on developments in the area of autonomy and control include satellite clusters, deep space exploration, air traffic control, and battlefield management with multiple uninhabited aerial vehicles (UAVs) and ground operations. For these applications, an autonomous architecture requires many online functions such as mission/trajectory planning, control reconfiguration, fault detection, and traditional lower-level control loops. In addition, effective human command of many vehicles requires information flow and fusion at several levels to facilitate (semi)autonomous operation.

The concept of autonomous control is typically described as a layered, feedback-based architecture with "degrees" of autonomy [1]. The approach here is to focus on two lower levels of this architecture, as shown in Figure 11.1. This typical future autonomous vehicle architecture includes the following: Low- and mid-level control including feedback compensation and trajectory generation; fault detection to locate and recover from typical failures within the system; and active state models that provide accurate model based information to the other software-enabled control (SEC) components so that they can perform adequately. SEC brings software and control algorithms together to enable new functionality within the autonomous control paradigm. Traditional feedback control techniques have few accommodations for maintaining stability and high performance while operating in highly uncertain environments and in the presence of failures or damage. SEC is designed to deliver enabling functions such as real-time fault detection and control customization, both of which will require models to accurately and reliably fulfill their goals.

Most control and fault detection methods are model-based; accurate and reliable modeling will allow these new autonomous control developments to move past the low/no-noise and full-state feedback restrictions on to a realistic implementation. For example, low-level reconfigurable control [2] typically requires *both* a point model and an uncertainty set to guarantee stability. Trajectory generation, such as those methods developed in the SEC progam [3], use online model predictive approaches that require nonlinear models with bounded uncertainties. This is formulated as a constrained optimization problem,

$$\min_{x,u} J = \min_{x,u} \int_0^T q(x,u,w,\upsilon,t)\, dt + V(x_F, u_F, t_F)$$

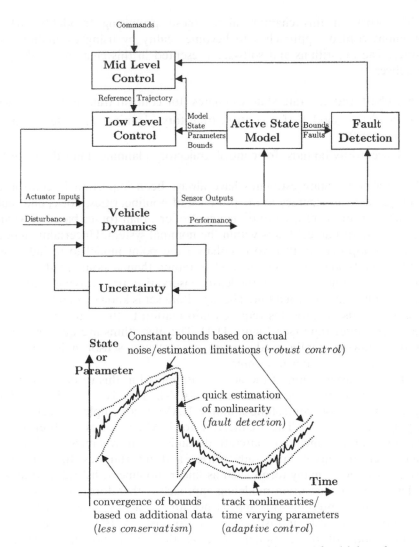

Figure 11.1. Overview of an autonomous control architecture for high-performance vehicles (top), and the active-state model concept (bottom), showing a state/parameter estimate and uncertainty bounds.

subject to the following constraints:

$$\dot{x} = f(x, u, w, t)$$
$$y = h(x, u, v, t)$$
$$g(x, u, t) \leq 0$$

where the uncertainties are written as constraints.

The work in this chapter allows these promising (model-based) autonomous control approaches to become reality by using estimation techniques. The algorithms and software presented in this chapter are designed to deliver

- Robust and accurate state estimates (for control, trajectory planning)
- Predictive models (for control, trajectory planning, fault detection and recovery)
- Uncertainty bounds (for control, trajectory planning, fault detection)

all in real time. State estimates have always been important in control; the challenge here is accuracy in the presence of nonlinearities, noise, and faults. Predictive models are used to design controllers, plan routes based on vehicle capabilities, and detect faults within the nominal system. Uncertainty models, a key development in this work, enhance the point model to include confidence bounds and/or a set of models rather than a point model. Ideally, uncertainty bounds should be the least conservative, time-varying bounds that the algorithm and system allow. The algorithm set is known as an "active-state model," because it provides required information to the autonomous system that adapts over time (see Figure 11.1). The algorithms are quite general in nature, allowing nonlinear dynamics, process and sensor noise, and even failures that are unknown *a priori*.

The focus application of the active state model in this work is a system of multiple uninhabited aerial vehicles (UAVs). This is a good test of the algorithms because controlling multiple UAVs is a complex, uncertain, and very dynamic problem. The architecture for UAVs requires a subset of many different models, including aircraft dynamics, aerodynamics, environment, engine, terrain, threat, faults, adversaries, and uncertainties. In addition, the models may be of many forms such as linear, nonlinear, hybrid, and look-up tables. A flexible, accurate modeling approach is therefore required.

This chapter presents two approaches for the active-state model in order to support the SEC infrastructure: a stochastic version and a bounded version. Both of these models have at their core nonlinear estimation algorithms that are then used to develop online state and parameter estimates and uncertainties. The algorithms are designed for online usage, interface well with control and fault detection methods, and are quite robust. After a brief introduction to each active-state model concept, the results from their implementation on a nonlinear F-15-like simulation are presented.

11.2. PRELIMINARIES: JOINT AND DUAL ESTIMATION

Consider the general discrete, nonlinear state-space system:

$$x_{k+1} = f(x_k, u_k, w_k, t_k) \tag{11.1}$$

$$y_k = h(x_k, u_k, v_k, t_k) \tag{11.2}$$

where $x_k \in \mathbb{R}^n$ is the state vector, $w_k \in \mathbb{R}^n$ is the disturbance vector, $y_{k+1} \in \mathbb{R}^{n_y}$ is the output vector, and $v_{k+1} \in \mathbb{R}^{n_y}$ is the sensor noise vector. A key aspect of this chapter is that estimating a model of a nonlinear system is no different conceptually than estimating the state. To see this, the above equations of motion can be rewritten as

$$x_{k+1} = f(x_k, \theta_k, u_k, w_k, t_k)$$
$$\theta_{k+1} = f_\theta(x_k, \theta_k, u_k, w_{\theta,k}, t_k)$$
$$y_k = h(x_k, \theta_k, u_k, v_k, t_k)$$

where θ_k is a vector of model parameters at each time step. The estimation problem can be set up using two distinct approaches, termed joint and dual. In joint estimation, a new state vector is defined as a combination of the old state and parameter vector, $\tilde{x}_k = [x_k \quad \theta_k]$, and the equations of motion can be written as

$$\tilde{x}_{k+1} = \tilde{f}(\tilde{x}_k, u_k, \tilde{w}_k, t_k)$$
$$y_k = \tilde{h}(\tilde{x}_k, u_k, v_k, t_k)$$

(11.3)

Thus, the problem is still a nonlinear estimation problem. In fact, even when the dynamics in Eq. (11.1) are linear, the joint state and model estimator in Eq. (11.3) is typically nonlinear.

Dual estimation is when the state and parameter estimation problems are solved separately, usually with two distinct estimators that feed each other. While joint estimation is primarily considered in this chapter, a dual formulation will also briefly be presented in the simulations. Note that as a notational simplification, Eq. (11.1) and x_k are used in the derivations to follow. The vector x_k can refer to the state or combined state/parameter estimation problems without loss of generality.

11.3. ROBUST NONLINEAR STOCHASTIC ESTIMATION

The stochastic active-state model requires a nonlinear estimation algorithm that performs well online, does not diverge if there are jumps in the sensor signal (such as damage), and is robust to various parameters such as tuning and initial conditions. Traditional state estimation for linear systems has a rich, elegant theory, namely the stochastic Kalman filter. Extension of Kalman filtering theory, especially guarantees, to nonlinear systems is nontrivial but important, considering such motivating applications as aircraft state estimation, global positioning system (GPS), and parameter estimation of linear models. The extended Kalman filter (EKF) is an approximate extension of Kalman filtering techniques to nonlinear systems by linearizing the state dynamics at each time step. The EKF does not require linearity, stability, or time-invariance of the system. It does require that the state dynamics are

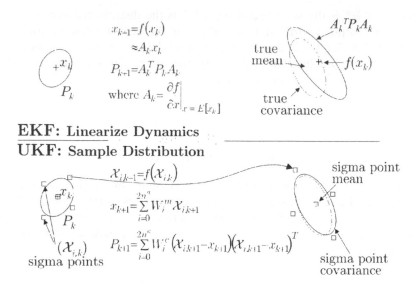

EKF: Linearize Dynamics

UKF: Sample Distribution

Figure 11.2. Comparison of the prediction step for the UKF and EKF nonlinear estimators.

sufficiently differentiable (i.e., the Jacobian exists). EKF has been used in a wide variety of applications, including radar tracking and aircraft parameter estimation. One major disadvantage of the EKF is that it requires *a priori* information regarding the initial state of the system and noise environment. Furthermore, sensitivities to this a priori information can lead to bias and even divergence [4].

It is understood that one cause of the poor performance in the EKF is due to the linearization of the model dynamics [5]. A new nonlinear filter has been introduced that is much simpler than the EKF, yet achieves better performance. Instead of truncating nonlinear dynamics to first order as with the EKF, the unscented Kalman filter (UKF) [6] approximates the distribution of the state with a finite set of points. These points, called sigma points, are then propagated through the nonlinear dynamics. The mean and covariance of the distribution are then calculated as a weighted sum and outer product of these propagated points. Because nonlinear dynamics are used without approximation (i.e., no derivatives are calculated as in the EKF), it is much simpler to implement and better results are expected. The performance of the EKF has been shown to be analytically similar to a truncated second-order EKF, but without the need to calculate Jacobians or Hessians of the dynamics [6]. A schematic comparison of these two approaches is given in Figure 11.2.

The EKF algorithm, including the numerically robust square-root formulation, is well known in the literature [7] and is not repeated here. Because the UKF algorithm utilizes the square root of the covariance matrix at each time step, the square-root implementation is a very elegant, as well as numerically robust, solution. A square-root formulation of the UKF for general nonlinear systems [8] is presented here, and it is compared with the square-root EKF to form the underlying algorithm for the stochastic active-state model.

Consider the nonlinear state-space system described in Eq. (11.1). An augmented state vector is defined as

$$x_k^a = \begin{bmatrix} x_k \\ w_k \\ v_k \end{bmatrix}$$

which has dimension $x_k^a \in \mathbb{R}^{n_a}$, where $n_a = n + n_w + n_v$. The initial state estimate and covariance are known *a priori* for the augmented state vector,

$$\hat{x}_0^a = E[x_0^a] = \begin{bmatrix} x_0 \\ 0 \\ 0 \end{bmatrix}, \quad P_{xx,0}^a = E\left[(x_0^a - \hat{x}_0^a)(x_0^a - \hat{x}_0^a)^T\right] = \begin{bmatrix} P_{xx,0} & 0 & 0 \\ 0 & Q_0 & 0 \\ 0 & 0 & R_0 \end{bmatrix}$$

where Q_0 and R_0 are the initial covariance matrices for w and v, respectively. The square root of the initial covariance is defined as

$$S_{xx,0}^a = \sqrt{P_{xx,0}^a} = \begin{bmatrix} \sqrt{P_{xx,0}} & 0 & 0 \\ 0 & \sqrt{Q_0} & 0 \\ 0 & 0 & \sqrt{R_0} \end{bmatrix}$$

The next step is to define $2n_a + 1$ sigma points,

$$\mathcal{X}_k = \begin{bmatrix} \hat{x}_k^a & \hat{x}_k^a - \sigma_f \, S_{xx,0}^a & \hat{x}_k^a + \sigma_f \, S_{xx,0}^a \end{bmatrix}$$

$$= \begin{bmatrix} \mathcal{X}_k^x \\ \mathcal{X}_k^w \\ \mathcal{X}_k^v \end{bmatrix}$$

and a set of associated weights,

$$W_0^m = \frac{\sigma_f^2 - n}{\sigma_f^2}, \quad W_0^c = \frac{\sigma_f^2 - n}{\sigma_f^2} + 3 - \frac{\sigma_f^2}{n}$$

$$W_i^m = W_i^c = \frac{1}{2\sigma_f^2}, \quad i = 1, \ldots, 2n_a$$

where σ_f is a scaling for the distance of the sigma points from the mean, and m and c denote mean and covariance, respectively. A diagonal covariance weighting matrix of all but the initial weight, W_0^c, is also defined,

$$W^c = \begin{bmatrix} W_1^c & & 0 \\ & \ddots & \\ 0 & & W_{2n}^c \end{bmatrix}$$

The general square-root UKF algorithm is given as follows:

UKF Prediction

$$\mathcal{X}_{k+1}^{x-} = f(\mathcal{X}_k^x, u_k, \mathcal{X}_k^w)$$

$$\mathcal{Y}_{k+1}^- = h(\mathcal{X}_k^x, u_k, \mathcal{X}_k^v)$$

$$\hat{x}_{k+1}^- = \sum_{i=0}^{2n_a} W_i^m \mathcal{X}_{i,k+1}^{x-}$$

$$\hat{y}_{k+1}^- = \sum_{i=0}^{2n_a} W_i^m \mathcal{Y}_{i,k+1}^-$$

$$\left[\mathcal{X}_{c0,k+1}^{x-} \quad \mathcal{X}_{c,k+1}^{x-}\right] = \left[\mathcal{X}_{0,k+1}^{x-} - \hat{x}_{k+1}^- \;\middle|\; \mathcal{X}_{1,k+1}^{x-} - \hat{x}_{k+1}^- \cdots \mathcal{X}_{2n_a,k+1}^{x-} - \hat{x}_{k+1}^-\right]$$

$$\left[\mathcal{Y}_{c0,k+1}^- \quad \mathcal{Y}_{c,k+1}^-\right] = \left[\mathcal{Y}_{0,k+1}^- - \hat{y}_{k+1}^- \;\middle|\; \mathcal{Y}_{1,k+1}^- - \hat{y}_{k+1}^- \cdots \mathcal{Y}_{2n_a,k+1}^- - \hat{y}_{k+1}^-\right]$$

UKF Update

$$K_{k+1} = \left[(\mathcal{X}_{c,k+1}^{x-})(\mathcal{Y}_{c,k+1}^-)^T + \frac{W_0^c}{W_i^c}(\mathcal{X}_{c0,k+1}^{x-})(\mathcal{Y}_{c0,k+1}^-)^T\right]$$

$$\times \left[(\mathcal{Y}_{c,k+1}^-)(\mathcal{Y}_{c,k+1}^-)^T + \frac{W_0^c}{W_i^c}(\mathcal{Y}_{c0,k+1}^-)(\mathcal{Y}_{c0,k+1}^-)^T\right]^{-1}$$

$$\hat{x}_{k+1} - \hat{x}_{k+1}^- + K_{k+1}(y_{k+1} - \hat{y}_{k+1}^-)$$

$$S_{xx,k+1} = update\left\{orth\left\{(\mathcal{X}_{c,k+1}^{x-} - K_{k+1}\mathcal{Y}_{c,k+1}^-)\sqrt{W^c}\right\},\right.$$

$$\left.(\mathcal{X}_{c0,k+1}^- - K_{k+1}\mathcal{Y}_{c0,k+1}^-)\sqrt{|W_0^c|}, sign(W_0^c)\right\}$$

This square-root formulation of the UKF makes use of two functions [9]: (1) orthogonalization or triangularization of a rectangular set of sigma points into an $n \times n$ matrix, denoted as $S = orth\{R\}$ or a Cholesky factor ($A = SS^T = RR^T$, $R \in \mathbb{R}^{n \times m}$, $S \in \mathbb{R}^{n \times n}$, $m > n$), and (2) rank-one update of this Cholesky factor, denoted as $X_u = update\{X, v, \pm \sqrt{r}\}$ ($A = X_u X_u^T = XX^T \pm \sqrt{r} vv^T$). Note that a rank one update of the zero sigma point, W_0^c, is required in the prediction step when $W_0^c < 0$; if $W_0^c > 0$, a simple orthogonalization can be performed.

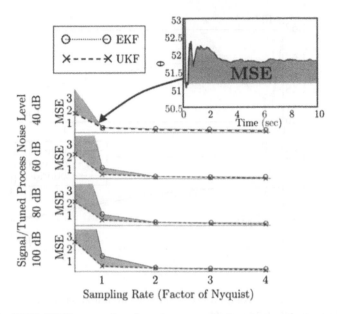

Figure 11.3. UKF–EKF comparison based on assumed process noise versus sampling rate for a wide variety of systems (averaged).

In order to select a baseline algorithm for the stochastic active state model, the following metrics are compared between the square-root versions of the EKF and UKF: accuracy, stability, tuning sensitivity, sampling rate sensitivity, and initialization sensitivity. These metrics are compared using past work, as well as through comparison of the UKF and EKF as applied to a large set of systems including: simple nonlinear systems (quadratics, jumps, sinusoids), parameter estimation in structural systems, and state/parameter estimation in nonlinear aircraft models [10]. Comparing accuracy, Julier et al. [6] showed that the UKF is equivalent to a second-order EKF in step-to-step accuracy, with a level of computation equivalent to a first-order EKF. Comparing tuning and sampling rate, Figure 11.3 shows an averaged set of results, where the mean-square error (MSE) of the joint state/parameter estimates is summarized (note that this is directly proportional to the cost minimized in the Kalman filter). Several trends are noted. First, as the sampling frequency decreases to near Nyquist, the UKF always has a lower MSE. Second, as the tuned process noise is decreased, the MSE for the EKF is larger than the UKF. Finally, as the tuned process noise changes, the MSE for the UKF changes only slightly, while the EKF gets better or worse. It is noted that the EKF also exhibited tuning sensitivity in nonlinear aircraft parameter estimation [11].

Comparing algorithm stability, Reif et al. [5] recently presented a stability analysis of the EKF, which is a direct function of the linearization error; obviously, because the UKF error is smaller (because it is equivalent to a second-order EKF in accuracy), it has more desirable stability properties. In

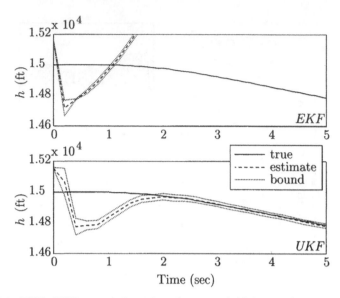

Figure 11.4. UKF–EKF comparison based on an initial covariance that is small compared to the initial-state error.

simulations, while both square root algorithms performed quite well, the EKF on occasion would diverge while the UKF never diverged. An example is shown in Figure 11.4. Here, the altitude for a UAV is estimated; initial covariance bounds are set such that the 3σ bounds do not include the initial condition. Notice that the UKF increases its bound in the early stages of estimation, and it converges to the estimate within a few seconds. The EKF, on the other hand, diverges quickly. Similar results occur with initial covariances that are too large, as well as with jumps in the data.

Past work and the study in reference 10 indicate that the UKF is a wise choice as the underlying nonlinear estimation algorithm for the stochastic active-state model because of several important facts/trends. The UKF is

- More accurate from one time step to the next
- Less susceptible to divergence
- Equivalent in numerical computation
- Less sensitive to tuning of process noise
- Less sensitive to slow sampling rates
- More robust to initial uncertainties (and jumps in the data)

compared to the EKF. The UKF also has the added benefit of not requiring Jacobians or Hessians of the dynamics to be calculated, and it is therefore simpler to derive and implement. These results support the conclusion that the UKF is better than the EKF not just in accuracy, but also in its online implementation characteristics for applications such as future UAVs.

11.4. NONLINEAR BOUNDED SET ESTIMATION

The stochastic algorithms, while excellent computationally for online implementation, have three shortcomings: (1) system noises must usually satisfy Gaussian noise assumptions, (2) the uncertainty bounds are *ad hoc*, and (3) the linearization error is ignored (or masked with tuned noise parameters). While this is fine for particular applications, it is not for those applications that require hard (guaranteed) uncertainty bounds such as many types of robust and optimized control customization approaches. Set membership methods assume (a) hard bounds on noise sources and states and (b) worst-case propagation. Sets are usually described with simpler mathematical approaches such as ellipsoids or polytopes. For linear systems, a recursive set membership filter (SMF) can be developed that looks strikingly similar to the Kalman filter [12]. Extensions of the linear SMF to nonlinear dynamics have been attempted for simple systems [13], but are much more difficult for estimating complex systems in real time such as aircraft applications.

The approach here is to cast the nonlinear dynamics in a way that is suitable for implementation within the linear SMF framework. Specifically, the nonlinear dynamics are linearized about the current estimate in a manner that is similar to the EKF. The remaining terms are then bounded using interval mathematics [14], and they are incorporated into the algorithm as additions to the process or sensor noise bounds. This allows the solution to be guaranteed for nonlinear systems so long as the bound on the nonlinear term is guaranteed. The proposed method is compatible with many current robust control methods that require hard bounds, as well as with planning algorithms that require guaranteed uncertainty information. It is also computationally efficient for online recursive implementation. The new algorithm is termed the Extended Set-Membership Filter (ESMF) [15].

Consider the discrete nonlinear system (removing u_k and using additive noise to simplify notation, with no loss in generality):

$$x_{k+1} = f(x_k) + w_k \tag{11.4}$$

with its nonlinear output equation:

$$y_k = h(x_k) + v_k \tag{11.5}$$

Both f and h are C^2 functions. The initial state, x_0, is known to be bounded by an ellipsoid given as

$$[x_0 - \hat{x}_0]^T \Sigma_{0,0}^{-1} [x_0 - \hat{x}_0] \leq 1 \tag{11.6}$$

where $\hat{x}_0$ is the center of the ellipsoid. The disturbance w_k and sensor noise v_k are also bounded at each time step k,

$$w_k^T Q_k^{-1} w_k \leq 1 \quad \forall\, k$$

$$v_k^T R_k^{-1} v_k \leq 1 \quad \forall\, k$$

where $\Sigma_{k,k} \in \mathbb{R}^{n \times n}$, $Q_k \in \mathbb{R}^{n \times n}$, and $R_k \in \mathbb{R}^{n_y \times n_y}$ are symmetric positive definite matrices. Note that no assumptions on the structure of the noise are made except that it is bounded; hence, many types of noises are included within this framework including random and biased signals.

Linearizing Eq. (11.4) about the current state estimate $\hat{x}_k$ yields

$$x_{k+1} = f(x_k)\big|_{x_k = \hat{x}_k} + \frac{\partial f(x_k, t_k)}{\partial x}\bigg|_{x_k = \hat{x}_k} (x_k - \hat{x}_k) + \text{H.O.T.} + w_k \quad (11.7)$$

where H.O.T. refers to the higher-order terms of the expansion. The H.O.T. and process noise are then bounded together such that Eq. (11.7) can be rewritten as

$$x_{k+1} = f(x_k)\big|_{x_k = \hat{x}_k} + \frac{\partial f(x_k, t_k)}{\partial x}\bigg|_{x_k = \hat{x}_k} (x_k - \hat{x}_k) + \hat{w}_k$$

where $\hat{w}_k$ is a new noise term that bounds both the original noise and the linearization remainder and is defined as

$$\hat{w}_k^T \hat{Q}_k^{-1} \hat{w}_k \leq 1 \quad \forall\, k \qquad (11.8)$$

The output equation [Eq. (11.5)] can be linearized in a similar fashion such that the nonlinearities are captured in the new measurement noise term:

$$y_k = h(x_k)\big|_{x_k = \hat{x}_k} + \frac{\partial h(x_k, t_k)}{\partial x}\bigg|_{x_k = \hat{x}_k} (x_k - \hat{x}_k) + \hat{v}_k$$

$$\hat{v}_k^T \hat{R}_k^{-1} \hat{v}_k \leq 1 \quad \forall\, k \qquad (11.9)$$

This approach, shown in Figure 11.5, is more amenable to on-line implementation as compared to similar approaches in references 16–18. Compared to the EKF, the ESMF now bounds the linearization error. In addition, the only restriction on the ESMF is that the Hessian and Jacobian must be continuous over the set of states.

In this work, interval mathematics are used to bound the H.O.T. Interval mathematics is an extension, or generalization, of real analysis, where intervals and interval arithmetic are applied instead of real numbers and real arithmetic. This field was primarily initiated by Moore [14], with many applications described by Hansen [19]. There are also several software packages for interval analysis [20]. As an example, consider the interval X given as $X = [a, b]$. Subtracting intervals now yields $X - X = [a, b] - [a, b] = [a - b, b - a]$, which bounds the worst-case subtraction.

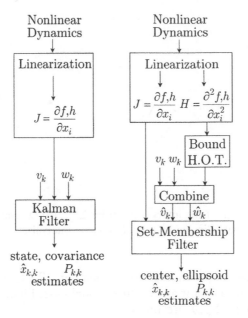

Figure 11.5. Simplified graphical representation of the EKF (*left*) and the nonlinear set-membership filter (*right*) at each time step.

Rewriting the linearized function with a remainder, one obtains

$$x_{k+1} = f(x_k)\big|_{x_k=\hat{x}_k} + \frac{\partial f(x_k, t_k)}{\partial x}\bigg|_{x_k=\hat{x}_k}(x_k - \hat{x}_k) + \cdots + \frac{f(x_k)^{(n)}}{n!}\bigg|_{x_k=\hat{x}_k}$$

$$\times (x_k - \hat{x}_k)^n + R_n(x_k - \hat{x}_k, X_k) + w_k \qquad (11.10)$$

where R_n is a remainder term, and $f^{(n)}$ is the nth derivative. The term X_k can take on any value over an interval for which $(x_k - \hat{x}_k)$ is defined. Thus, $R_n(x_k - \hat{x}_k, X_k)$ can be bounded by simply defining the interval X_k and evaluating $R_n(x_k - \hat{x}_k, X_k)$ using interval mathematics. Following Taylor's Theorem [19], the Lagrange remainder is written as

$$R_n(x_k - \hat{x}_k, X_k) = \frac{f^{(n+1)}(X_k)}{(n+1)!}(x_k - \hat{x}_k)^{n+1} \qquad (11.11)$$

For the ESMF proposed here, the remainder term, or equivalently the linearization error, is simply a function of the Hessian of the nonlinear dynamics [Eq. (11.11) with $n = 1$]. Equation (11.10) then simplifies to

$$x_{k+1} = f(x_k)\big|_{x_k=\hat{x}_k} + \frac{\partial f(x_k, t_k)}{\partial x}\bigg|_{x_k=\hat{x}_k}(x_k - \hat{x}_k) + \frac{f^{(2)}(X_k)}{2!}(x_k - \hat{x}_k)^2 + w_k$$

$$(11.12)$$

This is easily extended to the multidimensional case.

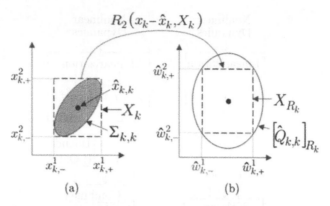

Figure 11.6. Derivation of the interval X_k from $\Sigma_{k,k}$. (b) Derivation of the bound on $R_2(x_k - \hat{x}_k, X_k)$ output ellipsoid by bounding using intervals.

The procedure for bounding the remainder, $R_2(x_k - \hat{x}_k, X_k)$, is shown pictorially in Figure 11.6 for a two-dimensional case. First, the extrema of the state ellipsoid, $S(\hat{x}_k, \Sigma_{k,k})$, are found as

$$x^i_{k,\pm} = x^i_k \pm \sqrt{\Sigma^{i,i}_{k,k}} \qquad (11.13)$$

where the superscript i denotes the ith state and the subscripts $+$ and $-$ denote the maximum and minimum values, respectively. The state interval bound, X_k, is then defined as

$$X^i_k = \left[x^i_{k,-}, x^i_{k,+} \right] \qquad (11.14)$$

The second step is to evaluate the interval of the remainder term, $R_2(x_k - \hat{x}_k, X_k)$ [Eq. (11.11)], using interval mathematics, which yields interval $X^i_{R_k}$. The third step is to bound $X^i_{R_k}$ using an ellipsoid. This outer bounding ellipsoid is not uniquely defined, but can be optimized by minimizing the volume of the ellipsoid. This yields a closed-form solution (see reference 13 for a derivation):

$$\left[\hat{Q}_{k,k} \right]^{i,i}_{R_k} = 2 \left(X^i_{R_k} \right)^2 \qquad (11.15)$$

The full process noise bound can then be found by adding the respective ellipsoids (H.O.T. and w_k), or

$$\hat{Q}_{k,k} = \frac{\left[\hat{Q}_{k,k} \right]_{R_k}}{1 - \beta_Q} + \frac{\left[Q_{k,k} \right]}{\beta_Q} \qquad (11.16)$$

where β_Q denotes the parameter that optimizes the mixture of the two matrices.

It is now possible to apply the linear SMF to any nonlinear system for which the Jacobian and the Hessian are well-defined. Given the system dynamics [Eqs. (11.4) and (11.5)], initial conditions [Eq. (11.6)], and noise assumptions [Eqs. (11.8) and (11.9)], the state ellipsoid and center, denoted $S(\hat{x}_{k,k}, \Sigma_{k,k})$, can now be recursively estimated. The ESMF is summarized as follows.

ESMF Prediction

$$\hat{x}_{k+1,k} = f(\hat{x}_{k,k})$$

$$\Sigma_{k+1,k} = A_k \frac{\Sigma_{k,k}}{1-\beta} A_k^T + \frac{\hat{Q}_k}{\beta} \qquad (11.17)$$

Update

$$\hat{x}_{k+1,k+1}$$

$$= \hat{x}_{k+1,k} + \Sigma_{k+1,k} C_{k+1}^T \left[C_{k+1} \frac{\Sigma_{k+1,k}}{1-\rho} C_{k+1}^T + \frac{\hat{R}_{k+1}}{\rho} \right]^{-1} (y_{k+1} - h(\hat{x}_{k+1,k}))$$

$$\overline{\Sigma}_{k+1,k+1}$$

$$= \frac{\Sigma_{k+1,k}}{1-\rho} - \frac{\Sigma_{k+1,k}}{(1-\rho)} C_{k+1}^T \left[C_{k+1} \frac{\Sigma_{k+1,k}}{1-\rho} C_{k+1}^T + \frac{\hat{R}_{k+1}}{\rho} \right]^{-1} C_{k+1} \frac{\Sigma_{k+1,k}}{(1-\rho)}$$

$$\Sigma_{k+1,k+1} = (1 - \delta_{k+1}) \cdot \overline{\Sigma}_{k+1,k+1} \qquad (11.18)$$

where

$$A_k = \left. \frac{\partial f(x_k)}{\partial x} \right|_{x_k = \hat{x}_{k,k}}, \quad C_{k+1} = \left. \frac{\partial h(x_k)}{\partial x} \right|_{x_k = \hat{x}_{k+1,k}}$$

The prediction step [Eq. (11.17)] is the addition of two ellipsoids, the noise ellipsoid $S(0, \hat{Q}_k)$, and the state uncertainty ellipsoid $S(\hat{x}_{k,k}, \Sigma_{k,k})$ rotated and scaled by A_k^T and A_k. In general, the addition of two ellipsoids has no closed-form solution and therefore requires a bounding ellipsoid. The parameter β indicates the mixing of the two ellipsoids to form the bounding ellipsoid. The update step is the intersection of two sets; the predicted state ellipsoid and the set described by the output equation. The intersection of these two sets also does not have a general closed-form solution. The parameter ρ indicates the mixing of the two ellipsoids to form the bounding ellipsoid. The scalar parameters β and ρ strictly lie between 0 and 1 and can be numerically optimized at each time step in an effort to find the smallest bounding ellipsoid. Note that several closed-form solutions exist for minimizing the trace of the bounding matrix [13].

The actual state x_k is then bounded by

$$[x_{k+1} - \hat{x}_{k+1}]^T [(1 - \delta_{k+1}) \Sigma_{k+1,k+1}]^{-1} [x_{k+1} - \hat{x}_{k+1}] \leq 1 \quad (11.19)$$

where δ_{k+1} is a recursively calculated value

$$\delta_{k+1} = [y_{k+1} - h(\hat{x}_{k+1,k})]^T \left[C_{k+1} \frac{\Sigma_{k+1,k+1}}{1-\rho} C_{k+1}^T + \frac{\hat{R}_{k+1}}{\rho} \right]^{-1}$$

$$[y_{k+1} - h(\hat{x}_{k+1,k})] \quad (11.20)$$

Obviously, if $\delta_{k+1} \geq 1$, the state uncertainty ellipsoid is not defined. This only occurs when the base algorithm assumptions have not been satisfied— that is, when the initial state and noise ellipsoids are not valid. Thus, δ_{k+1} is a good check of the health of the algorithm.

It is noted that the SMF described above is very similar to the KF, with the exception of the parameters ρ, β, and δ_{k+1}. Although the SMF and KF equations look similar, their physical interpretation is very different. Instead of propagating probability distributions, hard bounds are now propagated in time and guaranteed to contain the state. Therefore, the actual probability distribution of the state across the ellipsoid is uniform.

The full ESMF algorithm can be summarized by the following steps:

1. Calculate the state interval based on the ellipsoid maximums [Eqs. (11.13 and 11.14)].
2. Find the maximum interval for the H.O.T. using interval analysis [Eq. (11.11)].
3. Calculate the ellipsoid bounding the linearization error [Eq. (11.15)].
4. Calculate the final process/linearization error bound [Eq. (11.16)].
5. Calculate the predicted state ellipsoid using the linear SMF prediction step, and optimize the output state ellipsoid using β [Eq. (11.17)].
6. Calculate the updated state ellipsoid using the linear SMF update step, and optimize the output state ellipsoid using ρ [Eq. (11.18)].

The true state bound is then given by Eq. (11.19). For the nonlinear SMF, the estimator can be shown to be stable under certain conditions. Stable here is defined as the error between the true value and the center of the ellipsoidal set. The proof is derived in reference 15, and it is similar to proofs for the EKF and polytope set methods [5, 17]. The primary condition is that the linearized system is uniformly observable, which bounds the uncertainty ellipsoid.

11.5. SIMULATION RESULTS: F-15-LIKE SIMULATION

A simulation environment that contains many of the complexities of the future multiple UAV missions is required to validate the algorithms developed in the SEC program. The model used in this work is a Simulink simulation of a McDonnell Douglas F-15 Eagle aircraft, based on an AIAA challenge model with updated aerodynamics from Wright Patterson AFB [21]. It is a high fidelity nonlinear model which couples the equations of motion for the F-15 with aerodynamic, atmospheric, and engine performance models. The states are

$$x = \begin{bmatrix} V & \alpha & q & \theta & p & \phi & r & \psi & \beta & h & x_p & y_p \end{bmatrix}^T$$

which includes the vehicle velocity, angle of attack, pitch, yaw, and roll rates and angles, sideslip angle, altitude, and x, y position. The aerodynamic model is developed from multidimensional tables and linear interpolation. The Dryden model [9] is used to model the turbulence, and wind vectors are also integrated as velocity components $(u_{wind}, v_{wind}, w_{wind})$. The propulsion model, also based on flight data, integrates models of the core engine and afterburner sections, and it considers each engine independently. The atmospheric model is based on standard tables and the aircraft altitude. The mass and geometry characteristics of the aircraft are listed in Table 11.1. The control and sensor variables for this model are listed in Table 11.2. The sensors assumed to be available are typical of what may be assumed for future UAVs. Quoted accuracies are taken from the literature [22].

Both algorithms are developed such that they can be run in real time on today's flight computers. The UKF is presented in a square-root version, which can directly be used. The ESMF, while presented in a full form for clarity, can also be implemented in a similar square-root version. The ESMF can utilize online optimization tools to calculate the minimum ellipsoid sizes, or there are several closed-form solutions that also could be used [15], albeit with an increased level of conservatism. The sampling rates and hard/soft constraints of these algorithms are ultimately a function of their usage. For low-level control, the algorithms should be implemented using hard real-time

Table 11.1. Physical Characteristics of F-15-like Simulation

Parameter	Value	Units
Aircraft weight (W)	45,000	lb
Moments of inertia (I)	$\begin{bmatrix} 28,700 & & -520 \\ & 165,100 & \\ -520 & & 187,900 \end{bmatrix}$	slug-ft^2
Wing area (S)	608	ft^2
Wing span (b)	42.8	ft
Mean aerodynamic chord ($\bar{c}$)	15.95	ft

Table 11.2. Control Sensors and Surface Inputs

Quantity	Description	Sensor	Accuracy
p, q, r	Roll, pitch, yaw rates	Rate gyros	$0.004°/s$
ϕ, θ, ψ	Roll, pitch, yaw angles	Attitude gyros	$0.1°$
f_x, f_y, f_z	Force in x, y, z direction	Accelerometer	0.0328 ft/s^2
V	Airspeed	Airspeed sensor	10 ft/s
α	Angle of attack	Wind vane	$0.2°$
β	Sideslip angle	Wind vane	$0.2°$
u, v, w	Inertial velocity in x, y, z direction	GPS/INS	1.0 ft/s
x_p, y_p, h	Inertial x, y positions and altitude	GPS/INS	50 ft

Input	Input (degrees)	Range [21]	Trim
δ_e	Stabilator	$\pm 20°$	$-9.2°$
δ_r	Rudder	$\pm 30°$	$2.5°$
δ_a	Aileron	$\pm 20°$	$-1.8°$
T_L	Thrust left	$20-127°$	$54°$
T_R	Thrust right	$20-127°$	$55°$

coding; for higher-level planning, a priority basis could be used, although with adequate care.

11.5.1. Nonlinear Stochastic Estimation Simulation Results

The stochastic estimator results are based on an architecture where all 12 states and 6 aerodynamic forces and moments are estimated. This problem is particularly important for the general case of failure detection, because it allows the possibility of failure detection with no *a priori* information (an enabling SEC technology). Autonomous aircraft are envisioned with the capability to maintain a level of performance through system failures and/or physical damage. The performance capabilities of an aircraft are typically quantified by a "flight envelope," a multidimensional space describing the vehicle's capabilities, usually in terms of velocity, altitude, and load factor (the "g-force" an aircraft experiences). This three-dimensional flight envelope then describes a bounded volume, within which the vehicle's capabilities lie. Development of an aircraft's postdamage performance requires (1) tracking the aerodynamic force/moment environment and control surfaces in real time and (2) mapping these estimates to the aircraft's "flight envelope." The first step is demonstrated here.

Figure 11.7 shows a sample of the estimation results (altitude state and lift force) for the UKF on a benchmark simulation. The following benchmark trajectory is used with the F-15-like simulation: (1) hold heading while dropping 2000 ft in 45 s, (2) hold elevation while turning 180° in 45 s, and (3) hold heading while climbing 2000 ft in 30 s. Note that a 50% failure of the stabilator occurs at 75 s. The true value from the nonlinear simulation, state/parameter estimates, and 3σ bounds are plotted. Before the failure

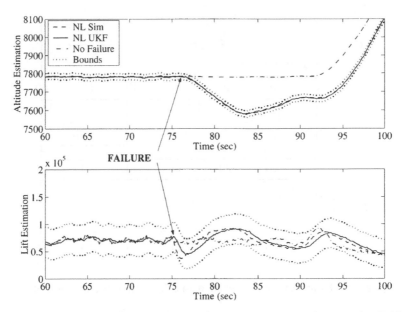

Figure 11.7. Simulation showing state and parameter estimation on the F-15-like simulation, where a 50% elevator failure occurs.

occurs, the estimator performs quite well on all states and parameters. After the failure, the lift force is difficult to estimate due to the instantaneous system change. The UKF reacts quickly (within 1 s) and follows the true parameters. In addition, the true value is still within the measurement error bounds. Eventually, all estimates do converge to their true values. These results are promising for the SEC objectives. The refined aerodynamic loads can then be mapped onto the flight envelope, which in turn gives a new set of constraints for controller design. In addition, the stochastic bounds can be used in conjunction with control and planning approaches as a good measure of the uncertainty.

A second scenario is to couple the active-state model with control customization in real time. Many control algorithms for aircraft use table models of the aerodynamics, based on expanding the highly nonlinear aerodynamic forces and moments in a Taylor-type series around a nominal operating condition. The coefficients of the series, called stability and control derivatives, are usually found using wind tunnel and flight tests. It is proposed here to estimate and track the derivatives online; thus, the aerodynamic model could be quickly updated if there is a failure. Two approaches were implemented: (1) a joint estimator, where all states, aerodynamic forces and moments, and stability and control derivatives are estimated, and (2) a dual estimator, where state and aerodynamic forces and moments are estimated (as in the previous results), and then a second (separate) UKF is used to estimate the 30 stability and control derivatives (with the aerodynamic loads as inputs).

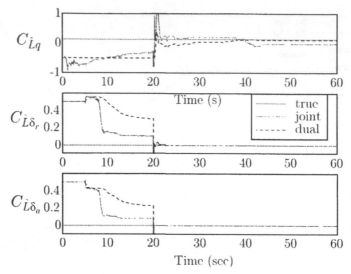

Figure 11.8. Estimation of the aerodynamic derivatives as a function of time.

The stability derivatives associated with the rolling moment $(\tilde{L})$ are illustrated in Figure 11.8. The rolling moment is a strong function of the aileron input, and its corresponding derivative $(C_{\tilde{L}\delta_a})$ is estimated well by both approaches. The estimate associated with the rudder command, $C_{\tilde{L}\delta_r}$, is also accurate. However, estimates for the rate influences on the rolling moment, such as $(C_{\tilde{L}q})$, are less accurate. These and other similar results support the observation that those derivatives with large observability on the dynamics are estimated well. While not plotted, the confidence bounds remain large for those parameters that are difficult to estimate (i.e., are not particularly observable). Overall, the dual approach appears to be slightly better for autonomous operations for several subtle reasons. First, separating state estimation from aerodynamic parameter estimation leads to a more robust algorithm because (a) the stability of the algorithms are not as directly tied together and (b) different aerodynamic models can be introduced at any time. Second, there is a lower computational cost because the joint formulation requires tracking the parameters and their coupling; this coupling is removed in the dual formulation.

11.5.2. Guaranteed Nonlinear Set Estimation Results

Validation of the ESMF can be done using any nonlinear estimation problem, such as those presented in the previous section. Perhaps a more appropriate validation, however, is the application of radar tracking because the output ellipsoid bounds can be used in constrained trajectory planning. More specifically, the ESMF has been developed here as a hybrid nonlinear estimator, used in the case of tracking of multiple vehicles in an uncertain environment, including noises, communication blackouts, and radar sensor losses. Consider the following concept. Many UAVs (friendly and foe) are flying in an uncertain environment. The friendly UAVs must coordinate in order to develop plans locally to achieve their goals. The estimation architecture must be capable of (1) tracking friendly positions (and uncertainties)

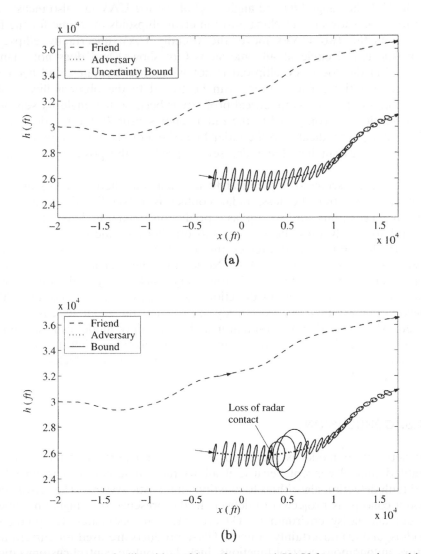

Figure 11.9. Guaranteed ellipsoid tracking of a second UAV from a moving vehicle during a failure or loss of line of sight. (a) Tracking during normal operations. (b) Tracking during a radar loss.

using cross-communication links, (2) tracking foe positions (and uncertainties) using radar, (3) tracking through communication and radar sensor losses, (4) coordinating in order to reduce conservatism, and (5) adapting in order to add or subtract entities (UAVs).

The first scenario is with two UAVs, a single friendly UAV and a single foe UAV; both vehicles are simulated using the F-15-like simulation. The friendly UAV must estimate its own position using the ESMF, as well as the position of the foe UAV using radar. In addition to its own sensors given in Table 11.2, the range (R) and angle (θ_R) of the foe UAV are also measured. Figure 11.9a shows the tracking position error ellipsoids over time for the foe UAV. As the two UAVs move, the friendly UAV uncertainty ellipse is approximately a circle at all time steps (not shown) and does not change shape. But the foe UAV ellipsoid is not symmetrical, rotates, and becomes smaller over time. These affects can be traced to the observability of the radar sensor. The nonsymmetrical property is because the angle θ_R sensor is more sensitive as compared to the radar range sensor R; the rotation of the ellipsoid is perpendicular to the radar line of sight (LOS). In addition, as the range becomes smaller, the radar sensor predicts the position more accurately.

The second scenario is identical to the first, but there is a loss of radar contact; soon after the loss, radar contact is reestablished. Because the filtering algorithms are set up as "prediction" and "update" steps, it is natural to simply turn the update step off (because there is no sensor to provide the update) and predict the foe UAV's reachable space. These results are shown in Figure 11.9b. Notice that the reachable space (uncertainty bounds) for the foe UAV grows very quickly. Physically, this implies that the UAV can turn in any direction and increase thrust to maximum. The tracking bounds then reduce quickly again when the radar lock is reestablished. This type of hybrid estimator and uncertainty model can be used in cases of mode changes, as well as in trajectory planning. Full details of this estimation approach, along with cooperation between the vehicles, are given in reference 23.

11.6. CONCLUSION

An active-state model is presented as an integral portion of the software-enabled control concept because it allows the online control customization and fault detection algorithms to function on realistic systems. The active-state model concept is projected to deliver, in the presence of a highly nonlinear model and noisy environment, (1) accurate state estimates, (2) predictive models, and (3) uncertainty bounds. These products are used for current and future autonomous control functions, including online control customization, trajectory planning, control design, and fault detection. The architecture presented makes use of a joint estimation problem, where states and model parameters are estimated together. A stochastic nonlinear estimation algorithm, called the UKF, is used to deliver each of the three products, along with confidence intervals. The UKF is more accurate and is more amenable to online implementation compared to the EKF. Simulations have shown that it can be used to track jumps in the data, which may occur as failures in the actual mission. For those components that require guaranteed uncertainty bounds, a guaranteed nonlinear set membership filter has been developed, called the ESMF, which combines set membership estimation, Taylor models, and interval mathematics. The ESMF delivers ellipsoid error bounds for nonlinear systems with dynamics that are continuously differentiable in the first two derivatives. Simulations of the ESMF have shown that it can be used to track position errors of multiple UAVs quite easily using a typical radar sensor. This work has been extended to include many UAVs, communication blackouts, reduced conservatism by sharing information, and integration with other components such as the trajectory planners.

Enabling SEC technologies include (1) online aerodynamic model identification, (2) online detection of vehicle capabilities after failure or damage, (3) soft (stochastic) and hard (guaranteed) state and model parameter uncertainty information for use in control design and trajectory planning, and (4) a guaranteed, hybrid estimator for multiple UAVs interacting in a highly nonlinear, noisy, and uncertain environment with communication/sensor blackouts.

REFERENCES

1. P. Antsaklis, and K. Passino, Towards intelligent autonomous control systems: Architecture and fundamental issues, *Journal of Intelligent and Robotic Systems* **1**(5):315–342, 1989.

2. H. E. Rauch, Autonomous control reconfiguration, *IEEE Control Systems Magazine* **15**:37–48, 1995.

3. R. M. Murray, J. Hauser, A. Jadbabaie, M. B. Milam, N. Petit, W. B. Dunbar, and R. Franz, Online control customization via optimization based control, *Software Enabled Control: Information Technology for Dynamical Systems*, T. Samad and G. Balas, editors, 2003.

4. L. Ljung, Asymptotic behavior of the extended Kalman filter as a parameter estimator for linear systems, *IEEE Transactions on Automatic Control* **AC-24** (1):36–50, February 1979.

5. K. Reif, S. Guenther, E. Yaz, and R. Unbehauen, Stochastic stability of the discrete-time extended Kalman filter, *IEEE Transactions on Automatic Control*, **44**(4):714–728, April 1999.

6. S. Julier, J. Uhlmann, and H. F. Durrant-Whyte, A new method for the nonlinear transformation of means and covariances in filters and estimators, *IEEE Transactions on Automatic Control* **45**(3):477–482 March 2000.

7. A. Jazwinski, *Stochastic Processes and Filtering Theory*, Academic Press, New York, 1970.

8. S. Brunke and M. Campbell, Estimation architecture for future autonomous vehicles, in *Proceedings of the American Control Conference*, 2002.

9. G. H. Golub. and C. F. V. Loan, *Matrix Computations*, 3rd edition, The John Hopkins University Press, Baltimore, MD, 1996.

10. S. Brunke, *Nonlinear Filtering and System Identification Algorithms for Autonomous Systems*, Ph.D. thesis, University of Washington, August 2001.

11. J. Garcia-Velo and B. Walker, Aerodynamic parameter estimation for higher performance aircraft using extended Kalman filtering, *Journal of Guidance, Control, and Dynamics* **20**:1257–1259, 1997..

12. F. C. Schweppe, Recursive state estimation: Unknown but bounded errors and system inputs, *IEEE Transactions on Automatic Control* **AC-13**(1):22–28, February 1968.

13. F. L. Chernousko, *State Estimation for Dynamic Systems*, 1st edition, CRC Press, Boca Raton, FL, 1994.

14. R. E. Moore, *Interval Analysis*, Prentice-Hall, Englewood Cliffs, NJ, 1966.

15. E. Scholte and M. E. Campbell, On-line nonlinear guaranteed estimation with application to a high performance aircraft, *Proceedings, American Control Conference*, 2002, to appear in the *International Journal pf Robust and Nonlinear Control*.

16. K. S. Tsakalis and L. Song, Set-membership estimation for weakly nonlinear models: An application to the adaptive control of semiconductor manufacturing processes, in *Proceedings of the 33rd Conference on Decision and Control* (Lake Buena Vista, FL), December 1994, pp. 1066–1071.

17. J. S. Shamma and K.-Y. Tu, Approximate set-valued observers for nonlinear systems, *IEEE Transactions on Automatic Control*, **42**(5):648–658, May 1997.

18. L. Jaulin and E. Walter, Set inversion via interval analysis for nonlinear bounded-error estimation, *Automatica* **29**(4):1053–1064, 1993.

19. E. Hansen, *Global Optimization Using Interval Analysis*, 1st edition, Monographs and Textbooks in Pure and Applied Mathematics, Marcel Dekker, New York, 1992.

20. J. Zemke, *b4m: A Free Interval Arithmetic Toolbox for Matlab*. http://www.ti3.tu-harburg.de/ zemke/b4m/index.html. version 1.0.2.004.

21. J. P. Dutton, Jr., *Development of a Nonlinear Simulation for the McDonnell Douglas F-15 Eagle with a Longitudinal TECS Control Law*, Master's thesis, Department of Aeronautics and Astronautics, University of Washington, 1993.

22. M. Laban and K. Masui, Total least squares estimation of aerodynamic model parameters from flight data, *Journal of Aircraft* **30**(1):150–152, 1993.

23. M. E. Campbell and J. Ousingsawat, On-line estimation and path planning for multiple vehicles in an uncertain environment, in *Proceedings, AIAA Guidance, Navigation and Control Conference*, Monterey, CA, 2002.

CHAPTER 12

AN INTELLIGENT METHODOLOGY FOR REAL-TIME ADAPTIVE MODE TRANSITIONING AND LIMIT AVOIDANCE OF UNMANNED AERIAL VEHICLES

GEORGE VACHTSEVANOS, FREEMAN RUFUS, J. V. R. PRASAD,
ILKAY YAVRUCUK, DANIEL SCHRAGE, BONNIE HECK and LINDA WILLS

Editors' Notes

This chapter describes intelligent control approaches for two related problems. The first application discussed is mode transitioning: ensuring the smooth and stable change of flight mode, such as from hover to forward flight. The controller proposed is a fuzzy neural network, an extension of the popular Takagi-Sugeno fuzzy model. In essence, the controller output is a nonlinear blending of different linear controllers, each of which is tailored for a particular region in the state space of the vehicle.

The mode transition controller is initially designed offline and then customized using an online adaptation scheme. An indirect adaptive control approach is proposed in which the controller modification is driven by the adaptation of a plant model. Simulation results are presented for a simplified helicopter model. (See also Chapter 16 for another multi-mode control application for autonomous helicopters.)

The second problem discussed in this chapter is limit detection and avoidance. Actuator control limits and controller inputs are redefined dynamically to allow high performance while ensuring that the vehicle's safety limits are not violated. The focus of this part of the chapter is on the adaptive limit detection subsystem. Normally, predictions of the asymptotic steady state under the current control setting and, similarly, of the control limits that would ensure that the vehicle outputs remain in the safe region, are done using a linearized model derived from an offline first principles analysis. The authors approach recomputes these two parameters based not only on the linearized model but also on a neural network that is adapted online to estimate the error of the linear model. These recomputed limits can therefore be expected to be more accurate. Simulation results are provided for a linear helicopter

Software-Enabled Control. Edited by Tariq Samad and Gary Balas
ISBN 0-471-23436-2 © 2003 IEEE

model and a nonlinear XV-15 tilt rotor aircraft. The adaptive limit detection scheme, integrated with limit avoidance control, is able to keep the vehicle within the safe region whereas a controller based on the linearized model would result in constraint violations.

12.1. INTRODUCTION

Under the Software Enabled Control program sponsored by DARPA, the Georgia Institute of Technology has been advancing technologies for unmanned aerial vehicles (UAVs) that address issues of mode transitioning control, limit checking and avoidance, and their implementation in an open-control platform (OCP) [1]. Special emphasis in this project is placed on the development of mid-level controllers combined with low-level flight controllers, as depicted in the mission intelligence flow chart of Figure 12.1.

In this context, flight modes are selected in the high- and mid-level modules of the control hierarchy and passed to the low-level controllers. The low-level controllers are used to carry out these flight modes in a smooth, stable, and safe way. The SEC program objectives include research products that can enable autonomous vehicles to perform *extreme maneuvers* in highly unstructured environments. Real-time mode transitioning and limit avoidance technologies are called upon in such extreme-performance flight scenarios if the vehicle is to meet mission objectives while maintaining its opera-

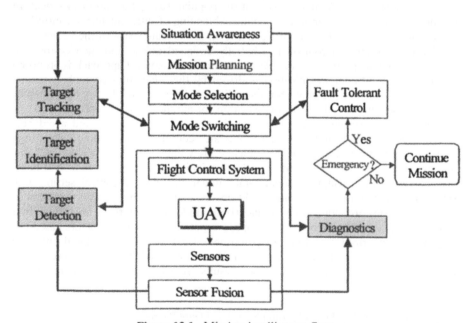

Figure 12.1. Mission intelligence flow.

tional integrity. Both mode transitioning and limit avoidance scenarios may be implemented in a complementary manner for a particular mission.

Complex large-scale systems such as unmanned aerial vehicles and industrial processes are demanded to possess the intelligence required to behave in an autonomous manner under uncertain environmental conditions. Typically, these systems are required to operate in a finite number of operational modes with robust, stable, and smooth transitions between them. A (local) operational mode (or mode of operation) is considered to be a region in the system's state space in which the system exhibits quasi-steady-state behavior. Thus, a *mode* is considered here to be a region of the state space around an operating state associated with a local controller such that the combined plant and controller maintains a stable behavior in the prescribed region. The local controller must ensure the robust stability of the combined closed-loop system for any trajectory within the mode. Different modes correspond to disjoint regions of the state space. For example, in the UAV case, we define such modes as *hover*, *forward flight*, and so on. A transition region between two modes is a region of the state space consisting of all the systems states not included in any local mode but in the trajectory connecting the modes. A mode transition (or mode to mode) controller refers to a controller that transitions a system from a start mode of operation to the goal mode. The problem of transitioning between two operational modes can be solved by nonadaptive techniques such as gain scheduling, sliding mode control, and the method of blending local mode controllers. However, when the system to be controlled differs significantly from the nominal system used in the design methods above, degraded tracking performance of the desired transition trajectory is to be expected.

Although there is no consistent theory that deals with dynamic transitions between various equilibria, gain scheduling has been used to design equilibrium to equilibrium controllers. The technique of gain scheduling constructs a nonlinear controller by combining the members of an appropriate family of linear controllers. In conventional gain scheduling, the transition between equilibria is governed normally by an auxiliary scheduling variable [2, 3] that should vary slowly with respect to the states. The disadvantages associated with gain scheduling include a reliance on a long trial-and-error design process, a lack of adaptability to online variations, and poor robustness to uncertainties [4]. The gain scheduling procedure is generally as follows [5]: (i) Parameterize the equilibrium operating points of the plant by a scheduling variable that involves some of the plant states; (ii) for a family of equilibrium operating points parameterized by a scheduling variable, linear models of the plant are created; linear controllers are obtained for each linear model; (iii) finally, an interpolation technique is used to interpolate between the linear controller gains for the equilibrium to equilibrium transition. Although gain-scheduled controllers are typically designed using plant linearizations at a number of equilibrium operating points, it is possible to apply gain scheduling to control design of linear time-varying systems obtained via linearizations relative to a trajectory [6, 7]. To overcome the restriction to near-equilibrium operation in traditional gain scheduling, a velocity-based gain-scheduling controller design has been developed [8]. This method uses plant dynamics at equilibrium and non-equilibrium operating points, which may lead to controller realizations that achieve better performance than classical gain-scheduling controllers.

Similar to gain scheduling, sliding mode control (SMC) uses more than one control law and is, in general, nonlinear. The performance index is specified as a manifold of the state space called the sliding surface. A sliding mode controller sends the system states onto the sliding surface and keeps them there. However, due to high control gain, SMC systems can suffer from the effects of actuator chattering due to switching and imperfect implementations [9]. When using SMC to track a desired trajectory, it may become possible for the closed-loop system to become unstable if the sliding surface changes faster than the SMC can follow it. In order to overcome the disadvantages of the SMC systems, the fuzzy sliding mode control (FSMC) has been introduced to provide better damping and reduced chattering [9]. The FSMC has been used for motion trajectory control in reference 10. In reference 11, FSMC is used with a fuzzy logic controller in order to track a prespecified position-velocity trajectory for an uncertain nonlinear system.

The basic objective of adaptive control is to maintain consistent performance of a system in the presence of uncertainty or unknown variation in plant parameters. Typically, adaptive control is developed for MIMO linear systems, SISO nonlinear systems, and certain classes of MIMO nonlinear systems. In reference 12, adaptive controllers were developed for a class of feedback linearizable nonlinear systems. In reference 13, an adaptive output feedback tracking control was proposed for a class of single-input single-output nonlinear systems with uncertain differentiable time-varying parameters. Recently, adaptive techniques based upon the one-step-ahead control strategy have been developed for more general nonlinear systems. In reference 14, neural-network-based one-step-ahead control strategies were proposed for a class of nonlinear SISO systems. In reference 15, a nonlinear one-step-ahead control scheme based upon a recurrent neural network model was proposed for nonlinear SISO processes. The neural network model was trained via a recursive least-squares (RLS) algorithm, and the gradient descent update rate for the control law was determined by stability considerations. Finally, for a general class of MISO nonlinear systems, an adaptive quasi-one-step-ahead control law was proposed in reference 16. The control law was derived using the sensitivity between the controlled system input and output and the quasi-one-step-ahead predictive output. The sensitivity of the plant was estimated using RLS and the predictive output was obtained by a recurrent neural network.

A continuous knowledge of "how far" the vehicle is from its structural limit boundaries is needed by the low-level controller for safe maneuvering of the vehicle. Controller commands need to be modified automatically, if a "fly safe" region violation in the flight envelope is foreseen. This will guarantee a safe flight, regardless of the commands from the high- and mid-level controller blocks in the chain. Neural networks have been shown to be promising tools for mapping of complex flight envelope limits as functions of flight condition and control inputs [17]. Similarly, piloted simulations of tactile cueing systems have been conducted using neural-network-based predictions for limit avoidance [17]. However the application of these techniques to adaptive flight control systems is a new area that needs to be fully exploited.

12.2. REAL-TIME ADAPTATION OF MODE TRANSITION CONTROLLERS

We propose a scheme for the real-time adaptation of mode transition controllers designed via the method of blending local mode controllers (BLMC) [18–20]. The control objective is to adapt the blending gains portion of the mode transition controllers such that the nonlinear plant state vector tracks the desired transition trajectory from a start mode of operation to a goal mode. The adaptation scheme is composed of a desired transition model, an active plant model and an active controller model which is the mode transition controller. The desired transition model, the active plant model and the blending gains portion of the active controller model are represented via a fuzzy neural network construct discussed in reference [12]. All three fuzzy neural models are trained offline, while the latter two models are adapted online. The active plant model that incorporates local model information is initially trained offline to capture the desired transition trajectory and controls. Afterwards, the active plant model is online adapted via structure and parameter learning to capture the input–output behavior of the nonlinear system to be controlled. Likewise, the blending gains portion of the mode transition controller is determined offline and is adapted online via structure and parameter learning to track the desired transition trajectory. The new blending gains to be developed by the mode transition controller are determined from the control sensitivity matrix and the one-step-ahead predictive output of the active plant model. The parameter learning for the active plant and active controller models is based upon Kaczmarz's algorithm [21].

12.2.1. The Fuzzy Neural Construct

The fuzzy neural network (FNN) proposed in reference 12 consists of a fuzzy rule base of Takagi–Sugeno fuzzy rules with the rule consequents being linear polynomials of the input premise variables. Both structure learning and parameter learning are used to adaptively develop the FNN construct. Structure learning inserts new membership functions, creates new fuzzy rules, and selects initial parameters of the new rules on the basis of the desired output data. Parameter learning updates the consequent weights via Kaczmarz's algorithm.

This FNN construct realizes the fuzzification, fuzzy reasoning, and defuzzification functionalities of a connectional fuzzy inference mechanism.

Let $x = [x_1, \ldots, x_m]^T$ and $y = [y, \ldots, y_p]^T$ denote the input and output vectors of the FNN, respectively. The fuzzy rule base of the FNN consists of a collection of N fuzzy rules of the form

$$R^j: \quad IF\ x_1\ is\ A_{1j}\ AND\ x_2\ is\ A_{2j}\ AND\ \cdots\ AND$$

$$x_m\ is\ A_{mj}$$

$$THEN\ f_j^1 = w_{0j}^1 + w_{1j}^1 x_1 + \cdots + w_{mj}^1 x_m\ AND\ \cdots$$

$$AND\ f_j^p = w_{0j}^p + w_{1j}^p x_1 + \cdots + w_{mj}^p x_m$$

where f_j^ℓ denotes the jth rule output associated with the ℓth output component y_ℓ. $w_{0j}^\ell, \ldots, w_{ij}^\ell, \ldots, w_{mj}^\ell$ are the polynomial coefficients connecting linearly the input variables to the f_j^ℓ consequent function. Finally, $A_{1j}, \ldots, A_{ij}, \ldots, A_{mj}$ are labels of the fuzzy sets in the premise space associated with the jth rule $R^{(j)}$. Each linguistic label A_{ij} is associated with a Gaussian membership function, $\mu_{A_{ij}}(x_i)$, which specifies the degree to which a given x_i satisfies the quantifier A_{ij}:

$$\mu_{A_{ij}}(x_i) = \exp\left[-\frac{1}{2} \frac{(x_i - m_{ij})^2}{\sigma_{ij}^2} \right]$$

where m_{ij} and σ_{ij} denote the mean and standard deviation of the Gaussian membership function, respectively. The degree of fulfillment (or the firing strength) of each rule $R^{(j)}$ is taken as

$$\mu_j(x_1, \ldots, x_m) = \mu_{A_{1j}}(x_1) \times \cdots \times \mu_{A_{mj}}(x_m)$$

Given an input vector x, the ℓth output component y_ℓ of the fuzzy system is inferred as follows:

$$y_\ell = \frac{\sum_{j=1}^N \mu_j \cdot f_j^\ell}{\sum_{j=1}^N \mu_j}, \qquad \ell = 1, \ldots, p$$

The inferred outputs result from the application of the weighted-average defuzzification method.

12.2.2. Learning Algorithm Using Local Model Information

The learning algorithm employs both structure learning and parameter learning to determine the premise part structure and the appropriate premise/consequent parameters of a fuzzy neural network. Structure learning determines the appropriate fuzzy neural structure by performing membership function insertion and parameter setting of initial premise/consequent parameters of newly established rules. Parameter learning updates consequent weights via a local least-squares estimation technique [22].

Parameter learning involves the computation of N locally weighted least-squares regressions, one for each rule, using only the training data within the rule's receptive field. The consequent weights are not updated for rules having zero training data within their receptive field.

The learning procedure follows the steps below:

1. Use a subset of the training data set to perform network initialization. This network initialization is conducted offline using structure learning.
2. Perform structure learning if necessary.
3. Perform parameter learning.
4. Repeat steps 2 and 3 for all training data entries.

12.2.3. FNN Linear Incremental Model

A linear incremental model is derived, as a basis for the offline design of mode transition controllers, by considering first-order input–output sensitivity terms of the original FNN model. For each fuzzy rule, the input–output relation is expressed as a function of the firing strength of the rule premise and consequent parameters, the latter being denoted as the mean and standard deviation of Gaussian functions.

12.2.4. Offline Design of Mode Transition Controllers

12.2.4.1. Mode Transition Problem. Given a large-scale dynamical system represented by the following state equation:

$$\dot{x} = F(x, u), \qquad x \in R^n, \quad u \in R^m$$

it is assumed that the system is composed of N_s subsystems S_i, $i = 1, 2, \ldots, N_s$ where each subsystem represents an operational mode of the system. The state equation for the ith subsystem is

$$\dot{x}_i = f_i(x_i, u_i), \qquad x_i \in R^{n_i}, \quad u_i \in R^{m_i}$$

Let $mode_p$ and $mode_q$ denote the pth and qth subsystem, respectively. How do we design a controller that stably and smoothly transitions a system from $mode_p$ to $mode_q$? It should be noted that the term "mode" refers to such operational modes of a helicopter as "hover," "forward flight," "turn left," and so on.

12.2.4.2. Offline Control Design. In reference 19, an offline design methodology, known as the BLMC approach, was developed to design mode transition controllers. This approach uses the aggregated states of the start and goal modes, while the output vector of the mode transition controller is determined by blending the individual output vectors of the start and goal mode controllers. The following is an outline of the BLMC approach:

- *Step 1:* Design regulators for the start and goal modes such that initial states are driven to the equilibria of the respective modes.
- *Step 2:* Model the dynamics that correspond to the aggregated states and controls of the start and goal mode so that a transitional path from the start mode to the goal mode can be determined.
- *Step 3:* Determine an optimal transitional path from the equilibrium state of the start mode to the equilibrium state of the goal mode by solving a nonlinear optimal control problem.
- *Step 4:* Determine the desired blending gains using the desired state and control trajectory determined from step 3.
- *Step 5:* Realize the blending gains via a fuzzy neural network construct proposed in reference 12.

12.2.5. Online Adaptation of Mode Transition Controllers

In this section, an adaptation scheme is proposed for the online customization of mode transition controllers designed offline via the method of blending local mode controllers. The control objective is to adapt the blending matrices such that the plant output vector tracks the output vector of a desired transition model. In order to apply the discrete-time adaptation scheme to the continuous-time system, it is assumed that the sample rate has been appropriately selected. Figure 12.2 shows the configuration for indirect adaptive mode transition control. The adaptation scheme is composed of five components: the *desired transition model*, the *active plant model*, the *plant adaptation mechanism*, the *active controller model*, and the *controller adaptation mechanism*.

Desired Transition Model
An offline trained fuzzy neural model of $x_{pq}^d(t_k) \rightarrow x_{pq}^d(t_{k+1})$ is determined for $k = 0, \ldots, N_1$, where $t_{k+1} - t_k = (t_f - t_0)/N_1$. Note when $k \geq N_1$, then $x_{pq}^d(t_k) = x_{pq}^d(t_{k+1}) = x_{pq}^d(t_f)$.

Active Plant Model
Given $x_{pq}^d(t_k)$ and $u_{pq}^d(t_k)$ for $k = 0, \ldots, N_1$, where $t_{k+1} - t_k = (t_f - t_0)/N_1$, a fuzzy neural model of the mapping $\{(x_{pq}^d(t_k), u_{pq}^d(t_k)) \rightarrow x_{pq}^d(t_{k+1})\}$ for $k = 0, \ldots, N_1$ is determined by offline training. Also, the linear model information $\partial x_{pq}(t_{k+1})/\partial x_{pq}(t_k)$ and $\partial x_{pq}(t_{k+1})/\partial u_{pq}(t_k)$ defined at

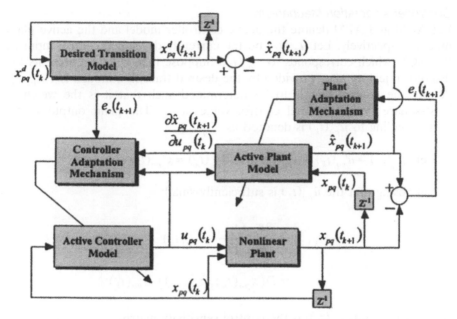

Figure 12.2. Configuration for indirect adaptive mode transition control.

$(x_{pq}^{d}(t_{k}), u_{pq}^{d}(t_{k}))$ is incorporated into the consequent part of the fuzzy neural model. Afterwards, the *active plant model* is adapted online via the plant adaptation mechanism.

Plant Adaptation Mechanism
The active plant is adapted online to account for plant variations on a real-time basis. At time instant t_{k}, the adaptation of the active plant model is accomplished by performing structure/parameter learning on the basis of the current input/output data $\{(x_{pq}(t_{k}), u_{pq}(t_{k})) \rightarrow x_{pq}(t_{k+1})\}$. Since a desired output is not known ahead of time when performing structure learning on the incoming input $(x_{pq}(t_{k}), u_{pq}(t_{k}))$, the strongest fired rule's consequence is used. Likewise, the strongest fired rule's consequent parameters are used to initialize the consequent parameters of the newly formed rule since the required linear model information is not known ahead of time.

Active Controller Model
The active controller model is the mode$_{p}$ to mode$_{q}$ controller. Given $x_{pq}^{d}(t_{k})$ and $k_{pq}(t_{k})$ for $k = 0, \ldots, N_{1}$ where $t_{k+1} - t_{k} = (t_{f} - t_{0})/N_{1}$, a fuzzy neural model of the mapping $x_{pq}^{d}(t_{k}) \rightarrow k_{pq}(t_{k})$ for $k = 0, \ldots, N_{1}$ is determined by offline training as suggested in Section 2.4. Afterward, the blending weights of the active controller model are adapted online using the *controller adaptation mechanism*.

Controller Adaptation Mechanism

Let ACM and APM denote the active controller model and the active plant model, respectively. Let $u_{pq}(t_k)$ be the currently developed control input by the ACM which corresponds to $x_{pq}(t_k)$. Suppose that $x_{pq}^d(t_k)$ represents the desired trajectory at t_k provided by the desired transition model. Let $u'_{pq}(t_k)$ denote the control that is to be determined such that it is the weighted least-square (WLS) optimal control value at t_k. The plant output vector corresponding to $u'_{pq}(t_k)$ is denoted as $x'_{pq}(t_k)$.

Let $\tilde{u}_{pq}(t_k) = u'_{pq}(t_k) - u_{pq}(t_k)$ and $\tilde{x}_{pq}(t_k) = x'_{pq}(t_k) - x_{pq}(t_k)$.

Supposing that the $\tilde{u}_{pq}(t_k)$ is sufficiently small:

$$\tilde{x}_{pq}(t_{k+1}) = \frac{\partial x_{pq}(t_{k+1})}{\partial u_{pq}(t_k)} \tilde{u}_{pq}(t_k)$$

$$= D\big(x_{pq}(t_k), u_{pq}(t_k)\big) \cdot \tilde{u}_{pq}(t_k)$$

where $D(x_{pq}(t_k), u_{pq}(t_k))$ is the control sensitivity matrix.

Let $\tilde{x}_{pq}(t_{k+1}) = x_{pq}^d(t_{k+1}) - x_{pq}(t_{k+1})$ and $\tilde{x}_{pq}(t_{k+1}) = x_{pq}^d(t_{k+1}) - x'_{pq}(t_{k+1})$. Therefore, $\tilde{x}_{pq}^d(t_{k+1}) = D \cdot \tilde{u}_{pq}(t_k) + \tilde{x}_{pq}(t_{k+1})$.

The optimal control input increments $\tilde{u}_{pq}(t_k)$ are determined such that the following performance index is minimized:

$$J = \tfrac{1}{2}\tilde{x}_{pq}(t_{k+1}) \cdot Q \cdot \tilde{x}_{pq}(t_{k+1}), \qquad Q > 0$$

The following steps implement the controller adaptation algorithm:

1. Apply ACM to $x_{pq}(t_k)$ and produce the current initial estimate of the control input to $u_{pq}(t_k)$. Since it is possible for the fuzzy neural model of the blending weights not to be sufficiently activated by $x_{pq}(t_k)$, structure learning with local model information is performed at this stage.
2. Input $u_{pq}(t_k)$ and $x_{pq}(t_k)$ to APM and produce $\hat{x}_{pq}(t_{k+1})$. Calculate $\tilde{x}_{pq}^d(t_{k+1})$ using the predictive one-step-ahead output $\hat{x}_{pq}(t_{k+1})$ in place of the unavailable output $x_{pq}(t_{k+1})$.
3. The true control sensitivity matrix $D(x_{pq}(t_k), u_{pq}(t_k))$ is approximated via the APM's incremental control matrix. When the APM is not to be sufficiently activated by $(x_{pq}(t_k) u_{pq}(t_k))$, the control sensitivity information contained in the strongest fired rule's consequence is used.
4. Compute the adjusted control law, $u'_{pq}(t_k) = u_{pq}(t_k) + [\hat{D}^T \cdot Q \cdot \hat{D}]^{-1} \hat{D} \cdot Q \cdot \tilde{x}_{pq}^d(t_{k+1})$. Afterwards, calculate the desired blending weights $k'_{pq}(t_k)$.

5. Train ACM to capture desired blending weights $k'_{pq}(t_k)$ given current input $x_{pq}(t_k)$. Note that parameter learning with local model information is used to train the ACM.

6. Put $t_k \leftarrow t_{k+1}$ and perform the same procedure at the next time t_{k+1}.

12.3. HOVER TO FORWARD FLIGHT EXAMPLE

12.3.1. Model of Helicopter's Forward Dynamics

The proposed adaptation scheme is illustrated on the following model representing the longitudinal channel dynamics of an Apache helicopter constrained to have no vertical motion; only longitudinal and pitch rotation motions are allowed [18]:

$$X = X_{trim} + X_{\dot{x}}(\dot{x} - \dot{x}_{trim}) + X_{\dot{\theta}}(\dot{\theta} - \dot{\theta}_{trim}) + X_{\delta_e}(\delta_e - \delta_{e,trim})$$

$$M = M_{trim} + M_{\dot{x}}(\dot{x} - \dot{x}_{trim}) + M_{\dot{\theta}}(\dot{\theta} - \dot{\theta}_{trim}) + M_{\delta_e}(\delta_e - \delta_{e,trim})$$

$$\ddot{x} = \frac{X}{m \cdot \cos(\theta)} - g \cdot \tan(\theta)$$

$$\ddot{\theta} = \frac{M}{I_Y}$$

where $\ddot{x}$, $\ddot{\theta}$, and δ_e represent the forward acceleration (ft/s^2), pitch angle acceleration (rad/s^2), and longitudinal cyclic input (deg), respectively. X represents the aerodynamic force along the "X axis," and M represents the pitching moment about the "Y axis." The parameters, X_{trim}, $X_{\dot{x}}$, $X_{\dot{\theta}}$, X_{δ_e}, M_{trim}, $M_{\dot{x}}$, $M_{\dot{\theta}}$, M_{δ_e}, x_{trim}, $\dot{\theta}_{trim}$, and $\delta_{e,trim}$ are functions of $\dot{x}$. X_{trim} and M_{trim} are the trim values of the aerodynamic force X and the pitching moment M, respectively. The variables, $X_{\dot{x}}$, $X_{\dot{\theta}}$, X_{δ_e}, M_{trim}, $M_{\dot{x}}$, and $M_{\dot{\theta}}$ are the partial derivatives of X and M with respect to $\dot{x}$, $\dot{\theta}$, and δ_e, respectively. The physical constants m and I_Y are the mass of the helicopter and the moment of inertia along the Y axis. The state vector of the helicopter model is $[x_1 \; x_2 \; x_3 \; x_4]^T = [\dot{x} \; \ddot{x} \; \theta \; \dot{\theta}]^T$. It is assumed that the output vector of the model is the same as the state vector.

12.3.1.1. Blending Local Mode Controllers Approach

Linear quadratic regulators are designed that regulate the vehicle about specified operating points. Scalar gains for the hover and forward flight controllers are calculated first such that the closed-loop system transitions from one state to the next in minimum time while satisfying physical constraints on the vehicle's velocity and acceleration profiles.

Afterwards, the blending gains $K_{hov}(\dot{x}, \ddot{x}, \theta, \dot{\theta})$ and $K_{FF}(\dot{x}, \ddot{x}, \theta, \dot{\theta})$ are realized via a fuzzy neural network construct described in a previous section.

12.3.1.2. Gain Scheduling Approach

A gain scheduled controller is designed via dynamic linearization about the minimum time trajectory determined in the previous section:

- 67 equidistant frozen times $t_1, \ldots, t_{67}$ are chosen where $t_{k+1} - t_k = (t_f - t_0)/66$.
- 67 linear autonomous open loop systems are obtained via Lyapunov linearization of the helicopter model about the minimum time state and control trajectories at frozen times $t_1, \ldots, t_{67}$.
- 67 linear quadratic regulators are designed for each time-frozen linear model of the helicopter created in the previous step.
- The 67 linear control laws are blended according to how near the current state is to each of the frozen operating states determined in the first step. The interpolated control law is applied to the system to be controlled.

12.3.1.3. Least Squares Adaptation Scheme.

A sample time of $T_S = 0.05$ s is chosen for the adaptation scheme. The desired minimum time trajectory and control are resampled such that they occur every T_S:

$$\bar{x}^d(t_k) \quad \text{and} \quad \bar{u}^d(t_k) \quad \text{for} \quad k = 0, \ldots, N$$

where $\bar{x}^d(t_k) = [\dot{x}^d(t_k) \quad \ddot{x}_d(t_k) \quad \theta^d(t_k) \quad \dot{\theta}^d(t_k)]^T$, $\bar{u}^d(t_k) = [\delta_c^d(t_k)]$ and $t_{k+1} - t_k = T_S$.

$$t_N \geq t_f \quad \text{and} \quad t_N - t_f < T_S,$$

Afterwards, the *desired transition model* of the following mapping is determined offline:

$$\bar{x}^d(t_k) \rightarrow \bar{x}^d(t_{k+1}), \quad k = 0, \ldots, N$$

The *active plant model* is initially determined offline for the following mapping:

$$\{(\bar{x}^d(t_k), \bar{u}^d(t_k)) \rightarrow \bar{x}^d(t_{k+1})\} \quad \text{for } k = 0, \ldots, N$$

Also, the linear model information defined at $(\bar{x}^d(t_k), \bar{u}^d(t_k))$ for $k = 0, \ldots, N$

$$\frac{\partial \bar{x}(t_{k+1})}{\partial \bar{x}(t_k)} \quad \text{and} \quad \frac{\partial \bar{x}(t_{k+1})}{\partial \bar{u}(t_k)}$$

is incorporated into the consequent part of the *active plant model*.

The *plant adaptation mechanism* adapts the active plant model with the following parameters:

$$\delta = 0.2, \quad \beta = 0.5, \quad \text{and} \quad \sigma^U = [\sigma_1^U \cdots \sigma_5^U] = [0.20 \quad 0.20 \quad 0.05 \quad 0.05 \quad 0.20]$$

where δ, β, and σ^U are lower thresholds for membership value, the desired overlap degree between membership functions, and the upper limit of the width of each membership function, respectively.

The *active controller model* is the hover to forward flight mode controller determined in Section 12.3.1.1.

The *controller adaptation mechanism* adapts the blending weights of the active controller model with the following parameters:

$$\delta = 0.2, \quad \beta = 0.5, \quad \text{and} \quad \sigma^U = \left[\sigma_1^U \cdots \sigma_4^U\right] = \begin{bmatrix} 0.20 & 0.20 & 0.05 & 0.05 \end{bmatrix}$$

12.3.2. Simulation Results

Figures 12.3 and 12.4 show the desired $\dot{x}$, $\ddot{x}$, θ, and $\dot{\theta}$ trajectories. For small parametric changes and wind disturbances, the controller exhibits good tracking performance of the desired transition trajectory. However, as the magnitude of the parametric changes and wind disturbances increases, the tracking performance of the controller degrades. As expected, if the approximate plant accurately captures the local model information and the input/output behavior of the system to be controlled, the adapted controller exhibits excellent tracking performance when encountering parametric changes and wind disturbances.

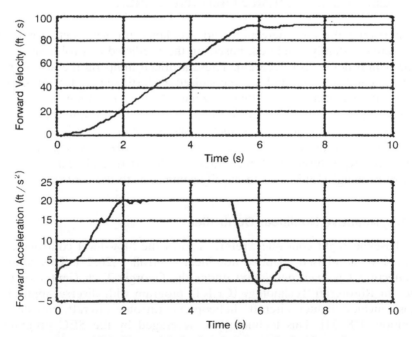

Figure 12.3. Plots of desired $\dot{x}$ and $\ddot{x}$ nominal trajectories.

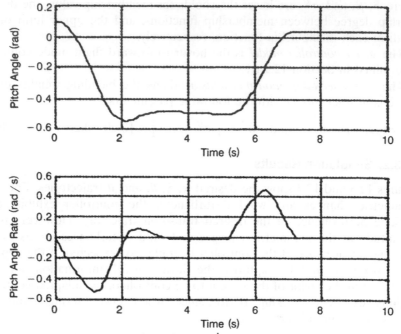

Figure 12.4. Plots of θ and $\dot{\theta}$ nominal trajectories.

12.4. LIMIT DETECTION AND LIMIT AVOIDANCE

In addressing the limit detection and avoidance problem, actuator control limits are constantly redefined, based on the predicted control limit margins using neural networks [17, 23, 24]. These artificial limits on the actuators are seen as "control limits" and are recognized as "artificial control saturations" in case of a violation of the "fly safe zone." At the same time, commands to the controller are modified accordingly in order to prevent the controller from "seeing" and reacting to vehicle response errors. This is also essential for controllers with adaptive features, in order to avoid "wrong adaptation" due to control saturation. A technique developed recently called "pseudo-control hedging" [25] is used to modify commands to the controller.

12.4.1. Adaptive Nonlinear Controller

The baseline flight controller, referred to as the "low-level controller," is a model-inversion-based adaptive nonlinear controller [26, 27]. This controller architecture has been developed under the Georgia Tech Center of Excellence in Rotorcraft Technology (CERT) program with diverse applications. Those include fighter aircraft, helicopters, tiltrotor aircraft, missiles, and munitions [28–31]. This technology is leveraged by the SEC program for further improvements and for integration with mid- and high-level controllers. The low-level controller can be used for trajectory or velocity tracking, but also can be reconfigured to receive angular rate commands for aggressive maneuvering [32]. An adaptive neural net block is used in the feedback path in the inner loop to account for inversion errors and to guarantee closed loop stability. More details on the inner- and outer-loop controllers along with control law derivations are given in references 28 and 29. Additional design details, derivation of the neural network update law, and a proof of closed-loop stability can be found in references 30–32.

12.4.2. Neural-Net-Based Limit Detection and Avoidance

The functional relationship between the set of limit parameters and the set of measurable state variables and control variables is modeled through a neural network. Data sets corresponding to dynamic trim solutions of the full model are used to train the neural network. In dynamic trim the states of the aircraft are divided into fast and slow variables. The slow states include flight parameters that vary slowly with time. The fast states include states that vary quickly with time and reach a steady-state value during a maneuver. The dynamic trim condition is defined as a quasi-steady condition in which the fast states have reached their steady state [17]. Estimates of limit parameters in dynamic trim can effectively predict a limit exceedance as soon as the control inputs are applied. The time where the transient dynamics occur provides a time buffer to take precautions to avoid exceedance of a selected limit. This technique is equally applicable for a vehicle with multiple control inputs with multiple limits as has been demonstrated in reference 17. The dynamic trim predictions are used for calculating the control margins between the current control positions and the control deflections that will result in a limit exceedance. The most critical control margin is then used as a variable control limit inside the flight controller for limit avoidance.

12.5. ADAPTIVE LIMIT DETECTION

As the method described above relies on accurate dynamic trim data, an online technique for predicting dynamic trim parameters and calculating the corresponding control margins is essential for developing robust automatic limit detection and avoidance algorithms. The online technique utilizes an observer-type adaptive neural network loop for the estimation of the correct aircraft model. The constructed aircraft model is then used to predict the quasi-steady response behavior of the limit parameters and the corresponding control margins using a second adaptive neural network loop. Though the approach does not require any offline training of the neural networks, existing offline trained neural network data maps can be accommodated in the procedure. Only standard sensor measurements are used for adaptation.

A detailed development of the method along with a simulation evaluation are included in reference 24. As offline training of the network weights is not required, the system has the advantage of adapting to varying flight condition and different vehicle configurations.

The proposed approach consists of two loops. The first loop is similar to an observer loop utilizing adaptive neural networks (ANN). This is used to capture the uncertain plant dynamics. Similar adaptive observers have been introduced previously [33], mainly for control purposes [34, 35]. The observer in our case is used to approximate the local behavior of the limit parameter. The second loop is used to obtain a dynamic trim solution using the dynamic model generated in the first loop. A neural network is also utilized in the second loop to speed up the solution convergence. In addition to finding the dynamic trim solution for the limit parameter, the second loop is used for finding the corresponding limit control margin as well. The limit control margin predictions can be used as artificial control limits to limit control commands from the low-level controller as described in the previous section. Simulation evaluations of the proposed adaptive limit detection and avoidance algorithms are carried out using a linearized helicopter model and the nonlinear 6-DOF Generic Tilt-Rotor Simulation, GTRSIM. High-g pull-up and push-over maneuvers are included as examples, to demonstrate satisfactory performance of the proposed technique.

12.5.1. Methodology

The adaptive limit detection and avoidance algorithm can be divided into two steps. The first step is to establish the functional relationship between the quasi-steady response of the limit parameter and a set of measurable state and control variables. The quasi-steady response of a limit parameter also corresponds to a dynamic trim condition. The second step is then to calculate the control deflections that would cause the vehicle to reach a given limit boundary. Let the following nonlinear state equation represent the equations of motion of an aircraft:

$$\dot{x} = f(x, u)$$
$$x \in \Re^n, \qquad u \in \Re^i \tag{12.1}$$

The state x can be divided into fast and slow variables, such that

$$\dot{x}_f = f_1(x_f, x_s, u)$$
$$\dot{x}_s = f_2(x_f, x_s, u) \tag{12.2}$$

where

$$x = [x_s \quad x_f]^T, \qquad x_f \in \Re^l, \ x_s \in \Re^{n-l} \tag{12.3}$$

The slow states include flight parameters such as forward speed and the Euler angles. The fast states are, for example, the angular rates, angle of attack, and sideslip. The dynamic trim condition corresponds to

$$x_f = 0 \qquad (12.4)$$

Assume a vector y_p consisting of measurable limit parameters as

$$y_p = g(x_f, x_s, u), \qquad y_p \in \Re^m \qquad (12.5)$$

Since the quasi-steady response of the limit parameters y_p will also correspond to the dynamic trim condition, the dynamic trim values of the limit parameters may be obtained using

$$y_{p_{DT}} = g(x_{f_{DT}}, x_s, u) \qquad (12.6)$$

The subscript "DT" denotes dynamic trim. Thus the dynamic trim response of the limit parameter can be calculated using the dynamic trim estimates of the necessary fast states. Differentiating the algebraic equation of y_p in Eq. (12.5), we obtain

$$\dot{y}_p = g_{s_f}\dot{x}_f + g_{x_s}\dot{x}_s + g_u\dot{u}$$
$$= g_{x_f}f_1 + g_{x_s}f_2 + g_u\dot{u} \qquad (12.7)$$

Now, let

$$\dot{u} = (u(t) - u(t - \Delta t))/\Delta t \qquad (12.8)$$

Since $u(t - \Delta t)$ gives rise to the current value of y_p, and assuming a unique mapping between the current values of the fast states and the limit parameters, Eq. (12.7) may be written in functional form as

$$\dot{y}_p = \phi(y_p, x_s, u) \qquad (12.9)$$

Now, using a linear approximation for the function ϕ, Eq. (12.9) can be rewritten as

$$\dot{y}_p = A_1 y_p + B_1 u + \xi_1(y_p, x_s, u) \qquad (12.10)$$

where the modeling error is represented as ξ_1. Using a neural network Δ_1 as an approximation of the modeling error, we obtain

$$\dot{\hat{y}}_p = A_1 \hat{y}_p + B_1 u + \Delta_1(\hat{y}_p, x_s, u) \qquad (12.11)$$

In this case, the quasi-steady predictions for the limit parameters are obtained by setting $\dot{\hat{y}}_p = 0$, which implicitly corresponds to the dynamic trim condition. Then in dynamic trim

$$A_1 \hat{y}_{p_{DT}} + B_1 u + \Delta_1(\hat{y}_{p_{DT}}, x_s, u) = 0 \qquad (12.12)$$

Also, for known limit parameter boundaries, $y_{p_{\lim}}$, an estimate of the limit control vector can be obtained using

$$A_1 y_{p_{\lim}} + B_1 \hat{u}_{\lim} + \Delta_1 \left(y_{p_{\lim}}, x_s, \hat{u}_{\lim} \right) = 0 \qquad (12.13)$$

Solutions of Eqs. (12.12) and (12.13) with respect to $\hat{y}_{p_{DT}}$ and $\hat{u}_{\lim}$ will provide the dynamic trim estimate and limit control vector estimate, respectively.

12.5.2. Adaptive Architecture

The proposed architecture involves two adaptive loops. The first adaptive loop generates approximate model dynamics of the necessary variables, in the form of Eq (12.11). This is similar to an observer loop and is graphically shown in Figure 12.5 for the limit parameter y_p. In Figure 12.5, the error between the measured y_p and the model generated estimate $\hat{y}_p$ is used to update an adaptive neural network, representing the approximation Δ_1 of Eq. (12.11). A proper observer gain matrix $\mathbf{K} = [k_1 \ k_2 \ \dots \ k_l]^T$ is chosen to ensure proper convergence of network weights.

The second adaptive loop (see Figure 12.6) is used to calculate the dynamic trim solution of the model obtained from the first adaptive loop (Figure 12.5). The condition for dynamic trim of the limit parameter is given in Eq. (12.12). The initial guess of the dynamic trim is updated by an adaptive neural network, ANN2, such that Eq. (12.12) is satisfied. The sum of the approximate linear model and Δ_1 is used to update the adaptive neural network ANN2. By forcing the error term e_2 to go to zero, the proposed architecture finds a solution to Eq. (12.12). At each time step, the function

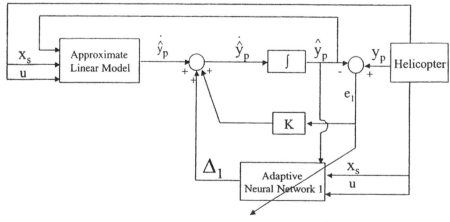

Figure 12.5. Observer loop with adaptive neural network for limit parameter dynamics.

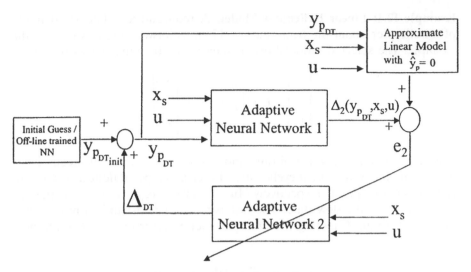

Figure 12.6. Dynamic trim solution for limit parameters.

Δ_1 is estimated by ANN1 from the first loop. The weights are "frozen'" for that instant, and they are used in the dynamic trim solution process in the second loop.

The second loop in Figure 12.6 can be structured to obtain the limit control vector estimations $\hat{u}_{\lim}$ of Eq. (12.13) by solving for $\hat{u}_{\lim}$, given the limiting parameter boundaries $y_{p_{\lim}}$. The control margins can then be obtained using

$$\hat{u}_{marg} = \hat{u}_{\lim} - u \tag{12.14}$$

A Lyapunov-based update law can be derived for both networks [24]. Although it is possible to use the approximate model to obtain initial guesses for the dynamic trim solutions, faster convergence may be obtained using a pretrained neural network. A similar approach can also be followed to calculate the dynamic trim solutions of fast variables [24]. An alternative method to calculate dynamic trim solutions and control margins from Eq. (12.11), without the use of a second network (ANN2) can be found in reference 36.

12.5.3. Simulation Results

In this section, two example simulations are used to demonstrate the effectiveness of the proposed architectures for adaptive limit prediction. In both examples, Sigma-Pi neural networks are used in the adaptation loops. Initial guesses needed for dynamic trim solutions are obtained using linear approximations of Eq. (12.12).

Example 12.1: Linear Helicopter Model. A reduced-order linearized model of a helicopter about the trim point at 85 knots forward flight is used in this example. The selected reduced-order linearized model may be written as

$$
\begin{bmatrix} \dot{q} \\ \dot{p} \\ \dot{V} \end{bmatrix} = [A] \begin{bmatrix} q \\ p \\ V \end{bmatrix} + [B] \begin{bmatrix} \delta_e \\ \delta_a \\ \delta_p \end{bmatrix}
$$

where q is pitch rate, p is roll rate, and V is change in forward speed. δ_e, δ_a, and δ_p are the longitudinal cyclic, lateral cyclic, and pedal deflections around the initial trim point, respectively. In this example, p and q are the fast states, and V is the slow state. The task is to estimate dynamic trim responses on the pitch rate. Choosing the limit parameter vector to consist of the pitch rate q, that is,

$$
y_p = [q]
$$

and using the sensor measurement for pitch, the architectures presented in Figures 12.5 and 12.6 can be used for adaptive limit prediction of pitch rate. Similar to Eq. (12.11), the following equation is used in the first adaptive loop:

$$
\dot{\hat{y}}_p = \hat{a}_{11} \hat{y}_p + \begin{bmatrix} \hat{b}_{1j} \end{bmatrix} \begin{bmatrix} \delta_e \\ \delta_a \\ \delta_p \end{bmatrix} + \Delta_1(\hat{y}_p, V, \delta_e, \delta_a, \delta_p)
$$

where $j = 1, 3$, and $\hat{a}_{11}$ and $\hat{b}_{1j}$ are known approximate estimate values of a_{11} and b_{1j}. The learning rate for the adaptive neural network (ANN1) was set at 300. The observability gain was chosen as $K = 2.5$. The dynamic trim solution is obtained using

$$
\hat{a}_{11} \hat{y}_{P_{DT}} + \begin{bmatrix} \hat{b}_{1j} \end{bmatrix} \begin{bmatrix} \delta_c \\ \delta_a \\ \delta_p \end{bmatrix} + \Delta(\hat{y}_{P_{DT}}, V, \delta_e, \delta_a, \delta_p) = 0
$$

The network gain for the second adaptive neural network (ANN2) was set at 15. In this example, the adaptive loops were turned on at 14 seconds into the simulation. Figure 12.7 shows the pitch rate response, along with the estimated response, and the dynamic trim predictions. The neural network weights are initially set to zero and remain constant until the adaptive algorithms are switched on. When adaptation is off, the observer follows its own approximate dynamics, which at this phase is a linear model with the approximate values $\hat{a}_{11}$ and $\hat{b}_{1j}$ as parameters. It is seen in Figure 12.7 that,

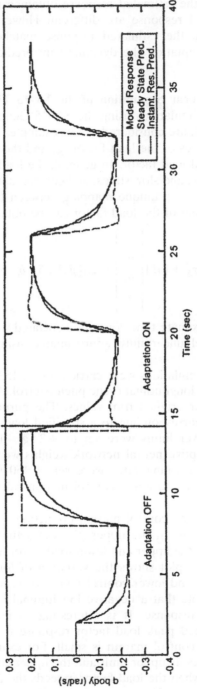

Figure 12.7. Dynamic trim estimation of pitch rate.

prior to adaptation, the estimated response obtained using the approximate model and the actual response are different. However, once the adaptive loops are turned on, the estimated response matches very well with the actual. Also, with adaptation the dynamic trim predictions are significantly improved.

Example 12.2: Nonlinear Simulation of the XV-15 Tiltrotor. The push-up, pull-over maneuver simulations using the XV-15 Generic Tilt-Rotor Simulation model, GTRSIM, are considered in this example. The task is to estimate the dynamic trim values of the load factor, g, and the corresponding control margin for a specified load factor. In general, the load factor can be written as a function of all fast and slow states, and control inputs. However, with the assumption that there is a unique mapping between the fast states and the load factor, an estimate of the load factor can be obtained using

$$
\dot{\hat{g}} = \tau \hat{g} + [B] \begin{bmatrix} \delta_e \\ \delta_c \\ \delta_a \\ \delta_p \end{bmatrix} + \Delta_1 \left(\hat{g}, V, \delta_e, \delta_c, \delta_a, \delta_p \right)
$$

where τ is the approximate time constant associated with load factor dynamics and Δ_1 is model uncertainty approximated using an adaptive neural network.

In this example, simulations were carried out at 153 knots forward speed, for step inputs in the longitudinal cyclic pitch control. All other controls were held constant at their original trim values. The purpose is to estimate the limit control vector corresponding to a specified load factor limit. The load factor upper and lower limits were set to $+2.5g$, $-0.5g$, respectively (see Figure 12.8). All adaptive neural network weights were initially set to zero. The neural network learning rates were set at 150 for ANN1 and 25 for ANN2 (both for upper and lower control margin calculations). The observability gain was $K = 10$.

Figure 12.8 shows the control margin predictions with predefined upper and lower limit margins. In the upper plot the actual load factor variation (solid line) along with the upper and lower load factor limits ($[-0.5g; +2.5g]$) are shown. The lower plot shows the variation of longitudinal cyclic input along with the upper and lower control limits corresponding to the selected load factor limits. Note that a negative longitudinal input corresponds to a positive load factor response. It is interesting to note that, as expected, whenever the predicted peak load factor response is close to the specified limits, the predicted control margin is small. For example, during 16–25 s, the load factor reaches its upper limit and the upper control margin becomes nearly zero. In fact, when the load factor exceeds the $2.5g$ limit around 37 s, the upper control margin becomes negative. Note that the control margin predictions have a similar lead time with respect to reaching a limit parameter boundary.

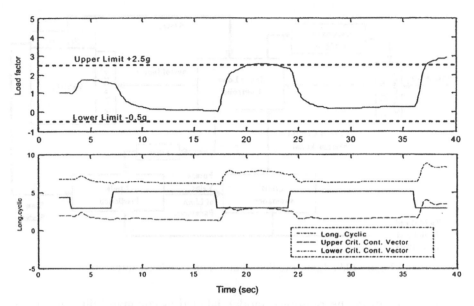

Figure 12.8. Load factor response and longitudinal cyclic input with upper and lower control margins, 153 knots case.

12.6. AUTOMATIC LIMIT AVOIDANCE FOR UAVs

Figure 12.9 is a block diagram representation of the integration of the limit detection and limit avoidance algorithms with the low-level controller. The limit detection block receives the current values of the slow states and the current control inputs, which are fed into the limit detection algorithms. The neural net predictions, which lead the actual response, are used to calculate control margin vector (for details, see reference 17).

The control margin vector ($\Delta\delta$) provides the maximum/minimum allowable control input travel before a limit is encountered. The maximum and minimum allowable control inputs therefore are computed using the limit avoidance algorithm:

$$\delta_{max\ soft\ lim} = \delta_{cmd} + \Delta\delta_{max\ estim}$$

$$\delta_{min\ soft\ lim} = \delta_{cmd} - \Delta\delta_{min\ estim}$$

$\delta_{max\ soft\ lim}$ and $\delta_{min\ soft\ lim}$ can be viewed as the upper and lower bounds of "control soft limits," in order to protect the vehicle from exceeding a performance limit. In order to reduce the undesirable effects associated with control saturation in the context of an adaptive controller, a technique called "pseudo-control hedging" is implemented. The pseudo-control hedging avoids the false adaptation that could result from artificially "saturating" the controls, while protecting the vehicle from exceeding a limit. In order to prevent

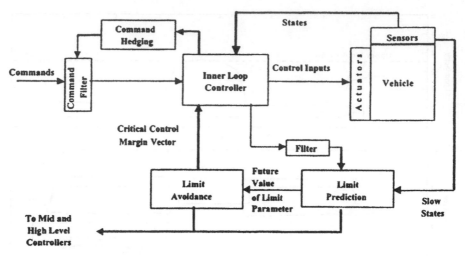

Figure 12.9. Limit prediction and limit avoidance integration.

wrong adaptation, the reference model, labeled as command filter in Figure 12.9, is "hedged," so that the adaptive law would not see these system characteristics [23].

The performance of the automatic limit avoidance algorithms is tested in reference 23 through a series of simulations using the TMAN helicopter model. High-g pull-up and push-over maneuver simulations are used to demonstrate satisfactory performance of the automatic limit avoidance system. In the example presented here, the unmanned helicopter is commanded to perform high-g push-up and pull-over maneuvers. The controller is in rate command controller mode. It receives a pitch rate command of 0.4 rad/s at 85 knots forward flight. Figure 12.10 shows the variation of the load factor

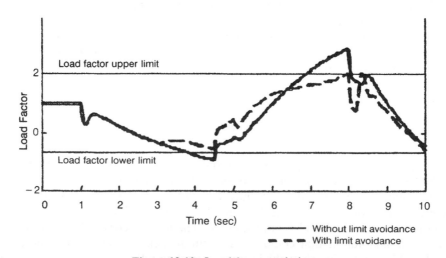

Figure 12.10. Load factor variation.

response with and without limit avoidance. The load factor limits for this aircraft were set as $+2g$ and $-0.5g$. The load factor clearly exceeds the prescribed limit boundaries for this maneuver without a limit avoidance system. However, with the limit avoidance system, the actuator commands are modified, forcing the vehicle to stay within the limit boundaries.

12.7. PERFORMANCE ASSESSMENT AND IMPLEMENTATION ISSUES

A performance assessment methodology has been developed to measure the stability and robustness of mode transition controllers with respect to initial condition uncertainty and/or parametric changes of the system to be controlled. Details of this approach can be found in reference 37. Mode transitioning and limit avoidance algorithms are implemented on a hardware-in-the-loop simulation of an autonomous UAV using the open-control platform (OCP) developed by a consortium of participating contractors in DARPA's Software Enabled Control program that includes Boeing, Georgia Tech, Honeywell, and U.C. Berkeley [38]. More recent developments in online adaptation of mode transition controllers and limit avoidance algorithms are implemented on a YAMAHA R-MAX UAV using the OCP. Georgia Tech intends to perform a series of flight demonstrations with this vehicle that will integrate adaptive mid-level controls, the OCP, and the vehicles instrumentation suite. Flight testing, initiated in 2002, will demonstrate extreme performance of such vehicles when advanced software-enabled controls are augmenting the vehicle's conventional capabilities.

12.8. CONCLUSION

An adaptation scheme is proposed for the online adaptation of mode transition controllers designed via the blending local mode controllers approach. When the active plant model of the adaptation scheme is a good approximation of the system to be controlled, then it is expected that the adapted controller will track the desired trajectory very well in the presence of parametric changes and disturbances. However, if the active plant model does not capture the local model and the input/output behavior of the systems to be controlled, poor tracking performance and unstable tracking can result. Further investigation is needed to add robustness capability to the adaptation scheme so that it can avoid controller faults due to poor approximation of the nonlinear plant. One way of preventing controller faults could be to decrease the aggressiveness of the scheme when approximation errors are significant and increase the aggressiveness when the errors are small.

Adaptive neural-net-based limit prediction and avoidance algorithms are integrated with a flight controller for avoiding performance limits during aggressive maneuvers of an unmanned helicopter. The limit avoidance provides "artificial limits" for the control. A pseudo-control hedging technique is used to eliminate the undesirable effects of control saturation in the context of controller adaptation.

ACKNOWLEDGMENTS

This research has been supported by the DARPA Software Enabled Control program. The support, assistance, and contributions of the DARPA program management and AFRL personnel are gratefully acknowledged.

REFERENCES

1. D. P. Schrage and G. Vachtsevanos, Software-enabled control for intelligent UAV's, in *Proceedings of the IEEE Control Applications Conference*, Hawaii, August 1999.

2. J. S. Shamma and M. Athans, Analysis of gain scheduled control for nonlinear plants, *IEEE Transactions in Automatic Control* **35**:898–907, 1990.

3. J. S. Shamma and M. Athans, Gain scheduling: Potential hazards and possible remedies, *IEEE Control Systems Magazine* **12**:101–107, 1992.

4. P. G. Gonsalves and G. L. Zacharias, Fuzzy logic gain scheduling for flight control, *International Journal of Control* **49**:952–957, 1994.

5. W. J. Rugh, Analytical framework for gain scheduling, *IEEE Control Systems Magazine* **11**:79–84, 1991.

6. R. Ravi, K. Nagpal, and P. Khargonekar, H_∞ control of linear time-varying systems: A state space approach, *SIAM Journal on Control & Optimization* **29**:1394–1413, 1991.

7. D. Dimiter, R. Palm, and U. Rehfuess, A Takagi–Sugeno fuzzy gain-scheduler, *IEEE International Conference on Fuzzy Systems* **2**:1053–1059, 1996.

8. D. J. Leith and W. E. Leithead, Gain-scheduled and nonlinear systems: Dynamic analysis by velocity-based linearization families, *International Journal of Control* **70**:289–317, 1998.

9. K. C. Ng, Y. Li, D. J. M. Smith, and K. C. Sharman, Genetic algorithms applied to fuzzy sliding mode controller design, Technical Report, Center for Systems and Control, University of Glasgow, 1994.

10. J. E. Slotine, Sliding controller design for nonlinear systems, *International Journal of Control* **40**:421–434, 1984.

11. W. Wang and H. Lin, Fuzzy control design for the trajectory tracking on uncertain nonlinear systems, *IEEE Transactions on Fuzzy Systems* **7**:53–62, 1999.

12. J. Theocharis and G. Vachtsevanos, Adaptive fuzzy neural networks as identifiers of discrete-time nonlinear dynamic systems, *Journal of Intelligent and Robotic Systems* **17**:119–168, 1996.

13. X. Ma and N. K. Loh, One-step-ahead controller design using neural networks, *Proceedings of the American Control Conference* **2**:958–962, 1992.

14. X. Ma, and R. N. K. Loh, Neural network-based successive one-step-ahead control of nonlinear systems, *Proceedings American Control Conference* **4**:2129–2133, 1994.

15. L. Mingzhong L and W. Fuli, Adaptive control of black-box nonlinear systems using recurrent neural networks, *Proceedings of the IEEE Conference on Decision and Control*, San Diego, California, pp. 4165–4170, 1997.

16. Y. Tan and A. V. Cauwenberghe, Nonlinear one-step-ahead control using neural networks: Control strategy and stability design, *Automatica* **32**(12):1701–1706, 1996.

17. J. Horn, A. J. Calise, and J. V. R. Prasad, Flight envelope cueing on a tilt-rotor aircraft using neural network limit prediction, *Journal of the American Helicopter Society* **46**(1):23–31, January 2001. .

18. F. Rufus, S. Clements, S. Sander, B. Heck, L. Wills, and G. Vachtsevanos, Software-enabled control technologies for autonomous aerial vehicles, in *18th Digital Avionics Systems Conference*, Vol. 2, pp. 6.A.5-1–6.A.5-8, 1999.

19. F. Rufus and G Vachtsevanos, Design of mode-to-mode fuzzy controllers, *International Journal of Intelligent Systems* **15**:657–685, June 7, 2000.

20. F. Rufus, G. Vachtsevanos and B. Heck, Real-time adaptation of mode transition controllers, *Journal of Guidance, Control and Dynamics*, 2001, accepted for publication.

21. T. Zhang, S. S. Ge, and Hang C.C. Neural-based direct adaptive control for a class of general nonlinear systems, *International Journal of Systems Science* **28**(10):1011–1020, 1997.

22. K. J. Hunt, G. R. Irwin, and K. Warwick, *Neural Network Engineering in Dynamic Control Systems*, Springer-Verlag, New York, 1991, pp. 42–45.

23. I. Yavrucuk and J. V. R. Prasad, Automatic limit detection and avoidance for unmanned helicopters, *Proceedings of the American Helicopter Society Forum*, 2001.

24. I. Yavrucuk, J. V. R. Prasad, and J. A. Calise, Adaptive limit detection and avoidance for carefree maneuvering, in *AIAA Atmospheric Flight Mechanics Conference Proceedings*, Montreal, August 2001.

25. E. N. Johnson, A. J. Calise, Hesham El-Shirbiny, and R. T. Rysdyk, Feedback linearization with neural network augmentation applied to X-33 attitude control, in *Proceedings of the AIAA Guidance and Navigation Conference*, Paper No. AIAA2000-4157, Denver, CO, August 2000.

26. J. E. Corban, A. J. Calise, and J. V. R. Prasad, Implementation of an adaptive nonlinear controller for flight test on an unmanned helicopter, *Proceedings of the 37th IEEE Conference on Decision and Control*, December 1998.

27. J. V. R. Prasad, A. J. Calise, Y. Pei, and J. E. Corban, Adaptive nonlinear controller synthesis and flight test evaluation on an unmanned helicopter, *Proceedings of the IEEE Control Applications Conference*, Hawaii, August 1999.

28. B. S. Kim and A. J. Calise, Nonlinear flight control using neural networks, *AIAA Journal of Guidance, Control and Dynamics* **20**(1) 1977.

29. J. Leitner, A. J. Calise, and J. V. R. Prasad, Analysis of adaptive neural networks for helicopter flight control, *AIAA Journal of Guidance, Control and Dynamics* **20**(5): September–October, 1997.

30. A. J. Calise and R. T. Rysdyk, Nonlinear adaptive flight control using neural networks, *IEEE Control Systems Magazine*, **18**(6):14–25, December 1998.

31. M. B. McFarland and A. J. Calise, Multilayer neural networks and adaptive nonlinear control of agile anti-air missiles, in *Proceedings of the AIAA Guidance, Navigation and Control Conference*, Paper No. AIAA 97-3540, August 1997.

32. I. Yavrucuk and J. V. R. Prasad, Simulation evaluation of a reconfigurable flight controller of a heli-UAV for extreme maneuvers, in *Proceedings of the AIAA Modelling and Simulation Technologies Conference*, Paper No. AIAA2000-4189, Denver, CO, August 2000.

33. Y. H. Kim, F. L. Lewis, and C. T. Abdallah, Nonlinear observer design using dynamic recurrent neural networks, in *IEEE 35th Conference on Decision and Control Proceedings*, Kore, Japan, December 1996, pp. 949–954.

34. F. L. Lewis, A. Yesildirek, and K. Liu, Multilayer neural-net robot controller with guaranteed tracking performance, in *IEEE Transactions on Neural Networks Proceedings*, March 1996, pp. 338–399.

35. D. Strobl, U. Lenz, and D. Schroder, Systematic design of stable neural observers for a class of nonlinear systems, in *IEEE International Conference on Control Applications Proceedings*, October 1997, pp. 377–382.

36. I. Yavrucuk, J. V. R. Prasad, A. J. Calise, and S. Unnikrishnan, Adaptive limit control margin prediction and limit avoidance, *Proceedings of the American Helicopter Society Forum*, 2002, Montreal, Canada.

37. G. Vachtsevanos and F. Rufus, Robust stability of mode to mode fuzzy controllers, American Institute of Aeronautics and Astronautics, *Journal of Guidance, Control, and Dynamics* **22**(6):823–832, *November–December* 1999.

38. L. Wills, S. Kannan, S. Sander, M. Guler, B. Heck, J. V. R. Prasad, D. Schrage, and G. Vachtsevanos, An Open Platform for Reconfigurable Control, *IEEE Control Systems Magazine* **21**(3):49–64, June, 2001.

CHAPTER 13

IMPLEMENTATION OF ONLINE CONTROL CUSTOMIZATION WITHIN THE OPEN CONTROL PLATFORM

RAKTIM BHATTACHARYA and GARY J. BALAS

Editors' Notes

This concluding chapter of Part III discusses several flight control implementation examples that integrate Matlab/Simulink within the open control platform (OCP) framework described in Chapters 4 and 5. A public domain version of an F-16 aircraft model, available as a Simulink block with OCP distribution 1.0, is used for illustration.

The first example is a linear quadratic regulator (LQR). The LQR controller was written as a Matlab "mex" function in C. A Matlab process behaves as an "LQR server." The "LQR client" is an OCP component in a Simulink block; when Simulink invokes the block, the client sends the simulated aircraft's state vector to the LQR server. The server generates the control signal and returns it to the client, which then writes it to the output port of the Simulink block. In this example, Simulink drives the simulation, which is therefore completely synchronous.

The second example demonstrates the use of Matlab/Simulink as a visualization package within OCP. A linear parameter varying (LPV) controller was designed, initially in Simulink. Both this controller and the F-16 dynamics were implemented as OCP component objects (this required translating the Simulink implementation to C++). The publish-subscribe execution model between the two OCP blocks in this example permits event-driven interaction. The development of this example uncovered limitations in the OCP from the point of view of a control engineer. In particular, it highlights the need for a tighter integration between Matlab/Simulink and the OCP.

The final example is a waypoint-tracking problem. A waypoint generator is integrated with an inner-loop tracking controller and can be replaced with a manual joystick by setting a switch. The aircraft model is the only OCP component in this example. The other components are implemented in Simulink and interfaced with the OCP via S functions.

Software-Enabled Control. Edited by Tariq Samad and Gary Balas
ISBN 0-471-23436-2 © 2003 IEEE

The chapter also describes the "sim2ocp" tool that is available to help users create OCP components from Simulink diagrams. At present only the I/O interface is handled by sim2ocp. Complications related to asynchronous systems—in particular the potential for differences between the ordering of execution steps in Simulink and OCP—are also noted.

13.1. INTRODUCTION

Today, the role of a control algorithm is evolving from static designs, synthesized offline, to dynamic algorithms that adapt in real-time to changes in the controlled system and its environment. The paradigm for control system design and implementation is also shifting from a centralized, single processor framework to a decentralized, distributed processor implementation framework. Distribution and decentralization of services and components is driven by the falling cost of hardware, increasing computational power, increasingly complex control algorithms, and development of new, low-cost micro sensors and actuators. A distributed, modular hardware architecture offers the potential benefit of being highly reconfigurable, fault-tolerant, and inexpensive. Modularity can also accelerate development time of products, since groups can work in parallel on individual system components. These benefits come with a price—that is, the need for sophisticated, reliable software to manage the distributed collection of components and tasks.

Communication within a distributed, decentralized environment becomes a significant issue. Hardware components and software processes operate in synchronous and asynchronous modes. These processes have to communicate with one another with a well-defined protocol to effectively control the system. Software tools needed for a distributed, real-time control architecture include: real-time execution, adaptive task scheduling, task synchronization, communication protocols and adaptive resource management. Hence software and its interaction with the controlled system will play a significantly larger role in the control of real-time systems. This drive toward distributed, dynamic control systems requires establishment of tighter ties between the controls and computer science communities for these systems to be successful. DARPA initiated the Software Enabled Control (SEC) program, in part, to address these issues.

A central theme of the DARPA SEC program is to develop software-based control technologies that use dynamic information about the controlled system to adapt, in real time, to changes in the subsystems and its operating environment. This software should be flexible to reconfiguration, allocate sensor capability and actuator control authority, achieve stable and robust operation across complex mode transitions, integrate and coordinate subsystem operations, and enable large-scale distribution of control [1]. The open-controls platform (OCP) [2], being developed by Boeing St. Louis, Honeywell Laboratories, and the Georgia Institute of Technology under the DARPA SEC program, provides a software infrastructure to enable control engineers to work seamlessly, in real time and simulation time, within a distributed control environment. The OCP software is built upon RT-CORBA and is an extension to the Bold Stroke software architecture developed by Boeing St. Louis to support aircraft avionics system integration. The OCP is middleware that consists of a set of services that allow multiple processes running on one or more machines interacting across a network.

The initial implementation of the OCP required users to have significant computer science background to implement a controller, which most control engineers do not have. Iteration of design and implementation through interaction of control designers and embedded system specialists can result in long product development times. The SEC program was designed to bridge the gap between control science and computer science so that control engineers can take their design from paper to hardware without having to deal with the details of implementation. Since the OCP is the proposed solution to bridge this gap, it is obvious that the success of the OCP in the controls community is strongly governed by level of its integration with Matlab, a popular control design software. In this chapter we demonstrate the integration of Matlab with the OCP with the help of three examples, based on flight control design for an F-16 aircraft model.

The chapter is organized as follows. A brief description of the OCP is provided in Section 13.2, followed by details of the F-16 model used in the examples. Three examples present different levels of integration of control algorithm within the OCP environment. A discussion follows the examples on a proposed C++ code generation tool in Matlab that will translate the simulation setup in Simulink to the OCP environment. Issues involved in creating asynchronous real-time systems from Simulink are also discussed.

13.2. WHAT IS THE OCP?

The OCP is an object-oriented software infrastructure that allows seamless integration of cross-platform software and hardware components in any control system architecture. It is designed to facilitate analysis, development, and testing of control algorithms in a distributed computing framework. The current application focus of the OCP is in the domain of uninhabited aerial vehicles (UAVs).

The OCP is a middleware that is based on the real-time common object request broker architecture (RT-CORBA). Middleware is connectivity software that consists of a set of services, allowing multiple processes running on one or more machines to interact across a network. In computer science jargon, middleware can take on the following different forms:

- *Transaction processing (TP) monitors*, which provide tools and an environment for developing and deploying distributed applications.
- *Remote procedure calls (RPCs)* which enable the logic of an application to be distributed across the network. Program logic on remote systems can be executed as simply as calling a local routine.
- *Message-oriented middleware (MOM)*, which provides program-to-program data exchange, enabling the creation of distributed applications. MOM is analogous to email in the sense that it is asynchronous and requires the recipients of messages to interpret their meaning and to take appropriate action.
- *Object request brokers (ORBs)*, which enable the objects that comprise an application to be distributed and shared across heterogeneous networks.

The essential feature of the OCP, from a user's point of view, is the notion of an OCP component. For control systems, an OCP component is a subsystem that is realized in the distributed computing framework. It could either be a pure software entity or a hardware entity with necessary software interface with the OCP. Interaction of the various OCP components results in the functioning of the designed control system. Intercomponent data are transferred via the publish–subscribe message passing model. When multiple applications need to receive the same data or message, publish–subscribe messaging is used. The central concept in a publish–subscribe messaging system is the topic. Multiple publishers may send messages to a topic, and all subscribers to that topic receive all the messages sent to that topic. From a control systems point of view, a topic could be the output port of a subsystem or an OCP component. All components, whose input ports are connected to this output port, subscribe to this topic. Hence, whenever there are data available at the output, they reach the input of those who subscribe. This model is extremely useful when applications want to notify each other of a particular occurrence.

13.3. F-16 AIRCRAFT MODEL

The nonlinear F-16 model used in this chapter, which is available in reference 3, can be represented as

$$\dot{x} = f(x, u) \tag{13.1}$$

The mathematical model uses the wind-tunnel data from the NASA–Langley wind-tunnel tests on a scale model of an F-16 aircraft [4]. The aerodynamic data are valid up to Mach 0.6, angle of attack range of $-10° \leq \alpha \leq 45°$, and side slip angle range $-30° \leq \beta \leq 30°$. The wind-tunnel tests were conducted on sufficiently close points to capture the nonlinear behavior of the aerodynamic force and moment coefficient. Aerodynamic data for the intermediate points are linearly interpolated. The states $x \in \mathcal{R}^{12}$ and controls $u \in \mathcal{R}^4$ in the model are defined as

npos = North position (ft)	T = Thrust (lb)
epos = East position (ft)	δ_e = Elevator deflection (degrees)
h = Altitude(ft)	δ_a = Aileron deflection (degrees)
ϕ = Bank angle (rad)	δ_r = Rudder deflection (degrees)
θ = Pitch angle (rad)	
ψ = Yaw angle (rad)	
V = Velocity (ft/s)	
α = Angle of attack (rad)	
β = Side slip angle (rad)	
p = Roll rate (rad/s)	
q = Pitch rate (rad/s)	
r = Yaw rate (rad/s)	

Actuators for the control surfaces and engine are modeled as first-order systems, details of which are available in references 3 and 5. This model is an example of a high-performance UAV that will be used to demonstrate the capabilities of the OCP. The F-16 model is available as a Simulink block, as an ANSI C function, as a C++ object, and as an OCP component. The Simulink implementation is available with OCP distribution 1.0. It serves as the baseline fixed-wing UAV model for the SEC program.

13.4. INTEGRATION OF MATLAB WITH OCP

It is clear that for the open-control platform to be useful for the controls community, it must be well integrated with Matlab. Most control engineers test new control algorithms in the Simulink environment. The convenience of rewiring interconnections, as well as adding, deleting, and modifying blocks within the Matlab environment, has made Simulink a popular simulation tool.

Recent versions of Matlab have features that enable control engineers to design control algorithms in Simulink and implement them on off-the-shelf DSP boards from Motorola and Texas Instruments. Several third-party vendors like dSPACE and RTSim also support the real-time implementation of Simulink. Matlab's Real-Time Workshop toolbox provides tools that can convert Simulink subsystem definition into intermediate ANSI C code. The C code can subsequently be compiled to run in some of the off-the-shelf DSP boards. Thus Matlab/Simulink not only provides an environment for system level design, it also has the necessary infrastructure for rapid real-time prototyping.

Hence, it is imperative to integrate Simulink with the OCP and take advantage of such an established and popular software package. The following subsections describe in detail the exact nature of Matlab and OCP integration demonstrated in the three examples.

13.4.1. Linear Quadratic Regulation of F-16

The first interface of Matlab with the OCP was developed using a client–server architecture. A linear quadratic regulation (LQR) of the F-16 model was demonstrated. The entire setup was created in Simulink as shown in Figure 13.1. A six-state model, describing the longitudinal motion of the F-16, was used to demonstrate the Matlab-OCP integration. The LQR controller was designed to regulate state perturbations to zero, using thrust and elevator control. This example was developed for the initial OCP version 0.1 to demonstrate the capability of integrating Matlab/Simulink with the OCP middleware.

In the setup shown in Figure 13.1, the block labeled *Remote Controller* is the linear quadratic regulator. It is implemented as a mex function written in C which interfaces with the OCP. Prior to starting the simulation, a Matlab process has to be started, which behaves as an LQR server. The LQR server is like any other server program in the computer science context. Typically, it is a program that is *bound* to some port of a computer and waits for some client program to connect. It is activated when a client connects to the designated port to establish a connection. The LQR server is designed to read in an array of numeric data (representing the state vector x), multiply it with a matrix K to produce a control vector $u = Kx$, and write it to the port for the client to read. This is illustrated in Figure 13.2. The LQR controller is implemented in Matlab. A set of socket interfaces was written to enable basic network programming in Matlab.

The client, an OCP component sitting inside the Simulink block labeled *Remote Controller*, connects to the Matlab based LQR server. The LQR server can be configured with the controller K during startup. The OCP client connects to the LQR server and establishes a connection. At every time step when Simulink invokes the *Remote Controller*, the OCP client sends the input vector signal to the LQR server. The LQR server generates the control signal and sends it to the OCP client. The OCP client then writes it to the output port of *Remote Controller*. Note that this form of message passing is very different from the publish–subscribe model of message passing. The *Remote Controller* OCP component essentially had no active role in the control of the aircraft. It simply served as a socket connection to the Matlab process.

There are obvious drawbacks to this form of interface. This client–server architecture has serious potential for a deadlock condition. In this setup, once the client sends the state vector to the LQR server, it waits for the server to respond with the control signal. On the other hand, the server

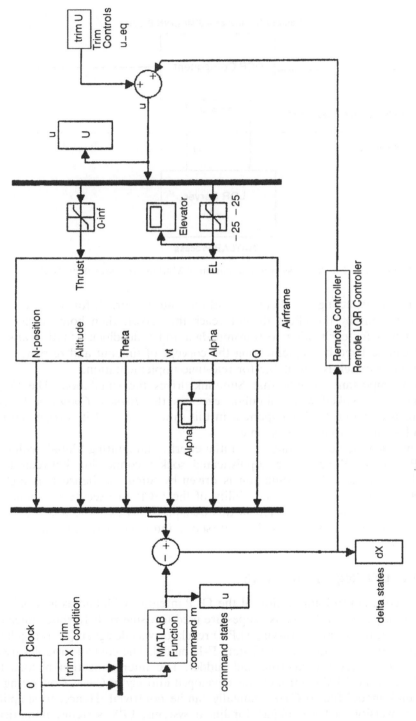

Figure 13.1. LQR Control in Simulink-OCP framework.

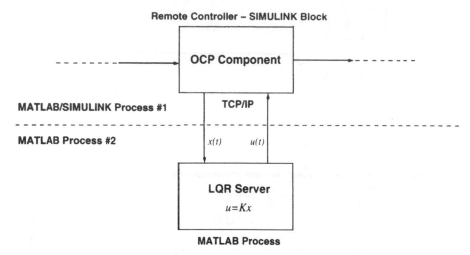

Figure 13.2. Interprocess communication of Matlab processes over TCP/IP.

program is expecting the client to send the state vector. If, for some reason, the data sent by the client do not reach the server, then both client and server wait for each other to transmit data and a deadlock situation arises. This can be fixed by implementing the concept of timeout and retrying, but such an approach is unsuitable for real-time implementation.

It is important to note that Simulink drives the simulation. The OCP manager is invoked when Simulink executes the *Remote Controller* block. There is only one OCP component in this simulation, and it is triggered by Simulink at fixed intervals of time.

This example demonstrates the initial effort in integrating Matlab with the OCP. The OCP acts as a sophisticated socket connection between two Matlab processes. The simulation is driven by Simulink; hence it is highly synchronous in execution. The ability of the OCP to execute components asynchronously is not exploited in this example. This example was designed to demonstrate a socket-based client–server interface between Matlab and the OCP.

13.4.2. LPV Regulation of F-16

The second control application of the OCP integrates Matlab as a visualization package, where Simulink scopes are used to visualize F-16 state trajectories. A linear parameter varying (LPV) regulator was designed [5] to regulate the perturbation states of the 12-state F-16 model. The control variables used were thrust, elevator, aileron, and rudder. The interconnection needed to implement the LPV controller was developed entirely in C++. Converting a Simulink model file to C++ manually can be nontrivial. Hence, to simplify this conversion, a C++ object for linear systems, LPV systems, numerical integration modules, and the full nonlinear F-16 model was developed.

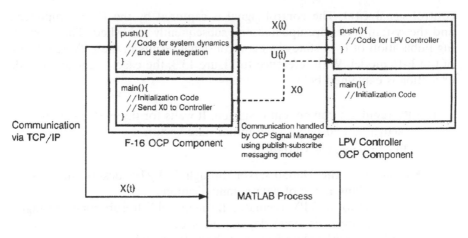

Figure 13.3. LPV Regulation in the OCP framework.

The simulation setup in the OCP framework is shown in Figure 13.3. There are two OCP components in this example, namely the F-16 component and the LPV controller component. Both of these components are implemented using the object oriented design philosophy of software engineering. From the application point of view, each of the two OCP components consists of a main(){ ... } function and a push(){ ... } function. The main() function, like any ANSI C program, is the entry point of the code. The push() function is the event handling code. In this example, the F-16 component subscribes to the output of the LPV controller and the LPV component subscribes to the output of the F-16 component. The push() function acts as an event handling routine that reads in the signal from the input port, computes the output signal, and publishes it. Note that this function is only called when an event (i.e., arrival of appropriate input data for the component) occurs.

To explain the inner workings of this example, let us assume that the executable code for the F-16 is a.out.F16 and for the LPV controller is a.out.LPV. To start the simulation, these two programs will have to be started in two different processes or xterms. The order of invocation is not important. When both the components have been started, each is waiting for the other to publish a signal. Neither can publish any signal because the necessary input is missing. This results in a deadlock. This can be solved by first computing the control signal $U(t)$ from the initial state vector X_0 of the F-16 model. The push() function in the F-16 component can't publish this signal because it can only be invoked when the LPV controller generates its output and triggers an event. Thus, the initial state vector X_0 is published by the main() function of the F-16 component during startup. This published signal is captured by the controller, and it computes and publishes the control signal $U(t)$. The control signal is captured by the F-16 component, and the system states are updated and subsequently published. The alternating publication of F-16 states and control signal results in the LPV regulation of the F-16 model. With reference to Figure 13.3, the exact sequence of code execution is explained below:

1. LPV regulator component is started. It waits for the F-16 component to be activated and publish the initial state vector.
2. F-16 component is started. It

   - connects to the Matlab server through TCP/IP sockets and uses the handshaking mode of data communication
   - connects to an LPV regulator through publish–subscribe messaging and exchanges acknowledgment messages
   - sends the initial state vector of the F-16 model

3. LPV captures the state vector, computes the control signal and publishes it.
4. F-16 captures the control signal, updates states, and publishes the state vector. It also sends the state vector to the Matlab server for trajectory visualization. Note that the data flow from the F-16 component to the Matlab process is not through publish–subscribe but through TCP/IP compliant sockets.

The last two steps are repeated until the simulation is over. The F-16 component then sends a termination signal to the LPV controller and both the processes terminate.

This example demonstrates the first publish–subscribe-based flight control of the F-16 model using the OCP. In this example, Matlab plays no role in controlling the airplane. It is used as a visualization package to plot the aircraft state trajectories in real time. Development of these simple OCP components revealed some of the complexities involved in using the OCP. The LPV controller was first tested in the Simulink environment and then ported to the OCP environment. The convenience of Matlab tools and Simulink blocks was missing when everything had to be converted to C++. Since interfacing to the OCP required all Simulink blocks to be recreated as C++ objects, development and testing of these objects was a time-consuming endeavor. It was also revealed that C++ programming knowledge alone is not sufficient to successfully develop an OCP component. Knowledge of the operating system and OCP internals are also necessary. We discovered that improper file I/O in the program can potentially cause race conditions that halt the computer. A race condition occurs when two processes try to access a shared resource and neither can. This race condition was resolved by changing the sequence of code execution.

Thus, with this example we were able to highlight the need for tighter links between Matlab/Simulink and the OCP. These will not only reduce the simulation setup time, but will also improve reliability of results. It was also clear that a code generation tool was necessary to make the internals of the OCP transparent to the users of OCP. We were also able to demonstrate how an external visualization package can be linked to the OCP via a socket connection.

13.4.3. Waypoint Tracking Problem

The third example is a waypoint tracking problem. In this problem, we are interested in having the F-16 aircraft model reach coordinates in space at specific times and states of the aircraft. Waypoint tracking is implemented as an outer-loop command generator and an inner-loop tracking controller. The aircraft inner-loop controller consists of two parts, the lateral controller and the longitudinal controller. The lateral controller is a classical controller given in reference 4, and the longitudinal controller is an LPV controller designed to track velocity (V) and flight path angle (γ) commands. The outer-loop controller, developed by Honeywell Laboratories researchers, generates velocity $V(t)$, climb angle $\gamma(t)$, and roll rate $p(t)$ reference signals on the basis of specified waypoints and the current position of the airplane. The inner-loop controller was designed to track these signals. The combined effect of the inner- and outer-loop controller flies the airplane to the specified waypoints at the specified times. Implementation of the waypoint tracking framework is shown in Figure 13.4. In this example, a switch was added to allow the waypoint command generator to be disconnected and have a joystick generate the required inner-loop commands. We use a three-button analog joystick to generate $[V(t), \gamma(t), p(t)]$ reference signals. It is possible to select the waypoint generator or the joystick by clicking the manual switch in the Simulink diagram.

The outer-loop controller is shown in Figure 13.5. It involves two subsystems, namely the waypoint loop and the flight path loop. The waypoint loop determines where the aircraft is located and estimates the velocity command vector, $[\chi(t), V(t)\gamma(t)]$. The velocity command vector is converted to a reference signal, $r(t) = [V(t), \gamma(t), p(t)]$ by the flight path loop. There is also a waypoint manager, shown in Figure 13.6, that decides when to update the current waypoint. The waypoint generator requires three waypoints to determine the reference signal.

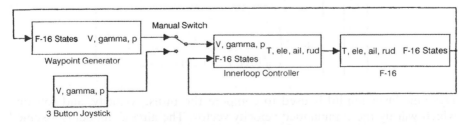

Figure 13.4. Waypoint tracking setup.

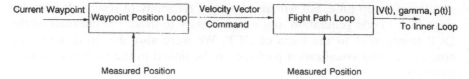

Figure 13.5. Honeywell outer-loop controller.

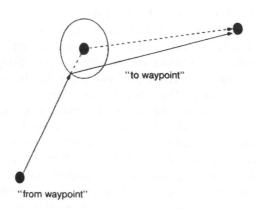

Figure 13.6. Honeywell waypoint generator.

The equations of motion used to determine the position of the aircraft are

$$
\begin{bmatrix} \dot{\xi} \\ \dot{\eta} \\ \dot{h} \end{bmatrix} = \begin{bmatrix} V\cos(\gamma)\cos(\chi) \\ V\cos(\gamma)\sin(\chi) \\ \sin(\gamma) \end{bmatrix} \tag{13.2}
$$

where ξ is the east position, η is the north position, and h is the altitude of the aircraft. From Eq. (13.2), we can obtain equations for χ, V and γ as

$$
\begin{bmatrix} \chi \\ V \\ \gamma \end{bmatrix} = \begin{bmatrix} \dfrac{\Delta\eta}{\Delta\xi} \\ \dfrac{\sqrt{(\Delta\xi)^2 + (\Delta\eta)^2 + (\Delta h)^2}}{\Delta t} \\ \sin^{-1}\left(\dfrac{\Delta h}{\sqrt{(\Delta\xi)^2 + (\Delta\eta)^2 + (\Delta h)^2}} \right) \end{bmatrix} \tag{13.3}
$$

Dynamic inversion [6] is used to compute the thrust, velocity, and roll rate which will fly the commanded velocity vector. The aircraft model is assumed to be a linear model given by

$$\dot{x} = Ax + Bu \tag{13.4}$$

$$y = Cx \tag{13.5}$$

and the control law is given by

$$\dot{y} = C\tilde{A}x + C\tilde{B}u \tag{13.6}$$

$$u = (C\tilde{B})^{-1}\left[\dot{y}^{des} - C\tilde{A}x\right] \tag{13.7}$$

This method is heavily model-based. $\tilde{A}$ and $\tilde{B}$ are models of the dynamics embedded in the control law given in Eq. (13.7). The nonlinear model of the F-16 aircraft is linearized at every timestep during simulation and used in dynamic inversion to compute the reference signals.

The longitudinal axis LPV controller was designed to track the $V(t)$ and $\gamma(t)$ reference signals. It also decouples the tracking, which provides two degrees of freedom in controlling the longitudinal flight path. The decoupling is essential because the outer loop is expected to generate commands where γ is fixed and velocity changes or vice versa. The controller was designed using a model matching technique. The ideal models for velocity and flight path angle tracking are given as

$$T_{ideal,V} = \frac{0.16}{s^2 + 0.8s + 0.16} \tag{13.8}$$

$$T_{ideal,\gamma} = \frac{-\frac{1}{3}s + 1}{s^2 + 2s + 1} \tag{13.9}$$

The LPV controller is scheduled on velocity and is valid for velocity ranging from 400 ft/s to 800 ft/s at an altitude of 10,000 ft.

This example is similar to the first example in the sense that the simulation is driven by Simulink. Here the plant model is an OCP component. Simulink is interfaced with the OCP via S-functions. In this example, communication between OCP and Simulink is achieved via publish–subscribe messaging. This differs from the first example where socket-based data communication is used. There is a minimal amount of integration with the OCP since there is only a single OCP component present. Deeper integration of the OCP with Simulink could have been achieved if all the subsystems in Figure 13.4 were converted to OCP components. Since the current version of the OCP allows components to be only defined in C++, converting the complicated interconnections in the subsystems would have required a very long development and validation time. Hence, it was avoided. This once again highlights the requirement of a code generation tool that will convert Simulink subsystems to OCP components.

One of the key elements in this example is the analog joystick. The joystick was included to enable tracking of human-generated reference signals. Since the joystick is an input device, inclusion of this hardware did not impose any real-time constraints on code execution. The joystick was implemented as a Simulink block using DirectX APIs in S-functions. The waypoint tracking example also integrates an external visualization software package, AVDS [7], for real-time visualization of F-16 state trajectories.

13.4.4. Summary

The three examples presented in this section demonstrate three different ways of integrating Matlab with the OCP. Each example depicts a more complex flight control scenario than the previous one. However, none of these examples fully demonstrate the capabilities of the OCP. The OCP components in the three examples execute in a highly synchronized manner. The ability of the middleware to handle asynchronous events like mode switching and other aspects of reconfigurability were not exemplified. Also, these examples were executed in a simulated environment. Hence, issues regarding schedulability and real-time CPU task allocation were not considered during both development and implementation phases. These simulations were initiated to explore the possible ways of integrating Matlab with the OCP. The experience obtained from developing these initial OCP-based examples helped us to recognize areas for improvement and avenues for further research that will make the OCP more user-friendly to control engineers.

13.5. SIMULINK TO OCP COMPONENTS

The current version of the OCP provides a tool called *sim2ocp* that helps users to create OCP components from Simulink diagrams. Currently, only the data flow specification of the simulation setup can be specified by the Simulink model file. The code generated by *sim2ocp* only handles the I/O interface of the components. The internal behavior of each component has to be manually coded. This is not trivial for most control engineers because it requires C++, CORBA, and operating-system-level knowledge.

The *sim2ocp* GUI has six tabs or pages where necessary information about the components, not expressible in Simulink, is entered. The different pages and the information entered are explained as follows.

- *Import Simulink Model File*. This requires the user to enter the Simulink model filename. The data flow information is obtained from the interconnection of the input and output ports.

- *Signal Definitions*. The names of subsystems and their ports are determined by *sim2ocp* from the Simulink diagram. The user defines the signal vector that connects these ports using a user-friendly graphical user interface (GUI). Each member of this user-defined `struct` defines an element of the signal vector. For example, for our F-16 example, we would need to define a signal vector that connects the output port of the controller to the input port of the F-16 component. This signal vector would contain thrust, elevator, rudder, and aileron control variables. Thus we would need to define a `struct` for this signal as follows:

```
typedef struct {
                double Thrust;
                double Elevator;
                double Rudder;
                double Aileron;
                } ControllerOutput;
```

- *Ports*. Once the necessary signals have been created, the signals need to be identified with the ports of the subsystems. The *sim2ocp* tool allows reusability of signal data structures wherever possible. This page also allows the user to specify scheduling information like criticality, execution rates, worst-case execution time, aperiodic deadline, and a choice to execute the code at a periodic interval. The periodic interval could be based on a software or hardware timer.

- *Components*. The Simulink block diagram entered in page one could have an interconnection of more than two subsystems. The user has the freedom to specify the number of OCP components generated and allocate the number of subsystems that will be contained in each component. Each OCP component behaves as an independent process, within which several subsystems can execute. Thus, the importance of scheduling comes into the picture when a single process has to manage the execution of more than one subsystem. In this page, the user can also choose to delay the execution of a subsystem until all its input dependencies have been satisfied.

- *Configuration States*. This page allows the user to specify different configuration states for mode switching scenarios. It is possible to create different configuration states where certain ports are activated or deactivated, criticality of the subsystems modified, and execution rates changed.

- *Finish*. This is the last page of the GUI. Once all the information has been entered, it is saved in a *Component Information File*. The C++ code for OCP components and specific I/O handling routines are generated by a `perl` script called `genopc.pl` based on the *Component Information File*.

The `perl` script generates Microsoft Visual C++ project files along with the code for the OCP components. The auto-generated wrapper code implements intercomponent communication, the component triggering mechanism, and support for advanced OCP features such as adaptive resource management and hybrid mode switching. The generated code, however is not ready to be compiled at this point. The code generator only generates the real-time asynchronous I/O interfaces. This tool makes the I/O interface transparent, but is still not very useful for control engineers. In particular, the code generator doesn't generate the subsystem definition. The auto-generated framework code has place-holders for code which implements the algorithms within the individual components. This allows heritage code which implements the algorithms, if it exists, to be plugged into the framework. If code implementing a component does not already exist, then code must be written, either by hand or with the aid of some additional auto-code functionality like that provided by the Mathworks Real-Time Workshop tool set. This generation of the code implementing the component algorithms is often nontrivial for complex systems.

13.6. ASYNCHRONOUS SYSTEMS AND SIMULINK MODELS

The Simulink diagram defines a data-flow specification of the system being simulated. Data-flow specifications can be used to automatically synthesize asynchronous systems [8]. The data-flow model, obtained from the Simulink diagram, represents many asynchronous concurrent computation events [9], where each event corresponds to the activation of a component. A component becomes active when all the inputs it requires are available. Thus in data-flow execution, many components may be ready to execute simultaneously, resulting in several asynchronous concurrent computation events.

In Simulink, subsystems are executed sequentially [10]. The order of execution is determined during the initialization phase of the simulation. A list is created such that any block with direct feedthrough is not updated until the blocks driving its inputs are updated. Such a list cannot be created if there are algebraic loops in the interconnection. Typically, algebraic loops are created when the output of a block with direct feedthrough is fed to its input, directly or by a feedback path through other blocks with direct feedthrough. Simulation in Simulink is a two step process. First, the output of each block is computed in the order determined during initialization. In the second pass, the system states are updated based on their current state and derivatives.

Once the Simulink subsystems are converted to OCP components and implemented within the OCP, their execution will occur in an asynchronous manner. Generally, the order of their execution in this environment may not be the same as that in Simulink. This difference in order of execution could lead to results that are different from that obtained from Simulink, though not necessarily erroneous.

It is clear that the simulation setup itself imposes certain synchronization of components. For instance, in the waypoint tracking example, the activation of the waypoint generator component is dependent on the availability of sensor data. The sampling rate of all the sensors may not be the same, hence the dependencies of the waypoint generator may not be satisfied all at once. Its activation has to be delayed until all the dependencies have been satisfied. Simulink satisfies this constraint when it defines the order in which the blocks are evaluated. In the asynchronous environment, this has to be explicitly satisfied. However, it may not be always necessary to delay the activation until all the dependencies are satisfied. For certain applications, it may be acceptable to use the input from previous updates to compute the current output.

Real-time deadlines for software elements can be hard or soft deadlines. However, if there are hardware elements in the interconnection, the real-time deadlines are hard deadlines. A hardware element will demand its input buffer to be filled with valid data at every sampling time. It is not possible to delay the execution of such an element until all its input dependencies have been satisfied. In case of violation of real-time deadlines, the system either propagates with partially current input signals or the sources of the input dependencies approximate their outputs to meet the real-time deadline. These errors will affect the overall performance of the control system. Simulation in Simulink does not reveal these issues. The sequential data flow (SDF) model of computation (MOC), implicitly enforced by Simulink, defines the ordering of firing mechanism. However, in the realization of the components in the OCP framework, the firing mechanism is replaced by asynchronous communication between components and the ordering is embedded in the priorities assigned for the components. Hence, a scientific investigation should be done in the direction of understanding the mapping from SDF MOC to ORB-like frameworks so that the semantics can be preserved.

13.7. CONCLUSION

A central theme of the DARPA SEC program is to use dynamic information about the controlled system to adapt, in real time, to changes in the system and its operating environment. A key enabling technology is the OCP middleware software. The OCP services will enable dynamic control algorithms to handle complex mode transitions and to integrate and coordinate subsystem operations and large-scale distribution of control. This chapter describes software tools developed to link the Matlab/Simulink environment and the OCP. These tools are key to allowing control engineers to seamlessly implement control algorithms within the OCP framework. To fully realize the benefits of the OCP in future military and commercial systems, the controls and computer science communities need to establish close ties and aggressively address issues at the real-time control/software interface.

REFERENCES

1. DARPA—Software Enabled Control, http://www.darpa.mil/ito/research/sec/index.html.
2. J. Paunicka, B. Mendel, and D. Corman, The OCP—An open middleware solution for embedded systems, *American Control Conference*, 2001.
3. B. Stevens and F. Lewis, *Aircraft Control and Simulation*, Wiley-Interscience Publication, John Wiley & Sons, New York, 1992.
4. L. Nguyen, M. Ogburn, W. Gilbert, K. Kibler, P. Brown, and P. Deal, Simulator Study of Stall/Post-Stall Characteristics of a Fighter Airplane with Relaxed Longitudinal Static Stability, Technical Report 1538, NASA Langley Research Center, Hampton, Virginia, 1979.
5. R. Bhattacharya, G. Balas, M. Kaya, and A. Packard, Nonlinear receding horizon control of F-16 aircraft, *American Control Conference* 1:3102–3105, 2001.
6. D. J. Bugajski and D. F. Enns, Nonlinear control law with application to high angle-of-attack flight, *Journal of Guidance, Control and Dynamics* **May-June**:761–767, 1992.
7. Aviator Visual Design Simulator, http://www.rassimtech.com/.
8. T. -Y. Wuu and S. B. K. Vrudhula, Synthesis of Asynchronous Systems from Data Flow Specifications, Technical Report, Information Sciences Institute, University of Southern California, ISI No. RR-93-366, NTIS No. ADA-278687, Dec. 1993, http://citeseer.nj.nec.com/wuu93synthesi.html.
9. W. A. Najjary, E. A. Lee, and G. R. Gao, Advances in the Dataflow Computational Model, http://citeseer.nj.nec.com/najjary99advance.html.
10. "SIMULINK Manual—Using Simulink," http://www.mathworks.com.

PART IV

HYBRID DYNAMICAL SYSTEMS

CHAPTER 14

HYBRID SYSTEMS: REVIEW AND RECENT PROGRESS

PANOS J. ANTSAKLIS and XENOFON D. KOUTSOUKOS

Editors' Notes

The last part of this volume focuses on hybrid dynamical systems, an area of research that has developed as a result of the increasing interaction, over the last several years, between the control engineering and computer science communities. This chapter provides a broad-based introduction to hybrid systems and discusses a number of topics of ongoing research.

Hybrid systems are systems that exhibit both continuous-time and discrete-event dynamics. In the former case, the dynamics can be defined by differential or difference equations. For the latter, common representations include finite state machines and Petri nets. A good example of hybrid dynamics is multimode behavior. For example, different continuous-time models may be useful for capturing the dynamics of an aircraft in take-off, landing, cruise, and other operational modes. Analogously, control laws are generally very different in these cases as well—both plant models and controllers can be hybrid systems. Given the compounded complexity of hybrid systems, new methods are needed for their design and analysis. Over the last few years, research in this area has matured to a point where a number of tools have been developed.

A formalism for hybrid systems that has proven productive is hybrid automata. A hybrid automaton is a finite-state machine extended with real-valued variables and differential equations that are defined differently for each (discrete) state of the FSM. An important special case is a linear hybrid automaton, in which the rates of change of the continuous variables are constant. Hybrid automata (especially in their simplified versions) have proven useful in addressing problems of verification in hybrid systems.

Software-Enabled Control. Edited by Tariq Samad and Gary Balas
ISBN 0-471-23436-2 © 2003 IEEE

14.1. HYBRID SYSTEM MODELS

A hybrid control system is a system in which the behavior of interest is determined by interacting processes of distinct characteristics—in particular, interacting continuous and discrete dynamics. Hybrid systems typically generate mixed signals that consist of combinations of continuous- and discrete-valued signals. Some of these signals take values from a continuous set (e.g., the set of real numbers) and others take values from a discrete, typically finite set (e.g., the set of symbols $\{a, b, c\}$). Furthermore, these continuous- or discrete-valued signals depend on independent variables such as time, which may also be continuous- or discrete-valued. Another distinction that can be made is that some of the signals can be time-driven, while others can be event-driven in an asynchronous manner.

The dynamic behavior of such hybrid systems is captured in hybrid models. In a manufacturing process, for example, parts may be processed in a particular machine, but only the arrival of a part triggers the process; that is, the manufacturing process is composed of the event-driven dynamics of the parts moving among different machines and the time-driven dynamics of the processes within particular machines. Frequently in hybrid systems in the past, the event-driven dynamics were studied separately from the time-driven dynamics, the former via automata or Petri net models (also via PLC, logic expressions, etc.) and the latter via differential or difference equations. To understand fully the system's behavior and meet high-performance specifications, one needs to model all dynamics together with their interactions. Only then may problems such as optimization of the whole manufacturing process be addressed in a meaningful manner. There are, of course, cases where the time-driven and event-driven dynamics are not tightly coupled or the demands on the system performance are not difficult to meet, and in those cases, considering simpler separate models for the distinct phenomena may be adequate. However, hybrid models must be used when there is significant interaction between the continuous and the discrete parts and high-performance specifications are to be met by the system.

Hybrid models may be used to significant advantage, for example, in automotive engine control, where there is a need for control algorithms with guaranteed properties, implemented via embedded controllers, that can substantially reduce emissions and gas consumption while maintaining the performance of the car. Note that an accurate model of a four-stroke gasoline engine has a natural hybrid representation, because from the engine control point of view, on one hand, the power train and air dynamics are continuous-time processes, while, on the other hand, the pistons have four modes of operation that correspond to the stroke each is in and so their behavior is represented as a discrete-event process described, say, via a finite-state machine model. These processes interact tightly, as the timing of the transitions between two phases of the pistons is determined by the continuous motion of the power train, which, in turn, depends on the torque produced by each piston. Note that in the past the practice has been to convert the discrete part of the engine behavior into a more familiar and easier-to-handle continuous model, where only the average values of the appropriate physical quantities are modeled. Using hybrid models, one may represent time- and event-based behaviors more accurately so as to meet challenging design requirements in the design of control systems for problems such as cutoff control and idle speed control of the engine [1]. For similar reasons—that is, tight interaction of continuous and discrete dynamics and demands for high performance of the system—hybrid models are important in chemical processes [2], robotic manufacturing systems [3, 4], transportation systems [5], and air traffic control systems [6], among many other applications.

There are other ways in which hybrid systems may arise. Hybrid systems arise from the interaction of discrete planning algorithms and continuous processes; and, as such, they provide the basic framework and methodology for the analysis and synthesis of autonomous and intelligent systems [7]. In fact, the study of hybrid systems is essential in designing sequential supervisory controllers for continuous systems, and it is central in designing intelligent control systems with a high degree of autonomy. Another important way in which hybrid systems arise is from the hierarchical organization of complex systems. In these systems, a hierarchical organization helps manage complexity and higher levels in the hierarchy require less detailed models (discrete abstractions) of the functioning of the lower levels, necessitating the interaction of discrete and continuous components [8].

In the control systems area, a very well-known instance of a hybrid system is a sampled-data or digital control system. Therein, a system described by differential equations, which involve continuous-valued variables that depend on continuous time, is controlled by a discrete-time controller described by difference equations, which involve continuous-valued variables that depend on discrete time. If one also considers quantization of the continuous-valued variables or signals, then the hybrid system contains not only continuous-valued variables that are driven by continuous and discrete times, but also discrete-valued signals. Another example of a hybrid control system is a switching system, where the dynamic behavior of interest can be adequately described by a finite number of dynamical models, which are typically sets of differential or difference equations, together with a set of rules for switching among these models [9, 10]. These switching rules are described by logic expressions or a discrete-event system with a finite automaton or a Petri net representation.

A familiar simple example of a practical hybrid control system is the heating and cooling system of a typical home. The furnace and air conditioner, along with the heat flow characteristics of the home, form a continuous-time system, which is to be controlled. The thermostat is a simple asynchronous discrete-event system (DES), which basically handles the symbols {*too hot, too cold*} and {*normal*}. The temperature of the room is translated into these representations in the thermostat and the thermostat's response is translated back to electrical currents, which control the furnace, air conditioner, blower, and so on.

There are several reasons for using hybrid models to represent the dynamic behavior of interest in addition to the ones already mentioned. Reducing complexity was and still is an important reason for dealing with hybrid systems. This is accomplished in hybrid systems by incorporating models of dynamic processes at different levels of abstraction; for example, the thermostat in the above example is a very simple model, but adequate for the task at hand, of the complex heat flow dynamics. For another example, to avoid dealing directly with a set of nonlinear equations, one may choose to work with sets of simpler equations (e.g., linear) and switch among these simpler models. This is a rather common approach in modeling physical phenomena. In control, switching among simpler dynamical systems has been used successfully in practice for many decades. Recent efforts in hybrid system research along these lines typically concentrate on the analysis of the dynamic behaviors and aim to design controllers with guaranteed stability and performance.

Hybrid systems have been important for a long time. The recent interest and activity in hybrid systems have been motivated in part by the development of research results on the control of DESs that occurred in the 1980s and on adaptive control in the 1970s and 1980s and by the renewed interest in optimal control formulations in sampled-data systems and digital control. In parallel developments, there has been growing interest in hybrid systems among computer scientists and logicians with an emphasis on verification of the design of computer software. Whenever the behavior of a computer program depends on values of continuous variables within that program (e.g., continuous-time clocks), one needs hybrid system methodologies to guarantee the correctness of the program. In fact the verification of such digital computer programs has been one of the main goals of several serious research efforts in the hybrid system literature. Note that efficient verification methodologies are essential for complex hybrid systems to be useful in applications. The advent of digital machines has made hybrid systems very common indeed. Whenever a digital device interacts with the continuous world, the behavior involves hybrid phenomena that need to be analyzed and understood. It should be noted that certain classes of hybrid systems have been studied in related research areas such as variable structure control, sliding mode control, and bang-bang control.

Hybrid systems represent a highly challenging area of research that encompasses a variety of challenging problems that may be approached at varied levels of detail and sophistication. Modeling of hybrid systems is very important, as modeling is in every scientific and engineering discipline. Different types of models are used, from detailed models that may include equations and lookup tables that are excellent for simulation but not easily amenable to analysis, to models that are also good for analysis but not easily amenable to synthesis, models for control, models for verification, and so on.

14.2. APPROACHES TO THE ANALYSIS AND DESIGN OF HYBRID SYSTEMS

Current approaches to hybrid systems differ with respect to the emphasis on or the complexity of the continuous and discrete dynamics and in whether they emphasize analysis and synthesis results, or analysis only, or simulation only. On one end of the spectrum there are approaches to hybrid systems that represent extensions of system theoretic ideas for systems (with continuous-valued variables and continuous time) that are described by ordinary differential equations to include discrete time and variables that exhibit jumps or that extend results to switching systems. Typically these approaches are able to deal with complex continuous dynamics. Their main emphasis has been on the stability of systems with discontinuities. On the other end of the spectrum there are approaches to hybrid systems embedded in computer science models and methods that represent extensions of verification methodologies from discrete systems to hybrid systems. Typically these approaches are able to deal with discrete dynamics described by finite automata and emphasize analysis results (verification) and simulation methodologies. There are additional methodologies spanning the rest of the spectrum that combine concepts from continuous control systems described by linear and nonlinear differential/difference equations, and from supervisory control of DESs that are described by finite automata and Petri nets to derive, with varying success, analysis and synthesis results.

It is very important to have good software tools for the simulation, analysis, and design of hybrid systems, which by their nature are complex systems. This need has been recognized by the hybrid system community, and several software tools have been developed. Here, we list some of the available software tools. It should be noted that the list of software tools dynamically changes to accommodate the progress in hybrid system research. Tools that have been traditionally used by engineers for simulation and design of continuous systems have been extended to provide various functionalities for hybrid systems. Modeling and simulation tools require the development of sophisticated algorithms that address various problems that arise in interfacing continuous and discrete dynamics. These issues have been studied in the literature; see, for example, references 11–13.

The Matlab/Simulink/Stateflow software environment [14] provides tools for visual modeling and simulation of hybrid systems that may include continuous-time, discrete-time, and event-driven dynamics. Ptolemy II [15] is a set of software packages supporting heterogeneous, concurrent modeling and design. It supports several models of computation including finite state machines, discrete event systems, and continuous-time systems as well as the appropriate interfaces that enable hybrid system modeling and simulation by orchestrating the interaction of these domains. It should be noted that Ptolemy II provides additional capabilities not necessarily related to hybrid systems; for details see reference 16. The object oriented language Modelica [4] has been developed for modeling physical systems and provides modeling paradigms for hybrid systems. Simulation engines and tools such as Dymola [17] and MathModelica [18] that support Modelica can be used for simulation of physical systems that exhibit hybrid phenomena. HCC [19] is a concurrent constrained-based object-oriented language that supports hybrid dynamics. An HCC compiler was initially developed at Xerox PARC and then extended at NASA Ames, where it has been used in various space applications. The Shift programming language [20] was developed for describing dynamic networks of hybrid automata. Although the primary motivation was the specification and analysis of automotive applications, Shift has been used in various application domains. OmSim is a software environment for modeling and simulation based on Omola, an object-oriented language for representing continuous-time and discrete-event dynamical systems [21]. Charon is a language for hierarchical and modular modeling of hybrid systems [22]. HyTech is a verification tool for hybrid systems based on the theory of linear hybrid automata [23]. Kronos [24] and UPPAAL [25] are verification tools for real-time systems modeled by timed automata. Hybrid system software tools have been developed in the chemical processing industry; for more details see reference 2. In addition, many research groups are developing software tools for supporting their work.

Several approaches to modeling, analysis, and synthesis of hybrid systems are described in this chapter. They are organized into three sections. First, approaches based on hybrid automata models are discussed. Then approaches that emphasize stability are presented. Finally, the supervisory control approach to the analysis and design of hybrid control systems is described. It should be noted that considerable research efforts have addressed additional topics such as optimal control, controllability, observability, and diagnosis. We believe that the research threads that were selected represent the most important topics with mature technical results.

14.3. HYBRID AUTOMATA

Hybrid automata were introduced in the study of hybrid systems in the early 1990s [26]. Hybrid automata provide a general modeling formalism for the formal specification and algorithmic analysis of hybrid systems. They are typically used to model dynamical systems that consist of both discrete and analog components that arise when computer programs interact with the physical world in real time. In the following, we review the hybrid automaton model and related approaches for analysis, verification, and synthesis of hybrid systems [26–29].

A hybrid automaton is a finite-state machine equipped with a set of real-valued variables. The state of the automaton changes either instantaneously through a discrete transition or through a continuous activity. The hybrid automaton in Figure 14.1 describes a thermostat and is used to introduce the modeling framework.

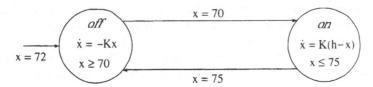

Figure 14.1. Hybrid automaton describing a thermostat.

Example 14.1. The hybrid automaton in Figure 14.1 models a thermostat controlling the temperature of a room by turning a heater on and off. The real-valued variable x denotes the temperature. The system has two control modes, *off* and *on*. When the heater is off, the temperature of the room falls according to the differential equation $\dot{x} = -Kx$. When the heater is on (control mode *on*), the temperature of the system rises according to the equation $\dot{x} = K(h - x)$, where h is a constant. Initially, the temperature is $x = 72$ and the heater is off. The heater will go on as soon as the falling temperature reaches 70°F; the discrete part of the state will then be in the position *on* (Figure 14.1), and the continuous part of the state will start at $x = 70$. When the heater is on, the temperature rises until it reaches 75°F. Then the heater will go off and the temperature will start falling again. This control policy guarantees that the temperature of the room will remain at between 70°F and 75°F.

A hybrid automaton consists of a finite set $X = \{x_1, \ldots, x_n\}$ of real-valued variables and a labeled directed graph (V, E). V is a finite set of vertices and E is a set of directed arcs or edges between vertices. The directed graph models the discrete (event) portion of the hybrid system. Directed graphs have a very convenient graphical representation. A circle is used to represent each vertex of the graph. An arrow starting at vertex v_i and terminating at vertex v_j represents the directed arc (v_i, v_j). The graph shown in Figure 14.1 consists of two vertices and two edges. Note that the arc labeled $x = 72$ is used for the initialization of the system.

The dynamics of the hybrid automaton are defined by labeling the vertices V and edges E of the graph with appropriate mathematical expressions involving the real-valued variables $X = \{x_1, \ldots, x_n\}$. The vertices represent continuous activities, and they are labeled with constraints on the derivatives of the variables in X. More specifically, a vertex $v \in V$ which is also called a (*control*) *mode* or *location* is equipped with the following labeling functions.

- A *flow condition* described by a differential equation in the variables in X. While the hybrid automaton is in control mode v, the variables x_i change according to the flow condition. For example, in the thermostat automaton, the flow condition $\dot{x} = K(h - x)$ of the control mode *on* ensures that the temperature is rising while the heater is on.

- An *invariant condition* $inv(v) \in \mathfrak{R}^n$ that assigns to each control mode a region of $\mathfrak{R}^n$. The hybrid automaton may reside in control mode v only while the invariant condition $inv(v)$ is true. For example, in the thermostat automaton, the invariant condition $x \leq 75$ of the control mode *on* ensures that the heater must go off when the temperature rises to 75°F.

An edge $e \in E$ is also called a *control switch* or *transition* and is labeled with an assignment of the variables in X called a guard. A transition is enabled when the associated guard is true, and its execution modifies the values of the variables according to the assignment. For example, the thermostat automaton has two control switches. The control switch from control mode *on* to *off* is described by the condition $x = 75$.

A *state* $\sigma = (v, x)$ of the hybrid automaton consists of a mode (control location) $v \in V$ and a particular value $x \in \mathfrak{R}^n$ of the variables in X. The state can change either by a discrete and instantaneous transition or by a time delay through the continuous flow. A discrete transition changes both the control location and the real-valued variables, while a time delay changes only the values of the variables in X according to the flow condition. A *run* of a hybrid automaton H is a finite or infinite sequence

$$\rho \colon \sigma_0 \xrightarrow[f_0]{t_0} \sigma_1 \xrightarrow[f_1]{t_1} \sigma_2 \xrightarrow[f_2]{t_2} \cdots$$

where $\sigma_i = (v_i, x_i)$ is the state of H and f_i is the flow condition for the vertex v_i such that (i) $f_i(0) = x_i$, (ii) $f_i(t) \in inv(v_i)$ for all $t \in \mathfrak{R} \colon 0 \leq t \leq t_i$, and (iii) σ_{i+1} is a transition successor of $\sigma_i' = (v_i, f_i(t_i))$, where σ_i' is a time successor of σ_i.

A hybrid automaton is said to be *nonzeno* when only finitely many transitions can be executed in every bounded time interval. Nonzenoness is an important notion for the realizability of the hybrid automaton.

Another labeling function assigns to each transition an event from a finite set of events Σ. The event labels are used to define the parallel composition of hybrid automata. Complex systems can be modeled by using the parallel composition of simple hybrid automata. The basic rule for the parallel composition is that two interacting hybrid automata synchronize the execution of transitions labeled with common events.

Example 14.2. A train-gate-controller system is used to illustrate modeling of hybrid systems using hybrid automata. The system consists of three components—the train, the gate, and the gate controller—as shown in Figure 14.2. A road crosses the train track, and it is guarded by a gate that must be *lowered* to stop the traffic when the train approaches and *raised* after the train has passed the road. The gate controller gets information from sensors located on the track and lowers or raises the gate.

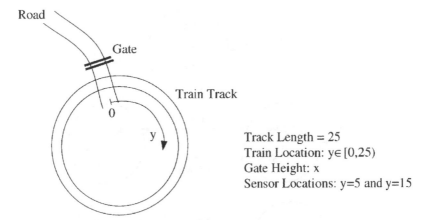

Road

Gate

Train Track

0

y

Track Length = 25
Train Location: y∈ [0,25)
Gate Height: x
Sensor Locations: y=5 and y=15

Figure 14.2. Train–gate–controller system.

The train moves clockwise on a circular track. The length of the track is $L = 25$. The location of the train is indicated by the state variable y, where $0 \leq y < 25$. The velocity of the train is described by the differential equation $\dot{y} = f(y)$, where $f(y)$ is an appropriate function of y. The gate is located at $y = 0$ on the train track, while the sensors are at $y = 5$ and $y = 15$. The train is modeled by a hybrid automaton with one control mode as shown in Figure 14.3.

The height of the gate is represented by the state variable x. When the gate is lowered, the height of the gate decreases according to the equation $\dot{x} = (1 - x)/2$. When the gate is raised, the height increases according to $\dot{x} = (10 - x)/2$. The hybrid automaton in Figure 14.3 is used to model the dynamic behavior of the gate. The automaton has two control modes, *RAISE* and *LOWER*. The transitions of the automaton are labeled with the events *UP* and *DOWN*, which are generated by the controller. The controller is also modeled as a hybrid automaton as shown in Figure 14.3. The controller receives information from the sensors and detects when the train reaches or moves away from the crossing. The controller automaton has two control locations, *DOWN* and *UP*, which trigger the transitions of the gate automaton. The hybrid automaton of the overall system is obtained by parallel composition and is shown in Figure 14.3.

The modeling formalism of hybrid automata is particularly useful in the case when the flow conditions, the invariants, and the transition relations are described by linear expressions in the variables in X. A hybrid automaton is *linear* if its flow conditions, invariants, and transition relations can be defined by linear expressions over the set X of variables. Note the special interpretation of the term linear in this context. More specifically, for the control modes the flow condition is defined by a differential equation of the form $\dot{x} = k$, where k is a constant, one for each variable in X, and the invariant

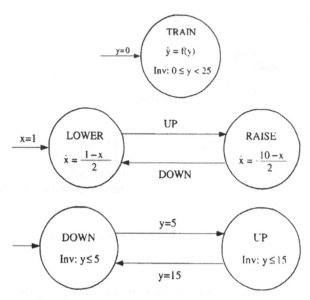

Figure 14.3. Hybrid automata modeling the train, gate, and controller.

$inv(v)$ is defined by linear equalities and inequalities (which correspond to a convex polyhedron) in X. Also, for each transition the set of guarded assignments consists of linear formulas in X, one for each variable. Note that the run of a linear hybrid automaton can be described by a piecewise linear function whose values at the points of first order discontinuity are finite sequences of discrete changes. An interesting special case of a linear hybrid automaton is a *timed automaton*. In a timed automaton, each continuous variable increases uniformly with time (with slope 1) and can be considered a *clock*. A discrete transition either resets the clock or leaves it unchanged.

Another interesting case of a linear hybrid automaton is a rectangular automaton. A hybrid automaton is rectangular if the flow conditions are independent of the control modes and the variables are pairwise independent. In a rectangular automaton, the flow condition has the form $\dot{x} = [a, b]$ for each variable $x \in X$. The invariant condition and the transition relation are described by linear predicates that also correspond to n-dimensional rectangles. Rectangular automata are interesting because the reachability and the controller synthesis problems are decidable under the assumption that the mode transitions occur using sampling at integer points in time [30]. A problem is decidable if there exists an algorithm that has as output the correct answer for every possible input. A problem is undecidable if there is no algorithm that takes as input an instance of the problem and determines whether the answer to that instance is "yes" or "no." *Semi decidable* procedures are often proposed to deal with undecidable problems. These algorithms produce the correct answer if they terminate, but their termination is not guaranteed.

The main decision problem concerning the analysis and verification of hybrid systems is the reachability problem, which is formulated as follows. Let σ and σ' be two states in the infinite state space S of a hybrid automaton H. Then, σ' is reachable from σ if there exists a run of H that starts in σ and ends in σ'.

While the reachability problem is undecidable even for very restricted classes of hybrid automata, two semidecision procedures, forward and backward analysis, have been proposed for the verification of safety specifications of linear hybrid automata. A *data region* R_v is a finite union of convex polyhedra in $\Re^n$ that can be defined using a logical formula with linear predicates [26]. A *region* $R = (v, R_v)$ consists of a location $v \in V$ and a data region R_v and is a set of states of the linear hybrid automaton. Given a region R, the precondition of R, denoted $pre(R)$, is the set of all states σ such that R can be reached from σ. The postcondition of R, denoted $post(R)$, is the set of all the reachable states from R. For linear hybrid automata, both $pre(R)$ and $post(R)$ are regions; that is, the corresponding data region is a finite union of convex polyhedra. Given a linear hybrid automaton H, an initial region R, and a target region T, the reachability problem is concerned with the existence of a run of H that drives a state from R to a state in T. Two approaches for solving the reachability problem have been proposed. The first one computes the region $post^*(R)$ of all states that can be reached from the initial state R and checks if $post^*(R) \cap T = \varnothing$ (forward reachability analysis). The second approach computes the region $pre^*(T)$ of the states from which one may reach T and checks if $pre^*(T) \cap R = \varnothing$ (backward reachability analysis). Since the reachability problem for linear hybrid automata is undecidable, these procedures may not terminate (semidecision procedures). They terminate with a positive answer if T is reachable from R, while they terminate with a negative answer if no new states can be added and T is not reachable from R. The crucial step in these approaches is the computation of the precondition or postcondition of a region.

The reachability problem is central to the verification of hybrid systems. The train–gate–controller example is used to illustrate the verification approach using hybrid automata.

Example 14.3. For the train–gate–controller example, the specification is that the gate must be lowered $(x < 5)$ whenever the train reaches the crossing. This is a safety specification that can be encoded as $y = 0 \Rightarrow x < 5$. This safety specification corresponds to a set S of safe states of the hybrid automaton shown in Figure 14.4, which consists of all four control locations. and the region of $\Re^2$ expressed by the set $\{(x, y): x < 5 \wedge y = 0\}$. To verify that the system satisfies the safety specification, we compute the set of all states R that can be reached from the initial conditions. If the reachable set

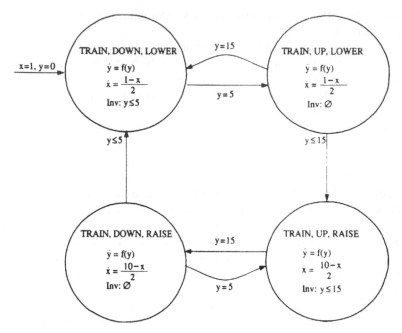

Figure 14.4. Hybrid automaton for the train–gate–controller system.

R is contained in the set of safe states $R \subset S$, then the gate is always down when the train reaches the crossing.

The undecidability of the reachability problem is a fundamental obstacle in the analysis and controller synthesis for linear hybrid automata. Nevertheless, considerable research effort has been focused on developing systematic procedures for synthesizing controllers for large classes of problems [8].

Control design algorithms have been developed for a class of hybrid systems with continuous dynamics described by pure integrators [31]. Although this class of hybrid systems is rather limited, these models are important for several applications including the control of batch processes. Note that even in the case where the continuous dynamics of the physical system are more complicated, it is sometimes useful to use low-level continuous controllers to impose linear ramp-like behavior around a set point.

Motivated by problems in aircraft conflict resolution, methodologies for synthesizing controllers for nonlinear hybrid automata based on a game theoretical framework have also been developed [32]. Another approach uses bisimulations to study the decidability of verification algorithms [8]. Bisimulations are quotient systems that preserve the reachability properties of the original hybrid system; therefore, problems related to the reachability of the original system can be solved by studying the quotient system. Quotient systems are simplified systems derived from the original system by aggregat-

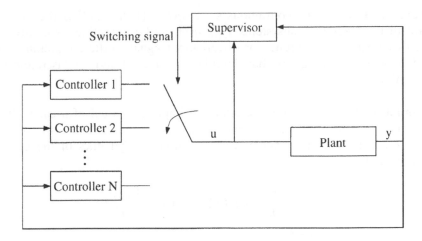

Figure 14.5. Switching controller feedback architecture.

ing the states in an appropriate manner. The idea of using finite bisimulations for the analysis and synthesis of hybrid systems is similar to the approximation of the continuous dynamics with DESs (see discussion in Section 14.5).

14.4. STABILITY AND DESIGN OF HYBRID SYSTEMS

In the area of control systems, powerful methodologies for analysis of properties such as stability and systematic methodologies for the design of controllers have been developed over the years. Some of the methodologies have been extended to hybrid systems, primarily to switched systems [33–35]; see also references 36–38 and the references therein. Switched systems are hybrid dynamical systems that consist of a family of continuous- or discrete-time subsystems and a rule that determines the switching between them. The switching behavior of these systems may be generated by the changing dynamics at different operating regions. Hybrid dynamical systems also arise when switching controllers are used to achieve stability and improve performance as shown in Figure 14.5. Typical examples of such systems are computer disk drives, constrained mechanical systems, switching power converters, and automotive power-train applications.

Mathematically, such hybrid systems can be modeled by the equations

$$\dot{x} = f(x(t), q(t), u(t))$$

$$q(t^+) = \delta(x(t), q(t))$$

where $x(t) \in \mathfrak{R}^n$ is the continuous state, $q(t) \in \{1, 2, \ldots, N\}$ is the discrete state that indexes the subsystems $f_{q(t)}$, $u(t)$ can be a continuous control input or an external (reference or disturbance) signal to the continuous part, and δ is the switching law that describes the logical and/or discrete-event dynamics.

Example 14.4. This example describes a simplified model of a car with an automatic transmission. Let m denote the mass of the car and v the velocity on a road inclined at an angle α. The simplified dynamics of the car are described by

$$\dot{v} = -\frac{k}{m} v^2 \, sign(v) - g \sin \alpha + \frac{G_{q(t)}}{m} T$$

$$\omega = G_{q(t)} v$$

where G_i, $i = 1, 2, 3, 4$, are the transmission gear ratios normalized by the wheel radius R, k is an appropriate constant, ω is the angular velocity of the motor, and T is the torque generated by the engine. The dynamic behavior of the car is indexed by the discrete state q. The discrete-state transition function that determines the switching between the gears is

$$q(t^+) = \begin{cases} i + 1 & \text{if } q(t) = i \neq 4 \text{ and } v = \dfrac{1}{G_i} \omega_{high} \\ i - 1 & \text{if } q(t) = i \geq 2 \text{ and } v = \dfrac{1}{G_i} \omega_{low} \end{cases}$$

where ω_{high} and ω_{low} are prescribed angular velocities of the engine.

Hybrid system stability analysis relies for the most part on classical Lyapunov stability theory. For conventional control systems, demonstrating stability depends on the existence of a continuous and differentiable Lyapunov (energy) function. In the hybrid system case, stability analysis is carried out using multiple Lyapunov functions (MLFs) to compose a single piecewise continuous and piecewise differentiable Lyapunov function that can be used to demonstrate stability. To illustrate the use of MLFs, we consider the autonomous form $[u(t) = 0]$ of the hybrid system model

$$\dot{x}(t) = f(x(t), q(t)) = f_{q(t)}(x(t)) \tag{14.1}$$

where $q(t) \in \{1, 2, \ldots, N\}$. It is also assumed that there are only a finite number of switchings in a bounded time interval. It should be noted that hybrid systems that exhibit infinitely many switchings in a finite interval are

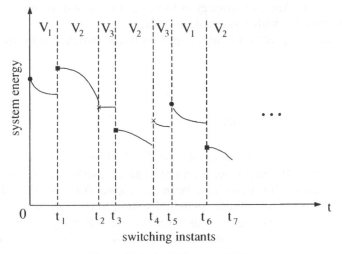

Figure 14.6. Stability condition.

called *Zeno* systems. Furthermore, it is assumed that the switchings occur instantaneously and they do not excite unmodeled high-frequency dynamics.

Consider the family of Lyapunov-like functions $\{V_i, i = 1, 2, \ldots, N\}$, where each V_i is associated with the subsystem $f_i(x)$. A *Lyapunov-like* function for the system $\dot{x} = f_i(x)$ and equilibrium point $\bar{x} \in \Omega_i \subset \Re^n$ is a real-valued function $V_i(x)$ defined over the region Ω_i which is *positive definite* ($V_i(\bar{x}) = 0$ and $V_i(x) > 0$ for $x \neq \bar{x}$, $x \in \Omega_i$) and has *negative semidefinite derivative* (for $x \in \Omega_i$ $\dot{V}_i(x) \leq 0$).

Given system (14.1), suppose that each subsystem f_i has an associate Lyapunov-like function V_i in the region Ω_i, each with equilibrium $\bar{x} = 0$, and suppose that $\bigcup_i \Omega_i = \Re^n$. Let $q(t)$ be a given switching sequence such that $q(t)$ can take on the value i only if $x(t) \in \Omega_i$; in addition,

$$V_i\big(x(t_{i,k})\big) \leq V_i\big(x(t_{i,k-1})\big) \tag{14.2}$$

where $t_{i,k}$ denotes the kth time the subsystem f_i is "switched in." Then system (14.1) is stable in the sense of Lyapunov. The stability condition (14.2) is illustrated in Figure 14.6. At every time instant the subsystem i becomes active, the corresponding energy function V_i decreases from the value it had the last time the subsystem i was switched in.

The general result presented above gives sufficient conditions for stability. Implicitly, this result provides a methodology for switching between subsystems to achieve a stable trajectory. One strategy that may stabilize a hybrid system is to pick that subsystem that causes maximal descent of a particular energy function. Another strategy is to select the subsystem according to the Lyapunov function with the smallest value.

In the following, the emphasis is put on linear switched systems described by

$$\dot{x} = A_{q(t)} x(t)$$

$$q(t^+) = \delta(x(t), q(t))$$

where $q(t) \in \{1, 2, \ldots, N\}$ and $A_{q(t)} \in \Re^{n \times n}$. For this restricted class of hybrid systems, stronger results and systematic methodologies to construct multiple Lyapunov functions have been developed. An important observation is that it is possible for a linear switched system to be unstable even when all the subsystems are stable as illustrated in the following example. On the other hand, it is possible to stabilize a linear switched system even when all the subsystems are unstable.

An important problem is to find conditions that guarantee that the switched system $\dot{x}(t) = A_{q(t)} x(t)$ is stable for any switching signal. This situation is of importance when a given plant is being controlled by switching among a family of stabilizing controllers, each of which is designed for a specific task. A supervisor determines which controller is to be connected in closed loop with the plant at each instant of time. Stability of the switched system can usually be ensured by keeping each controller in the loop long enough to allow the transient effects to dissipate. Another approach that can be used to demonstrate stability for any switching signals is to guarantee that the matrices A_i share a common quadratic Lyapunov function $V(x) = x^T P x$, such that $\dot{V}(x) \leq -x^T Q x, Q > 0$ [Q is positive definite ($Q > 0$) when $x^T Q x > 0$ for any $x \neq 0$]. These conditions on $V(x)$ are equivalent to finding matrices P and Q that satisfy the inequalities $A_i^T P + P A_i + Q \leq 0$ for all i. Note that the existence of a common Lyapunov function, although sufficient, is not necessary for stability.

The application of the theoretical results to practical hybrid systems is accomplished usually using a linear matrix inequality (LMI) problem formulation for constructing a set of quadratic Lyapunov-like functions [39]. The existence of a solution to the LMI problem is a sufficient condition and guarantees that the hybrid system is stable. The methodology begins with a partitioning of the state space into Ω-regions that are defined by quadratic forms. Physical insight, a good understanding of the LMI problem, and brute force are often required to choose an acceptable partitioning. Let Ω_i denote a region where one searches for a quadratic Lyapunov function $V_i = x^T P_i x, x \in \Omega_i$, that satisfies the condition

$$\dot{V}_i(x) = \left[\frac{\partial}{\partial x} V_i(x) \right] A_i x = x^T \left(A_i^T P_i + P_i A_i \right) x \leq 0 \qquad (14.3)$$

The goal is to find matrices $P_i > 0$ that satisfy the above condition. To constrain the stability conditions to local regions, two steps are involved. First, the region Ω_i must be expressed by the quadratic form $x^T Q_i x \geq 0$. Second, a technique called the S-procedure is applied to replace the constrained stability condition by a condition without constraints. By introducing a new unknown variable $\xi \geq 0$, the relaxed problem takes the unconstrained form

$$A_i^T P_i + P_i A + \xi Q_i \leq 0 \tag{14.4}$$

which can be solved using standard LMI software tools. A solution to the relaxed problem (14.4) is also a solution to the constrained problem (14.3). It should be noted that, in general, several subsystems A_i can be used in each Ω-region.

In addition, the LMI formulation requires that whenever there is movement to an adjacent region Ω_j with corresponding Lyapunov function V_j, then $V_j(x) \leq V_i(x)$. Using local quadratic Lyapunov-like functions, this condition can be written $x^T P_j x \leq x^T P_i x$. The states where this condition must be satisfied also have to be expressed by quadratic forms. The S-procedure is used to replace the constrained condition with an unconstrained LMI problem that can be solved very efficiently.

Nonquadratic Lyapunov functions have also been used for studying the stability of switched linear systems [40]. For example, Lyapunov functions defined by the infinity norm result in polyhedral partitions of the state space that can be used very efficiently for the stability analysis and verification of linear switched systems.

14.5. SUPERVISORY CONTROL OF HYBRID SYSTEMS

In the 1980s, systems with discrete dynamics such as manufacturing systems attracted the attention of the control research community, and models such as finite automata were used to describe such discrete-event dynamical systems. Important system properties such as controllability, observability, and stability were defined and studied for discrete-event systems, and methodologies for supervisory control design were developed [41–44]. In related developments, the relation between inherently discrete planning systems and continuous feedback control systems attracted attention [45]. In addition to finite automata, other modeling paradigms such as Petri nets gained the attention of control and automation system researchers in the last decade, primarily in Europe. Petri nets have been used in the supervisory control of DESs as an attractive alternative to methodologies based on finite automata [6, 47].

In this section, we review the supervisory control framework for hybrid systems [48–54]. One of the main characteristics of the supervisory control

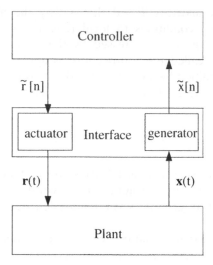

Figure 14.7. Hybrid system model in the supervisory control framework.

approach is that the system to be controlled is approximated by a DES, and the design is carried out in the discrete domain. The hybrid control systems in the supervisory control framework consist of a continuous (state, variable) system to be controlled, also called the plant, and a discrete-event controller connected to the plant via an interface in a feedback configuration as shown in Figure 14.7. It is generally assumed that the dynamic behavior of the plant is governed by a set of known nonlinear ordinary differential equations,

$$\dot{x}(t) = f(x(t), r(t))$$

where $x \in \Re^n$ is the continuous state of the system and $r \in \Re^m$ is the continuous control input. In the model shown in Figure 14.7, the plant contains all continuous components of the hybrid control system, such as any conventional continuous controllers that may have been developed, a clock if time and synchronous operations are to be modeled, and so on. The controller is an event-driven, asynchronous DES, described by a finite-state automaton. The hybrid control system also contains an interface that provides the means for communication between the continuous plant and the DES controller.

The interface consists of the generator and the actuator as shown in Figure 14.7. The generator has been chosen to be a partitioning of the state space (see Figure 14.8). The piecewise continuous command signal issued by the actuator is a staircase signal not unlike the output of a zero-order hold in a digital control system. The interface plays a key role in determining the dynamic behavior of the hybrid control system. Many times the partition of the state space is determined by physical constraints, and it is fixed and

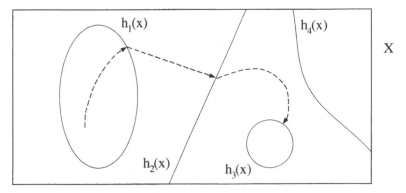

Figure 14.8. Partition of the continuous state space.

given. Methodologies for the computation of the partition based on the specifications have also been developed.

In such a hybrid control system, the plant, taken together with the actuator and generator, behaves like a DES; it accepts symbolic inputs via the actuator and produces symbolic outputs via the generator. This situation is somewhat analogous to the way a continuous-time plant, equipped with a zero-order hold and a sampler, "looks" like a discrete-time plant. The DES that models the plant, actuator, and generator is called the DES plant model. From the DES controller's point of view, it is the DES plant model which is controlled.

The DES plant model is an approximation of the actual system, and its behavior is an abstraction of the system's behavior. As a result, the future behavior of the actual continuous system cannot be determined uniquely, in general, from knowledge of the DES plant state and input. The approach taken in the supervisory control framework is to incorporate all the possible future behaviors of the continuous plant into the DES plant model. A conservative approximation of the behavior of the continuous plant is constructed and realized by a finite-state machine. From a control point of view, this means that if undesirable behaviors can be eliminated from the DES plant (through appropriate control policies), then these behaviors will be eliminated from the actual system. On the other hand, just because a control policy permits a given behavior in the DES plant is no guarantee that the behavior will occur in the actual system.

We briefly discuss the issues related to the approximation of the plant by a DES plant model. A *dynamical system* Σ can be described as a triple (T, W, B) with $T \subseteq \Re$ the *time axis*, W the *signal space*, and $B \subset W^T$ (denoting the set of all functions $f : T \to W$) the *behavior*. The behavior of the DES plant model consists of all the pairs of plant and control symbols that it can generate. The time axis T represents here the occurrences of events. A necessary condition for the DES plant model to be a valid approximation

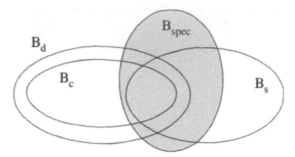

Figure 14.9. The DES plant model as an approximation.

of the continuous plant is that the behavior of the continuous plant model B_c is contained in the behavior of the DES plant model, that is, $B_c \subseteq B_d$.

The main objective of the controller is to restrict the behavior of the DES plant model in order to specify the control specifications. The specifications can be described by a behavior, B_{spec}. Supervisory control of hybrid systems is based on the fact that if undesirable behaviors can be eliminated from the DES plant, then these behaviors can likewise be eliminated from the actual system. This is described formally by the relation

$$B_d \cap B_s \subseteq B_{spec} \Rightarrow B_c \cap B_s \subseteq B_{spec}$$

and is depicted in Figure 14.9. The challenge is to find a discrete abstraction with behavior B_d which is a approximation of the behavior B_c of the continuous plant and for which it is possible to design a supervisor to guarantee that the behavior of the closed loop system satisfies the specifications B_{spec}. A more accurate approximation of the plant's behavior can be obtained by considering a finer partitioning of the state space for the extraction of the DES plant.

An interesting aspect of the DES plant's behavior is that it is distinctly nondeterministic. This fact is illustrated in Figure 14.10, which shows two trajectories generated by the same control symbol. Both trajectories originate in the same DES plant state $\tilde{p}_1$. Figure 14.10 shows that for a given control symbol, there are at least two possible DES plant states that can be reached from $\tilde{p}_1$. Transitions within a DES plant will usually be nondeterministic unless the boundaries of the partition sets are invariant manifolds with respect to the vector fields that describe the continuous plant [55].

There is an advantage to having a hybrid control system in which the DES plant model is deterministic. It allows the controller to drive the plant state through any desired sequence of regions, provided, of course, that the corresponding state transitions exist in the DES plant model. If the DES plant model is not deterministic, this will not always be possible. This is because even if the desired sequence of state transitions exists, the sequence of inputs which achieves it may also permit other sequences of state transi-

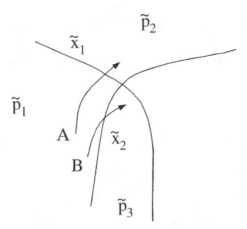

Figure 14.10. Nondeterminism of the DES plant model.

tions. Unfortunately, given a continuous-time plant, it may be difficult or even impossible to design an interface that leads to a DES plant model which is deterministic. Fortunately, it is not generally necessary to have a deterministic DES plant model in order to control it. The supervisory control problem for hybrid systems can be formulated and solved when the DES plant model is nondeterministic.

A language theoretic framework to describe performance specifications for hybrid systems and to formulate the supervisory control problem has been developed. Once the DES plant model of a hybrid system has been extracted, a supervisor can be designed using control synthesis techniques based on DES. The main differences are that the DES plant models of the hybrid control framework are nondeterministic and that the plant events cannot be disabled individually.

Example 14.5. The hybrid system in this example consists of a typical thermostat and furnace. Assuming that the thermostat is set at 70°F, the system behaves as follows. If the room temperature falls below 70°F, the furnace starts and remains on until the room temperature reaches 75°F. At 75°F, the furnace shuts off. For simplicity, we assume that when the furnace is on, it produces a constant amount of heat per unit time.

The plant in the thermostat/furnace hybrid control system is made up of the furnace and room. It can be modeled with the following differential equation

$$\dot{x} = 0.0042(T_0 - x) + 0.1r$$

where the plant state, x, is the temperature of the room in degrees Fahrenheit, the input, r, is the voltage on the furnace control circuit, and T_0 is the outside temperature. This model of the furnace is certainly a simplification, but it is adequate for this example.

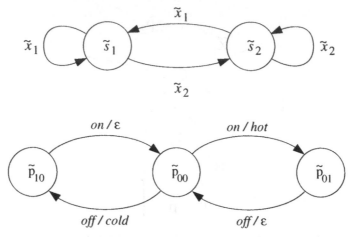

Figure 14.11. Controller and DES plant for the thermostat/furnace system.

The thermostat partitions the state space of the plant with the following hypersurfaces:

$$h_1(x) = x - 75$$
$$h_2(x) = 70 - x$$

The first hypersurface detects when the state exceeds 75°F, and the second detects when the state falls below 70°F. The associated functions with the interface that generate the plant events, α_1 and α_2, are very simple in this case:

$$\alpha_i(x) = \tilde{x}_i$$

So there are two plant symbols, $\tilde{x}_1$ and $\tilde{x}_2$. The DES controller is shown in Figure 14.11. The output function of the controller is defined as

$$\phi(\tilde{s}_1) = \tilde{r}_1 \Leftrightarrow \text{off}$$
$$\phi(\tilde{s}_2) = \tilde{r}_2 \Leftrightarrow \text{on}$$

and the actuator operates as

$$\gamma(\tilde{r}_1) = 0$$
$$\gamma(\tilde{r}_2) = 12$$

where the constants for the control inputs correspond to particular given data.

The thermostat/heater example has a simple DES plant model that is useful to illustrate how these models work. Figure 14.11 shows the DES plant model for the heater/thermostat. The convention for labeling the arcs is to list the controller symbols that enable the transition, followed by a slash and then the plant symbols that can be generated by the transition. Notice that two of the transitions are labeled with null symbols, ϵ. This reflects the fact that nothing actually happens in the system at these transitions. When the controller receives a null symbol, it remains in the same state and reissues the current controller symbol. This is equivalent to the controller doing nothing, but it serves to keep all the symbolic sequences, $\tilde{s}, \tilde{p}$, and so on, in phase with each other.

14.6. CONCLUSION

This chapter presents a brief overview of three research threads in hybrid systems, namely, hybrid-automata-based modeling and verification, stability analysis, and supervisory control. It should be noted that considerable progress has been achieved in related areas such as analysis and synthesis of piecewise linear systems, optimal control of switched systems, and hybrid system diagnosis, among others. The research efforts in the area of hybrid dynamical systems address many important challenges, such as real-time control reconfiguration, mode switching, safety, reachability, and liveness, and provide the theoretical background and the principles for the development of software-enabled control technologies. Although many important research developments were omitted, it is hoped that this chapter will provide a useful and representative description of main approaches to hybrid systems together with references, and as such it will be a useful resource to researchers. In summary, although many important problems related to hybrid systems are intrinsically difficult, there are efficient simulation, analysis, and synthesis algorithms for large classes of systems. Recent research efforts toward hybrid system design have shown that there are classes of hybrid systems for which computationally tractable procedures can be applied. Many practical applications can be modeled accurately enough by suitable hybrid models. Again, the choice of such models depends on their suitability for studying specific problems.

REFERENCES

1. A. Balluchi, L. Benvenuti, M. D. Benedetto, C. Pinello, and A. Sangiovanni-Vincentelli, Automotive engine control and hybrid systems: Challenges and opportunities, *Proceedings of IEEE* **88**(7):888–912, July 2000.
2. S. Engell, S. Kowalewski, C. Schulz, and O. Stursberg, Continuous–discrete interactions in chemical processing plants, *Proceedings of IEEE* **88**(7):1050–1068, July 2000.
3. M. Song, T. -J. Tarn, and N. Xi, Integration of task scheduling, action planning and control in robotic manufacturing systems, *Proceedings of IEEE* **88**(7):1097–1107, July 2000.

4. D. Pepyne and C. Cassandras, Optimal control of hybrid systems in manufacturing, *Proceedings of IEEE* **88**(7):1108–1123, July 2000.

5. R. Horowitz and P. Varaiya, Control design of an automated highway system, *Proceedings of IEEE* **88**(7):913–925, July 2000.

6. C. Livadas, J. Lygeros, and N. Lynch, High-level modeling and analysis of the traffic alert and collision avoidance system, *Proceedings of IEEE* **88**(7):926–948, July 2000.

7. P. Antsaklis, K. Passino, and S. Wang, Towards intelligent autonomous control systems: Architecture and fundamental issues, *Journal of Intelligent and Robotic Systems* **1**:315–342, 1989.

8. R. Alur, T. Henzinger, G. Lafferriere, and G. Pappas, Discrete abstractions of hybrid systems, *Proceedings of IEEE* **88**(7):971–984, July 2000.

9. A. Morse, Logic-based switching and control, in B. A. Francis and A. R. Tannenbaum, editors, *Feedback Control, Systems, and Complexity*, Springer-Verlag, New York, 1995, pp. 173–195.

10. A. S. Morse, editor, *Control Using Logic-Based Switching*, Lecture Notes in Control and Information Sciences Vol. 222, Springer-Verlag, New York, 1997.

11. J. Taylor and D. Kebede, Modeling and simulation of hybrid systems in Matlab, in *IFAC 13th Triennial World Congress*, Vol. J, San Francisco, CA, 1996, pp. 275–280,

12. P. Mosterman, An overview of hybrid simulation phenomena and their support by simulation packages, in *HSCC 99: Hybrid Systems—Computation and Control*, Lecture Notes in Computer Science, Vol. 1569, Springer-Verlag, New York, 1999, pp. 165–177.

13. J. Liu, X. Liu, T. Koo, J. Sinopoli, S. Sastry, and E. Lee, A hierarchical hybrid system model and its simulation, in *Proceedings of the 38th IEEE Conference on Decision and Control*, pp. 3508–3513, Phoenix, AZ, December 1999.

14. Matlab, The Mathworks, Inc., Homepage: http://www.mathworks.com.

15. Ptolemy II, Department of EECS, UC Berkeley, Homepage: http://ptolemy.eecs. berkeley.edu/ptolemyII/.

16. E. Lee, Overview of the ptolemy project, Technical Memorandum UCB/ERL M01/11, University of California, March 6, 2001.

17. Dymola, Dynasim AB, Homepage: http://www.dynasim.se/.

18. Mathmodelica, MathCore, Homepage: http://www.mathcore.com.

19. Hybrid CC, hybrid automata, and program verification, Homepage: http://www. parc.xerox.com/spl/projects/mbc/languages.html.

20. Shift, California PATH, UC Berkeley, Homepage: http://wwww.path.berkeley. edu/shift/.

21. OmSim, Department of Automatic Control, Lund Institute of Technology, Homepage: http://wwww.control.lth.se/cace/omsim.html.

22. Charon, Department of Computer and Information Science, University of Pennsylvania, Homepage: http://www.cis.upenn.edu/mobies/charon/.

23. HyTech, Homepage: http://wwww-cad.eecs.berkeley.edu/tah/HyTech/.

24. Kronos, VERIMAG, Homepage: http://www-verimag.imag.fr/ TEMPORISE/kronos/.

25. UPPAAL, Homepage: http://www.docs.uu.se/docs/rtmv/uppaal/.

26. R. Alur, C. Courcoubetis, N. Halbwachs, T. Henzinger, P.-H. Ho, X. Nicollin, A. Oliveiro, J. Sifakis, and S. Yovine, The algorithmic analysis of hybrid systems, *Theoretical and Computer Science* **138**:3–34, 1995.

27. T. Henzinger, P.-H. Ho, and H. Wong-Toi, A user guide to HyTech, in *First Workshop on Tools and Algorithms for the Construction and Analysis of Systems*, *TACAS95, Lecture Notes in Computer Science*, Vol. 1019, Springer-Verlag, New York, 1995, pp. 41–71.

28. N. Lynch, R. Segala, F. Vaandrager, and H. Weinberg, Hybrid I/O automata, in R. Alur, T. A. Henzinger, and E. D. Sontag, editors, *Hybrid Systems III, Verification and Control, Lecture Notes in Computer Science* Vol. 1066, Springer-Verlag, New York, 1996, pp. 496–510.

29. M. Lemmon, K. He, and I. Markovsky, Supervisory hybrid systems, *Control Systems Magazine* **19**(4):42–55, August 1999.

30. T. Henzinger and P. Kopke, Discrete-time control for rectangular hybrid automata, *Theoretical Computer Science* **221**:369–392, 1999.

31. M. Tittus and B. Egardt, Control design for integrator hybrid system, *IEEE Transactions on Automatic Control* **43**(4):491–500, 1998.

32. C. Tomlin, J. Lygeros, and S. Sastry, A game theoretic approach to controller design for hybrid systems, *Proceedings of IEEE* **88**(7):949–970, July 2000.

33. M. Branicky, Multiple Lyapunov functions and other analysis tools for switched and hybrid systems, *IEEE Transactions on Automatic Control* **43**(4):475–482, 1998.

34. H. Ye, A. Michel, and L. Hou, Stability theory for hybrid dynamical systems, *IEEE Transactions on Automatic Control* **43**(4):461–474, 1998.

35. A. Michel and B. Hu, Towards a stability theory of general hybrid dynamical systems, *Automatica* **35**(3):371–384, 1999.

36. D. Liberzon and A. Morse, Basic problems in stability and design of switched systems, *IEEE Control Systems Magazine* **19**(5):59–70, October 1999.

37. A. Michel, Recent trends in the stability analysis of hybrid dynamical systems, *IEEE Transactions on Circuits and Systems I* **46**(1):120–134, 1999.

38. R. DeCarlo, M. Branicky, S. Pettersson, and B. Lennartson, Perspectives and results on the stability and stabilizability of hybrid systems, *Proceedings of IEEE* **88**(7):1069–1082, July 2000.

39. M. Johansson and A. Rantzer, Computation of piecewise quadratic Lyapunov functions for hybrid systems, *IEEE Transactions on Automatic Control* **43**(4):555–559, 1998.

40. X. Koutsoukos and P. Antsaklis, Characterizing of switching stabilizing sequences in switched linear systems using piecewise linear Lyapunov functions, in M. D. Benedetto and A. Sangiovanni-Vincentelli, editors, *Hybrid Systems—Computation and Control 2001, Lecture Notes in Computer Science* Vol. 2034, Springer-Verlag, New York, 2001, pp. 347–360.

41. P. Ramadge and W. Wonham, The control of discrete event systems, *Proceedings of the IEEE*, **77**(1):81–89, January 1989.

42. C. Özveren and A. Willsky, Observability of discrete event dynamic systems, *IEEE Transactions on Automatic Control* **35**(7):797–806, 1990.

43. C. Özveren, A. Willsky, and P. Antsaklis, Stability and stabilizability of discrete event dynamic systems, *Journal of the ACM* **38**(3):730–752, 1991.

44. K. Passino, A. Michel, and P. Antsaklis, Lyapunov stability of a class of discrete event systems, *IEEE Transactions on Automatic Control* **39**(2):269–279, 1994.

45. P. Antsaklis and K. Passino, editors, *An Introduction to Intelligent and Autonomous Control*, Kluwer Academic Publishers, 1993.

46. T. Murata, Petri nets: Properties, analysis and applications, *Proceedings of IEEE* **77**(4):541–580, 1989.

47. K. Yamalidou, J. Moody, M. Lemmon, and P. Antsaklis, Feedback control of Petri nets based on place invariants, *Automatica* **32**(1):15–28, 1996.

48. A. Nerode and W. Kohn, Models for hybrid systems: Automata, topologies, controllability, observability, in R. L. Grossman, A. Nerode, A. P. Ravn, and H. Rischel, editors, *Hybrid Systems, Lecture Notes in Computer Science*, Vol 736, Springer-Verlag, New York, 1993, pp. 317–356.

49. P. Antsaklis, J. Stiver, and M. Lemmon, Hybrid system modeling and autonomous control systems, in R. L. Grossman, A. Nerode, A. P. Ravn, and H. Rischel, editors, *Hybrid Systems, Lecture Notes in Computer Science*, Vol. 736, Springer-Verlag, New York, 1993, pp. 366–392.

50. J. Lunze, Qualitative modelling of linear dynamical systems with quantised state measurements, *Automatica* **30**(3):417–431, 1994.

51. J. Stiver, P. Antsaklis, and M. Lemmon, A logical DES approach to the design of hybrid control systems. *Mathl. Comput. Modelling* **23**(11/12):55–76, 1996.

52. J. Raisch and S. O'Young, Discrete approximation and supervisory control of continuous systems, *IEEE Transactions on Automatic Control* **43**(4):568–573, 1998.

53. J. Lunze, B. Nixdorf, and J. Schroder, Deterministic discrete-event representations of linear continuous-variable systems, *Automatica* **35**(3):396–406, 1999.

54. X. Koutsoukos, P. Antsaklis, J. Stiver, and M. Lemmon, Supervisory control of hybrid systems, *Proceedings of IEEE* **88**(7):1026–1049, July 2000.

55. J. Stiver, X. Koutsoukos, and P. Antsaklis, An invariant based approach to the design of hybrid control systems, *International Journal of Robust and Nonlinear Control* **11**(5):453–478, 2001.

56. Modelica, Homepage: http://www.modelica.org.

CHAPTER 15

A MANEUVER-BASED HYBRID CONTROL ARCHITECTURE FOR AUTONOMOUS VEHICLE MOTION PLANNING

EMILIO FRAZZOLI, MUNTHER A. DAHLEH, and ERIC FERON

Editors' Notes

In the control community, motion planning for vehicles such as autonomous aircraft is usually considered a continuous design problem. We attempt to find trajectories over time for the vehicle state that satisfy certain constraints and possibly optimize a cost function. In this chapter, however, motion planning is seen as a hybrid system design problem. The authors' approach consists of first defining (offline) equilibrium trim trajectories and maneuvers that effect transitions between the trim trajectories, and then composing an overall motion plan by the online scheduling of appropriate trim trajectories and maneuvers.

This results in the development of a hybrid maneuver automaton, a dynamical system in which the states represent motion primitives and the control input corresponds to switches between primitives. The definition of the motion primitive library exploits the symmetry properties of the system and ensures that flight envelope and saturation constraints are not violated. A variety of methods can be used, including, for maneuvers, the analysis of experimental data, or the recording of a human pilot's actions.

The online planning problem can be formulated as an optimal control problem—a dynamic program in a hybrid state space. Solution methods that can be used include approximate dynamic programming algorithms. Thus the hybrid maneuver automaton architecture lends itself to computationally efficient online motion planning, although the solutions derived are not globally optimal.

The authors demonstrate their approach on an experimental three-degree-of-freedom tethered laboratory "helicopter." The online motion planning is performed in real-time on a 300 MHz Pentium II PC which also executes a low-level feedback controller.

Software-Enabled Control. Edited by Tariq Samad and Gary Balas
ISBN 0-471-23436-2 © 2003 IEEE

15.1. INTRODUCTION

This chapter focuses on one of the basic problems that must be faced and solved by autonomous vehicles—that is, the generation and execution of a motion plan, aimed at moving the vehicle from its initial location in space to a given target location, to accomplish a desired task. It is desired that the motion planning algorithm provide safety and possibly performance guarantees; moreover, in realistic situations, the motion planning problem must be solved in real time, using limited on-board computational resources. The fulfillment of the mission objectives might also require the exploitation of the full maneuvering capabilities of the vehicle.

The real-time interaction that occurs between the physical components and the software components of a system such as an autonomous vehicle creates a new set of challenging problems which have attracted the attention of both the computer science and the systems and control communities. In particular, we are concerned with systems which evolve on a state space that includes both continuous dynamics (i.e., the physical component of system) and a discrete logic component (i.e., the software component). These systems are commonly referred to as *hybrid systems*.

Hybrid control systems have been the object of a very intense and productive research effort in the recent years, which has resulted in the definition of very general frameworks (for example, see references 1–3 and references therein). Using these general frameworks, methods for analyzing the properties of hybrid systems, such as well-posedness and stability, and for solving optimal control problems have been developed. However, the applicability of such methods for nontrivial real-time applications is still limited. This can be seen as a consequence of the fact that very little structure is imposed on the system's dynamics, to preserve the generality of the model. The main limitation of such approaches is the rapid explosion of the computation requirements as the number of dimensions of the state space of the system increases. Stronger results have been obtained for limited classes of systems, which possess a relatively simple structure [4]. Unfortunately, these results cannot be readily extended to systems with the rich dynamics of aerospace vehicles or other complex systems, at least at the desired levels of performance.

On the other hand, it should be recognized that the dynamics of most vehicles are inherently continuous, and that the introduction of discrete logic is a *design choice* in the development of the flight control software. This choice is most often linked to hierarchical designs, in which a set of low-level control laws, driving directly the vehicle's actuators, is defined for each mode of operation of the vehicle; a higher level control logic is responsible for switching among the available modes, and the corresponding control laws [5–7]. The introduction of a hierarchical structure and discrete logic is aimed at a simplification of the control synthesis task, but often complicates the analysis of the behavior of the complete system. As a matter of fact, the verification of such systems is often based on extensive simulations.

An alternative approach is to exploit the degrees of freedom available in the control design process to build a hybrid system in such a way that it offers safety and performance guarantees *by design*, at least in an idealized situation, as defined by a nominal or clearly identified model of the vehicle and its environment. In other words, instead of imposing an arbitrary hierarchical structure on the system, which could result in a very difficult analysis problem, it is desirable to exploit the designer's insight into the nominal behavior of the system. Consequently, the analysis and verification problems of the hybrid system are translated to a robustness analysis problem, which can be addressed using the relevant tools from systems and control theory [8–10].

The technical approach that we propose is based on the fact that the dynamics of many systems of interest, including aerospace vehicles, exhibit a very rich structure: The definition of our control framework is based on the exploitation of such structure to develop new tools and methods for real-time software design, and of new tools to prove the correctness of the complete system, including its physical and software components.

In reference 11 an architecture for motion planning was introduced that is based on the offline definition of a library of "maneuvers," or more precisely motion primitives (and corresponding feedback control laws), and on the online scheduling of such primitives to achieve the mission objectives. This leads to the development of a hybrid "maneuver automaton," a new kind of dynamical system where the states represent the motion primitives being executed by the system, and the control input corresponds to switches between such primitives. The definition of such a maneuver automaton ensures that all the trajectories generated by such a dynamical system will satisfy all the constraints on the vehicle's behavior—including dynamical, flight envelope, and control saturation constraints.

The practical implementation of the maneuver automaton architecture is made possible by the capabilities of modern computers, in terms of both storage (e.g., maneuver library) and, to a lesser extent, computing power. The resulting motion planning algorithms, defined on what can be called a "maneuver space," are computationally efficient and of general applicability, and the architecture provides the means to obtain guarantees on the behavior of the complete software-driven dynamical system.

Due to space limitations, it will not be possible to discuss in this chapter all the details and the characteristics of a vehicle control system based on the maneuver automaton architecture, and we will limit ourselves to the presentation of (a) its main features and (b) its usage for motion planning in an obstacle-free environment. Furthermore, we will present the results of a simple experiment on a three-degree-of-freedom helicopter setup. More details on the maneuver automaton architecture and its system-theoretic characteristics, together with a more in-depth exploration of its applicability to aerospace vehicles of interest and a discussion of more general motion planning algorithms, are available in reference 12.

The chapter is organized as follows: In Section 15.2 we briefly introduce the class of dynamical systems we are interested in, and discuss some of their characteristics. In Section 15.3 we formulate the problem we want to address in this chapter. In Section 15.4 we finally introduce the maneuver automaton architecture, and in Section 15.5 we present corresponding algorithms for motion planning in a free environment. An example is presented next, including some experimental results. Some remarks on future work conclude the chapter.

15.2. SYSTEM DYNAMICS

The most usual representation of the dynamics of a vehicle is based on a set of ordinary differential equations (ODEs), of the form

$$\dot{x}(t) = f(x(t), u(t)) \tag{15.1}$$

where x is the state of the vehicle, u represents the control input, and the dot represents differentiation with respect to time. We will assume that $x(t)$ belongs to an n-dimensional manifold $\mathcal{X}$, and $u(t)$ belongs to a set $\mathcal{U} \subseteq \mathbb{R}^m$, where m is the number of available independent control inputs.

The form of the model (15.1) implies time-invariance of the system's dynamics. Under suitable technical conditions—usually satisfied in the case of vehicles—the ODE (15.1) can be integrated; starting from some initial condition $x(t_0) = x_0$, and under the action of a given input signal $u: t \mapsto u(t)$, the state at time t can be written as $x(t) = \phi_u(t, x_0)$. The function ϕ_u represents the *state flow* of the system and is also known as the *transition function*.

The dimension of the state space of aerospace vehicles of practical interest is at least twelve, when vehicles are modeled as simple rigid bodies. In this case, the vehicle configuration is described by an element g of the (six-dimensional) group of rigid displacements $SE(3)$ [13]. The rate of change of the vehicle's configuration, including linear velocities and angular rates, is described by a six-dimensional vector ξ, belonging to the Lie algebra $\mathfrak{se}(3)$. The full state of the vehicle, modeled as a rigid body, is then given by $x = (g, \xi)$. (From now on we will omit the explicit dependence of the state on time.) In several cases of interest, the rigid-body model is not enough for a detailed description of the vehicle's dynamics, and higher-dimensional models are needed. This is true, for example, when it is necessary to model the dynamics of the actuators, of the engine, or of the vehicle's structure; in this case we can append the required state variables to the vehicle's configuration and rates, to get $x = (g, \xi, z)$. Using the language of matrix Lie groups [13],

Eq. (15.1) can be rewritten as

$$\begin{cases} \dot{g} = g\xi \\ \dot{\xi} = \Xi(g, \xi, z, u) \\ \dot{z} = Z(g, \xi, z, u) \end{cases} \tag{15.2}$$

The model describing the system dynamics is usually high-dimensional and highly nonlinear. Nonlinearities cannot be neglected in many cases, especially when it is desired to exploit the vehicle's maneuvering capabilities, without restricting its operations to a neighborhood of an equilibrium condition. This makes the solution of general steering problems intractable from the computational point of view.

On the other hand, most vehicles exhibit a fundamental property, which can be exploited to simplify the motion planning problem. This is the property of symmetry: Roughly speaking, a symmetry is a group action on the state under which the dynamics are invariant. More precisely, consider a group $H \subseteq SE(3)$ and define its left action Ψ on the state $x = (g, \xi, z)$ in the following way: $\Psi: H \times \mathcal{X} \to \mathcal{X}$, $\Psi(h, (g, \xi, z)) = (hg, \xi, z)$. In the following, we will use the notation $\Psi_h(\cdot)$ as a shorthand for $\Psi(h, \cdot)$. We say that H is a (spatial) symmetry group if the state flow $\phi_u(t, \cdot)$ and the group action Ψ_h commute that is, if

$$\Psi_h \circ \phi_u(t, x_0) = \phi_u(t, \Psi_h(x_0))$$

for all $h \in H$, $x_0 \in \mathcal{X}$, $t \in \mathbb{R}$ and for all control signals $u: \mathbb{R} \to \mathcal{U}^*$.

The presence of symmetries in the system's dynamics yields powerful methods for simplifying the analysis of the system's behavior [14]. Nonetheless, it has been argued [15] that symmetries have not been used as much in control theory as they have in physics and mechanics (with some notable exceptions, e.g., references 16 and 17). The architecture we propose in this chapter is based on the exploitation of the symmetries and corresponding relative equilibria in the system's dynamics.

15.3. PROBLEM FORMULATION

Given the initial condition $x(0) = x_0$ and the desired final configuration x_f, the basic problem we need to solve is the generation of an input signal $u: t \mapsto u(t)$ and the corresponding state trajectory $x(t) = \phi_u(t, x_0)$, such that $x(t_f) = \phi_u(t_f, x_0) = x_f$ for some $t_f > 0$. For the purposes of this chapter, we assume that the system behaves exactly as described by the model (15.1), and there are no external disturbances.

---

*An additional symmetry group for time-invariant systems is translation in time; a system is said to be time-invariant if translation in time commutes with the state flow.

In many cases of interest, the choice of possible trajectories and input signals is limited. Constraints are imposed on the evolution of the state of the system—for example, to ensure safe operation of the vehicle (e.g., the flight envelope constraints for aircraft, speed limits for ground vehicles). Moreover, actuators are subject to saturation limits. Such constraints can typically be encoded by inequalities of the form

$$F(x(t), u(t)) \leq 0 \qquad (15.3)$$

where F can be a vector of constraints, and the inequality should be read componentwise.

Flight envelope constraints share some of the fundamental properties of the dynamical system, namely its symmetries; we make the assumption that the constraints (15.3) are invariant with respect to group actions on the symmetry group, that is,

$$F(x, u) = F(\Psi_h(x), u), \qquad \forall h \in H$$

Finally, for most practical applications, it is desirable to characterize the quality of the solution according to a meaningful performance measure, usually referred to as "cost" of the generated trajectory. In these cases, the nature of the motion planning problem moves from that of a simple *feasibility* problem to that of an *optimization* problem. A common way to measure the performance of a motion plan is based on the definition of a cost functional on the state-control trajectory, as follows:

$$J(x(\cdot), u(\cdot)) := \int_{t_0}^{t_f} \gamma(x(t), u(t)) \, dt \qquad (15.4)$$

The objective of an efficient motion planning algorithm will then be the minimization of the cost of the computed solution, according to (15.4). Hence we can state the problem:

Problem 15.1 (Optimal Motion Planning in a Free Environment). *Given a dynamical system described by the ODE* (15.1), *the initial and final conditions* $x_0, x_f \in \mathcal{X}$ *and a cost functional* $J(x(\cdot), u(\cdot))$, *find a control input signal* $u: t \mapsto u(t)$ *such that* $x_f = \phi_u(t_f, x_0)$, $J(\phi_u(\cdot, x_0), u(\cdot))$ *is a global minimum for J, and* $F(\phi_u(t, x_0), u(t)) \leq 0$ *for all* $t \in [t_0, t_f]$.

We will restrict our discussion to a particular class of cost functions, which share the symmetry properties of the system. Formally, we require that the incremental cost $\gamma(x, u)$ be invariant with respect to group actions on the symmetry group H, that is,

$$\gamma(x, u) = \gamma(\Psi_h(x), u), \qquad \forall h \in H \qquad (15.5)$$

Examples of optimization problems that meet this requirement include minimum time, minimum path length, and minimum control effort problems.

In this paper, we will concentrate on the solution of Problem 15.1. As shown in reference 18, a motion planning algorithm in a free environment can be used as the main building block to efficiently solve more complex problems involving time-varying obstacle avoidance and integral constraints.

15.4. MANEUVER AUTOMATON

The approach that we propose to reduce the computational complexity of the motion planning problem for a constrained, nonlinear, high-dimensional system is based on a quantization of the system dynamics. By quantization of the system dynamics we mean that we restrict the feasible nominal system trajectories to the family of time-parameterized curves that can be obtained by the interconnection of appropriately defined primitives. In this sense, we are not introducing a state, control, or time discretization, but rather we want to reduce the complexity of motion planning problems by quantizing (a) the possible choices to be made at each decision step and (b) the resulting trajectories. Such quantization should preserve the desirable characteristics of the system's dynamics, and possibly approximate the quality of the trajectories achievable through the solution to the full optimal control problem. The main task in developing such a quantization is to define and select its building blocks.

15.4.1. Motion Primitives

The class of systems to which our approach applies is characterized by the existence of (nontrivial) symmetries and relative equilibria. This class is very large and includes many systems of practical interest. As a matter of fact, the dynamics of most human-built vehicles, including bicycles, automobiles, aircraft, ships, and so on, under fairly reasonable assumptions—such as homogeneous and isotropic atmosphere, constant gravity acceleration—are invariant with respect to translations and with respect to rotation about a vertical axis (i.e., an axis parallel to the direction of the local gravitational field). In such a case the symmetry group H corresponds to $H = SE(2) \times \mathbb{R} \cong \mathbb{R}^3 \times S^1 \subset SE(3)$. An element $h \in H$ is completely described by a translation vector $p \in \mathbb{R}^3$ and a heading angle $\psi \in S^1$.

The property of symmetry induces an equivalence relation between trajectories:

Definition 15.2 (Trajectory Equivalence). *Consider two trajectories, π_1, π_2: $[t_0, t_f] \to \mathcal{X}$. The trajectories π_1 and π_2 are said to be equivalent on the interval $[t_0, t_f]$ if there exists an element $h \in H$ such that $\pi_2(t) = \Psi_h(\pi_1(t))$ for all $t \in [t_0, t_f]$.*

Two trajectories that are equivalent under the above definition can be transformed exactly to one another by the (left)-action of an element of the symmetry group. Hence we can formalize the notion of a motion primitive:

Definition 15.3 (Motion Primitive). *We define a motion primitive as an equivalence class of trajectories, under the trajectory equivalence relation.*

Since motion primitives are an equivalence class of trajectories, without loss of generality, each equivalence class can be represented by a prototype, starting at the identity on the symmetry group. The notion of symmetry thus becomes fundamental in the construction of a "library" of trajectories.

At this point we have to address the problem of selecting an appropriate set of motion primitives from among all possible trajectories—modulo a symmetry transformation—which can be executed by the system.

15.4.2. Relative Equilibria

A first class of motion primitives is composed of relative equilibria, or trim trajectories, defined in the following. Assume that a system (15.1) admits a symmetry group H, and consider the associated Lie algebra $\mathfrak{h}$. It might be possible to set the initial conditions $x_0 = (g_0, \xi_0, z_0)$ and find a constant control input u_0 in such a way that:

- The velocity expressed in the body frame ξ is constant: $\Xi(g_0, \xi_0, z_0, u_0) = 0$.
- The higher-order terms z are constant: $Z(g_0, \xi_0, z_0, u_0) = 0$.
- The velocity expressed in the inertial frame is an element of the Lie algebra of H (and hence its exponential is an element of H): $g_0 \xi_0 g_0^{-1} = \eta_0 \in \mathfrak{h}$.

In such a case, it is easy to verify that the evolution of the system is given by

$$\begin{cases} g(t) = \exp(\eta_0 t) g_0, \\ \xi(t) = \xi_0 \\ z(t) = z_0 \end{cases} \tag{15.6}$$

We call such a trajectory a *relative equilibrium* of the system. Along these steady-state trajectories the velocity in body axes ξ and the control inputs u are constant: In the aerospace literature such relative equilibria are also known as *trim trajectories*. (In the following, to simplify notation, we will consider just a rigid-body model; however, all the considerations made in the paper extend to higher-dimensional models.)

In order to apply our methodology, we will assume that the vehicle under consideration does indeed admit nontrivial relative equilibria. As in the case of symmetries, most human-built vehicles do indeed satisfy this assumption. From the definition of relative equilibria, it is apparent that these motion primitives are characterized by an infinite time horizon: Once a trim trajectory is initiated, it can be followed indefinitely (it is a steady-state condition).

Relative equilibria include trivially all equilibrium points of the system (e.g., points for which $\dot{g} = 0$, $\dot{\xi} = 0$). The simplest possible motion primitive is represented by equilibrium points. In a system with multiple equilibrium points, each equilibrium point can be chosen as a motion primitive. In the case in which the system admits only a trivial relative equilibrium, the framework we are considering in this chapter reduces to a version of the control quanta methodology [19].

In the general case of an unconstrained dynamical system, the collection of all possible trim trajectories defines, under some smoothness conditions on the system dynamics, an m-dimensional manifold in $\mathcal{X} \times \mathcal{U}$, known as the *trim surface*. When the flight envelope and control saturation constraints (15.3) are taken into account, the relative equilibria that are actually attainable by the system are typically contained in a compact set.

Trim trajectories are at the foundation of the vast majority of control design methods. Linear control is typically based on the Jacobian linearization of a nonlinear system's dynamics around equilibria, or relative equilibria (steady-state conditions). Moreover, the class of trim trajectories has been used widely to construct switching control systems, in which point stabilization is achieved by switching through a sequence of controllers progressively taking the system closer to the desired equilibrium [20, 21]. The ideas of gain scheduling and of linear parameter varying (LPV) system control can also be brought into this class [22], as well as other integrated guidance and control systems for UAV applications [23].

However, restricting the choice of motion primitives to trim trajectories alone has several drawbacks. First of all, there is a lack of hierarchical consistency [24]. A dynamical system cannot transition instantaneously from one trim trajectory to another one, since there will always be a transient between the two.

An additional reason for which trim trajectories alone could not be an appropriate set of motion primitives for some systems is due to the fact that in some cases there could be several disconnected regions on the trim surface. For example, it is not possible to steer a sailboat while staying always close to some equilibrium point. Transitions from a starboard tack to a port tack and vice versa, known as tacks and jibes, are very far from steady-state conditions.

15.4.3. Maneuvers

To address the above-mentioned issues, we must better characterize trajectories that move far from the trim surface—that is, what we will call *maneuvers*. The formal definition of maneuver needs to address the following points: (i) When do we say that a maneuver starts? (ii) When do we say that a maneuver ends? (iii) How can we build a consistent interface between maneuvers, so that we can sequence them efficiently?

Our answers are summarized in the following definition:

Definition 15.4 (Maneuver). *A maneuver is defined as a finite-time transition between two trim trajectories.*

The transition can also be from and to the same trajectory (e.g., in the case of aircraft, acrobatic maneuvers like loops and barrel rolls can be considered as transitions from and back to straight and level flight, and in the case of cars a lane change maneuver is a transition from and back to straight forward motion). The execution of a maneuver can also connect two trim trajectories belonging to disconnected components of the feasible trim set: This is what is customarily accomplished by sailors when tacking or jibing. Finally, the requirement that maneuvers start and end at trim trajectories does not mean that the system must remain at equilibria: it must go *through* equilibria when switching from one maneuver to another: this is required to provide a set of interfaces between maneuvers (i.e., a standard set of connections).

Each maneuver is characterized by the following:

- A time duration T.
- A control signal u: $[0, T] \rightarrow \mathcal{U}$.
- A function ϕ $[0, T] \rightarrow \mathcal{X}$, describing the evolution of the system over the maneuver.

Since a maneuver has to start and end at trim trajectories, the following boundary conditions must be satisfied:

$$u(0) = \bar{u}_{\text{start}}, \qquad\qquad u(T) = \bar{u}_{\text{end}}$$

$$\phi(0) = \left(h(0)\,\bar{g}_{\text{start}}, \xi_{\text{start}} \right), \qquad \phi(T) = \left(h(T)\,\bar{g}_{\text{end}}, \xi_{\text{end}} \right)$$

for some $h(0)$, $h(T) \in H$ and with the subscripts start and end referring to the trim trajectories connected by the maneuver. Because of the symmetry of the dynamics, the displacement on the symmetry group $h_m := h(0)^{-1}h(T)$ is invariant and is a constant characterizing the maneuver. Hence, one of the main properties of maneuvers is the fact that each maneuver is characterized by a well-defined time duration and results in a well-defined displacement. Even though the displacement on the symmetry group h_m can be derived from the function ϕ, it is convenient to store this displacement along with the other data on the maneuver.

We will not discuss here the details of how to generate the nominal state and control trajectories describing maneuvers: Several methods can be used depending on the application at hand, the desired performance, and the available computing, simulation, and experimental resources. Among these methods, we can mention actual tests or simulations with human pilots [25] or stabilizing controllers (as in our example), offline solutions to optimal control problems [26], or real-time trajectory generation methods [6].

Finally, we would like to remark that the design of the nominal trajectories can be carried out so as to ensure that the vehicle does not violate the internal constraints (3)—that is, flight envelope and control saturation constraints [27]. Thus in this sense the objective of envelope protection is ensured implicitly by the construction of the maneuvers (this can be extended to the case in which the behavior of the system is not nominal, or the system is subject to the action of external disturbances, through the introduction of appropriate feedback control laws [12]).

15.4.4. Maneuver Automaton Definition

To discuss the details of the control architecture, recall that we have identified two classes of motion primitives, namely: (i) relative equilibria (or trim trajectories) and (ii) maneuvers. The control framework we propose is based on the selection of a finite number of relative equilibria, connected by a finite number of maneuvers: Such motion primitives will form a *maneuver library*.

The control architecture will involve switching from one motion primitive to another, always alternating relative equilibria and maneuvers. Such a control system will include both continuous and discrete dynamics, thus belonging to the realm of hybrid control. The continuous dynamics correspond to the dynamics of the system along a trajectory primitive. The discrete dynamics correspond to the switching between motion primitives, and hence to the logical state of the controller.

The *discrete dynamics* of such a system are well represented by a finite-state automaton, and can be conveniently depicted by a directed graph, as in Figure 15.1. The nodes of the directed graph correspond to trim trajectories in the automaton, whereas the edges correspond to maneuvers. The directed graph representation is a convenient form of depicting the discrete dynamics

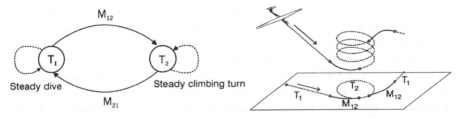

Figure 15.1. A deceptively simple maneuver automaton. These four primitives are enough to ensure important reachability properties [12].

of the system—that is, all the possible sequences of maneuvers/trim trajectories that the system is able to perform. Keep in mind, however, that a simple directed graph is not enough to convey all the information contained in the automaton, since it does not show its continuous evolution.

From the properties and definitions of trim trajectories and maneuvers, it follows that the discrete state can stay indefinitely in any node, but it must traverse the edges in a very definite time. Also, once the discrete state switches to an edge $q \in Q_M$ (i.e., the execution of a maneuver is started), the only possible outcome is for the system to switch to the next trim trajectory, after T_q seconds.

We will call such a control architecture a maneuver automaton:

Definition 15.5 (Maneuver Automaton). *A maneuver automaton MA over a dynamical system* (15.1), *with symmetry group H, is described by the following objects*:

- *A finite set of indices $Q = Q_T \cup Q_M \subset \mathbb{N}$, where the subscript T relates to trim trajectories, and the subscript M relates to maneuvers.*
- *A finite set of trim trajectory parameters $(\bar{g}, \xi, \bar{u})_q$, with $q \in Q_T$.*
- *A finite set of maneuver parameters, and state and control trajectories $(T, u, \phi)_q$, with $q \in Q_M$.*
- *The maps* Pre: $Q_M \to Q_T$, *and* Next: $Q_M \to Q_T$ *such that* Pre(q) *and* Next(q) *give, respectively, the index of the trim trajectories from which the maneuver q starts and ends.*
- *A discrete state $q \in Q$.*
- *A continuous state, denoting the position on the symmetry group, $h \in H$.*
- *A clock state $\theta \in \mathbb{R}$, which evolves according to $\dot{\theta} = 1$ and which is reset to zero after each switch on q.*

Decisions can be made about the future evolution of the system only when the system is executing a trim trajectory (that is, the discrete state is in one of the nodes in the graph). While executing a maneuver, the system is committed to it and must keep executing the maneuver until its completion.

This is admittedly a limiting assumption, especially for applications where the system might be required to react instantly to external events, such as failures, or the appearance of new obstacles or threats. However, maneuvers can be designed in such a way that their duration is very small, so that the execution of a maneuver does not block the system for too long. Moreover, often it is physically very challenging, if not impossible, to get out of a maneuver when initiated. For example, a human cannot easily abort a jump after leaving ground contact. Possible extensions to the concept of maneuver discussed in this chapter are the object of current research; however, in this chapter we will assume that maneuvers cannot be aborted. Hence, for motion planning and control design purposes, we can concentrate the study of the evolution of the system on and between nodes.

While on a trim trajectory, the state of the system is *completely* determined by the couple $(q, h) \in Q_T \times H$: The corresponding full continuous state will in fact be given by (hg_q, ξ_q). We will call the couple (q, h) the *hybrid state* of the system.

At any time while in trim trajectory q, it is possible to start a maneuver p, as long as $\text{Pre}(p) = q$. Once a maneuver is initiated, and executed to its completion, the hybrid state will undergo a net change, described by the following jump relations:

$$q^+ = \text{Next}(p)$$

$$h^+ = h^- h_{m,p}$$

$$\theta^+ = 0$$

with $h_{m,p}$ being the displacement on the symmetry group caused by maneuver p.

The decision variables, when executing a trim trajectory $q \in Q_T$, are the following:

- The duration of the permanence of the system in the current trim trajectory. We call this *coasting time*, and we indicate it with τ.
- The next trajectory to be executed. We indicate the next trajectory as $p \in \text{Pre}^{-1}(q)$.*

At this point, along with the hybrid state (q, h), we have a hybrid control variable $v := (p, \tau)$. The action of the hybrid control on the hybrid state (e.g., the evolution of the system between two arrivals at nodes) is given by

$$(q, h) \xrightarrow{(p,\tau)} \left(\text{Next}(p), h \exp(\overline{\eta}_q \tau) h_m \right)$$

In the following, we will indicate the action of the hybrid control with the maneuver-end map $\Phi \colon Q_T \times H \times Q_M \times \mathbb{R} \to Q_T \times H$. Moreover, we will use the notation $\Phi_v(q, h)$ as a shorthand for $\Phi(q, h, v)$.

15.4.5. The Maneuver Automaton as a Dynamical System

The maneuver automaton is in its own right a dynamical system, with well-defined state, control, and transition function. This system evolves on what we call the *maneuver space*.

In particular, the effect of the maneuver-end map, and the maneuver-trim trajectory switching, is that of giving rise to a new, discrete "time" system, where "time" now is an abstraction. We will indicate the state at different "times"—that is, after the execution of i maneuver-trim trajectory sequences, by square brackets:

$$(q, h)[i + 1] = \Phi((q, h)[i], v[i])$$

---

*Here Pre^{-1} is a set-valued function such that $p \in \text{Pre}^{-1}(q) \Leftrightarrow q = \text{Pre}(p)$.

Notice how one of the control inputs (namely, the coasting time τ) corresponds to a physical time duration.

The maneuver automaton can be seen as a consistent hierarchical abstraction of the continuous dynamics, in the sense outlined in reference 24: any sequence of motion primitives generated by the maneuver automaton results by construction in a trajectory which is executable by the full continuous system, and satisfies the constraints (3).

In this paper, the discussion is limited to the nominal dynamics of vehicles. However, the concepts introduced in this paper can be extended to more realistic cases, modelling the effects of external disturbances, or uncertainties in the plant and in the environment, through a process of robustification of the maneuver automaton framework [12]. The key requirement is that the exogenous inputs be modeled as isotropic disturbances, and that the feedback be designed in such a way that the invariance properties of the system are preserved. A robust maneuver automaton (RoMA) is obtained from a MA by (i) associating to each motion primitive an invariant feedback control law, that is, a control law which preserves the open-loop symmetries under closed-loop control [15] and (ii) verifying some "nesting conditions" on the invariant sets of the motion primitives. This process gives a deeper meaning to the concept of hierarchical consistency, taking into account external disturbances, in the sense that all trajectories generated by a RoMA, are not only executable by the system, but also trackable by the assigned control laws, to within a known and uniform (in space and time) bound.

It is out of the scope of this chapter to discuss these extensions, and other system-theoretic properties of the maneuver automaton as a dynamical system, such as well-posedness and controllability, which have been analyzed in reference 12.

15.5. MOTION PLANNING IN THE MANEUVER SPACE

The hybrid control architecture lends itself to computationally efficient solutions to many problems of interest for practical applications. The price that we must pay in using the maneuver automaton is the suboptimality of the computed solutions, owing to the fact that the stored motion primitives do not represent the whole dynamics of the system. However, the number of motion primitives stored in the automaton can be increased, depending on the available memory and computational resources, so the suboptimality gap can be reduced to a point where it is not noticeable for practical purposes. Moreover, very often a suboptimal solution which is computable online can be worth more than an optimal solution that requires computational resources only available for offline planning.

By the structure of the maneuver automaton, it is also evident that its applicability is limited to problems in which the target state (and, to a more limited extent, the initial conditions) corresponds to trim trajectories. This is not a very limiting requirement, since in fact most meaningful motion planning problems are stated exactly in these terms. For example, a typical requirement for a helicopter could be to stop or hover at a certain location, altitude, and heading. The target roll angle and pitch heading, and other state variables, are not directly specified, and will assume the values required to achieve the desired steady hovering conditions.

15.5.1. Optimal Control in a Free Environment

Consider again the cost functional (15.4), and remember that the incremental cost $\gamma(x, u)$ is assumed to be invariant with respect to actions in the symmetry group. We can derive the following expression for the cost increment along trajectories generated by the maneuver automaton. Along a trim trajectory q, we have (with a slight abuse of notation)

$$\frac{dJ}{dt} = \gamma(g, \xi, u) = \gamma\left(h_0 \exp(\bar{\eta}_q t)\bar{q}_q, \bar{\xi}_q, \bar{u}_q\right) = \gamma\left(\bar{g}_q, \bar{\xi}_q, \bar{u}_q\right) =: \gamma_q$$

In the above we used the fact that γ is invariant with respect to group actions on H. The cost increment along a trim trajectory is hence a constant, depending only on the trim trajectory being executed.

Similarly, after the execution of a maneuver p we have

$$\Delta J = \int_0^{T_p} \gamma\left(\Psi_{h_0} \phi_p(t), u_p(t)\right) dt = \int_0^{T_p} \gamma\left(\phi_p(t), u_p(t)\right) dt =: \Gamma_p$$

The (finite) cost increment after the execution of a maneuver is a constant, depending only on the maneuver index.

We can now formulate an expression for the cost functional (15.4) that is compatible with the maneuver automaton structure:

$$\tilde{J}(q[\cdot], h[\cdot], p[\cdot], \tau[\cdot]) = \sum_{i=1}^{n_f - 1} \gamma_{q[i]} \tau[i] + \Gamma_{p[i]} \tag{15.7}$$

with the constraint that $(q, h)[n_f] = (q_f, h_f)$ for some finite n_f. At this point, Eqs. [15.4] and [15.7] are equivalent, the only difference being the different parameterization of trajectories.

Since now we have a discrete "time" system, the application of the optimality principle leads us to a version of the Bellman's equation (as opposed to the Hamilton–Jacobi–Belman partial differential equation). Define the optimal cost function $\tilde{J}^*$, obtained by optimizing on the automaton structure (i.e., on the maneuver space). The optimality principle implies that the optimal cost function satisfies

$$\tilde{J}^*(q, h) = \min_{(p, \tau)} \left\{\gamma_q \tau + \Gamma_p + \tilde{J}^*\left[\Phi_{(p, \tau)}(q, h)\right]\right\} \tag{15.8}$$

Since the trajectories generated by the automaton are a subset of all trajectories executable by the full continuous system, the optimal costs computed on the maneuver automaton will be in general greater than the true optimal cost:

$$\tilde{J}^*(q,h) \geq J^*\left(h\bar{g}_q, \xi_q\right)$$

On the other hand, the solution to the Bellman's equation (15.8) is, from the computational point of view, much simpler than the solution of the Hamilton–Jacobi–Bellman partial differential equation for the original continuous system. The hybrid state space $Q_T \times H$ (the maneuver space) is much smaller than the continuous state space $\mathcal{X}$: For the general vehicle case the hybrid state has dimension four—consisting of three position and one heading coordinates—while the continuous state has dimension at least 12. Moreover, the decision variable set in the automaton case is $Q_M \times \mathbb{R}_+$, which is much smaller than the continuous control set $\mathcal{U}$ (in the case of aircraft, usually at least four).

If the dimension of the symmetry group is still too large for computing the optimal cost function, it can be possible to (a) further reduce the computational complexity of the problem by decomposing the group H into smaller symmetry groups (possibly of dimension one) and (b) solve the optimal control problem in each subgroup in sequence. Naturally, such an approach would reduce the optimality of the resulting trajectory. (Essentially, this amounts to working on one coordinate at a time.)

If the optimal cost function $\tilde{J}^*(q, h)$ is known, the optimal control can be computed by solving

$$(p, \tau)^* = \arg\min_{(p, \tau)} \left\{ \gamma_q \tau + \Gamma_p + \tilde{J}^*\left[\Phi_{(p, \tau)}(q, h)\right] \right\} \qquad (15.9)$$

which is an optimization problem on a discrete variable $p \in Q_M$ and a continuous, scalar variable $\tau \in \mathbb{R}_+$ (and as a consequence can be solved as a sequence of scalar optimization problems). Thus the formulation of the optimal control problem on the "Maneuver Space" can be seen as a trade-off between optimality and computational complexity.

An important property of (15.9) is that it is a stabilizing feedback control law, providing a degree of robustness with respect to the action of external disturbances and measurement errors. An additional level of robustness is provided by a low-level feedback control law which augments the maneuver automaton structure, as briefly stated in Section 15.4.5.

Since the dimension of the hybrid state is relatively small, it is possible to apply efficient computational methods developed for approximate dynamic programming, also known as neurodynamic programming [28], and reinforcement learning in the artificial intelligence community. The main approaches to the solution of dynamic programming problems are based on the computation of the optimal cost J^* (value iteration), of the optimal control policy μ^*: $(q, h) \mapsto (p, \tau)^*$ (policy iteration), or of the so-called *Q-factors*—that is, the functions to be minimized in Eqs. (15.9) and (15.8) (Q-learning). We refer the reader to reference 28 for more information and details on the value iteration algorithm and other computational methods for approximate dynamic programming.

15.6. EXAMPLE: THREE-DEGREE-OF-FREEDOM HELICOPTER

In this section we will address a minimum-time control problem for a three-degree-of-freedom helicopter experimental setup. The main motivation of this section is to show how a maneuver automaton can be built from the analysis of experimental data, together with insight in the fundamental properties of the system's dynamics. Even though a detailed model for the system could be derived (including all aerodynamic forces and rotor inflow modeling), the construction of the maneuver automaton for this system will be carried out with the assumption that the dynamics are not precisely known, and most of the information about the vehicle has to be derived from experiments.

15.6.1. Selection of Motion Primitives

Consider the three-degree-of-freedom (3-DOF) helicopter setup shown in Figure 15.2. The setup consists of a helicopter body carrying two propellers, which can be commanded independently. The body of the helicopter is attached to an arm and is free to rotate about the axis of the arm. The arm, in turn, is attached to the ground via a spherical joint. As can be recognized

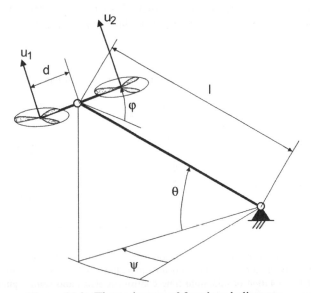

Figure 15.2. Three-degree-of-freedom helicopter.

from Figure 15.2, the dynamics are invariant with respect to rotations about a vertical axis. The symmetry group is in this case the group $S^1 \subset SO(3)$ of rotations about a vertical axis. Invariance with respect to rotation about a vertical axis can be established without even writing down the equations of the system: It is a fundamental property of the 3-DOF helicopter dynamics.

Relative equilibria correspond to constant heading rate rotations. For simplicity of exposition, we will only consider rotations with zero pitch angle (this measures the inclination of the arm with respect to the horizontal plane). Moreover, for safety reasons, the roll angle ϕ is limited to $\pm 30°$.

Experimentally, it was found that the maximum speed of the helicopter, while maintaining constant altitude and a roll angle of $\pm 30°$, is about $110°$ per second, which is reached asymptotically. All velocity settings in $(-110, 110)$ are achievable in steady state, and so the trim trajectory selection can be carried out at will in this set. As a practical maximum velocity setting we can choose $\pm 80°/s$, which is achieved in about 16.5 s and a total rotation of about $800°$, and then choose other velocity settings using a logarithmic scaling (for details on the motivation of this choice, see references 12 and 29). Six velocity settings other than zero were considered, namely, $(\pm 80, \pm 40, \pm 20)°/s$.

Since the dynamics are not well known, especially at high-velocity regimes, the problem of determining the full trim conditions (i.e., $(\bar{g}, \bar{u})_q$, other than the imposed $\bar{\eta}_q$) is solved through the use of a feedback controller, designed to track steps in $\dot{\psi}$; the feedback control law used is a simplification of the control law for small helicopters detailed in reference 30.[1] The results are reported in Table 15.1. Notice the asymmetry in the trim roll angle, due to a small imbalance in the experimental setup.

The maneuvers are also generated through the same feedback control law. Strictly speaking, since the controller provides only asymptotic tracking (even if at an exponential rate), these are not *finite-time* transitions from trim to

Table 15.1. Trim Settings for the Three Degree of Freedom Helicopter

Index $q \in Q_T$	Heading rate $\dot{\psi}$[deg/s]	Roll angle ϕ[deg]
0	0	-1
1	20	-5
2	40	-7.5
3	80	-15
4	-20	2
5	-40	4.5
6	-80	12

[1]It is important to remark that this feedback control law is only used for the offline generation of motion primitives. While it can also be used for online local stabilization of the system, it is not appropriate for the solution of minimum-time control problems and cannot guarantee satisfaction of the flight envelope constraints for arbitrary initial conditions.

Table 15.2. Example of Maneuver Data for the 3-DOF Helicopter

Index $p \in Q_M$	Pre(p)	Next(p)	Maneuver Duration T (s)	Maneuver Displacement $\Delta\psi$ (deg)
11	0	2	7.5	66.8
13	0	4	4	-40.7
20	1	5	10.5	-147.6
22	2	0	6	110.5
34	4	0	3.5	-30.5

trim. However, for all practical purposes it is possible to choose the maneuver termination time as the time at which the state of the system settles to within a given accuracy (depending for example on the noise level of the sensors).

Maneuvers were generated and recorded for each transition between trim trajectories (and averaged over several runs). Results in the form of the aggregate maneuver data $(T, \Delta\psi)$ are given in Table 15.2. Note that we "unroll" the heading coordinate (we keep track of the number of revolutions:

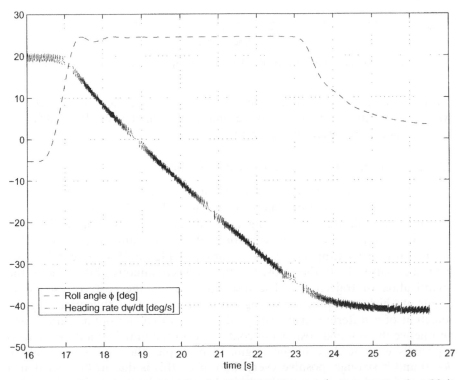

Figure 15.3. Example maneuver for the 3-DOF helicopter (maneuver number 20 in Table 15.2). The "heading rate" is obtained through high-pass filtering of the digital position measurements.

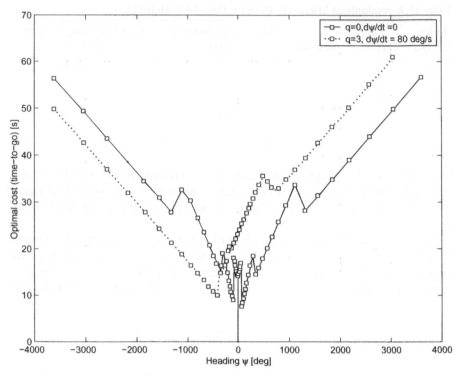

Figure 15.4. Optimal cost for the 3-DOF helicopter maneuver automaton.

a rotation of 360° is different from a zero rotation); as a consequence our symmetry group is actually $\mathbb{R}$. An example of a maneuver, from 20°/s to −40°/s, is also reported in Figure 15.3.

15.6.2. Real-time Optimal Maneuver Selection

Once the maneuver library is available, it is desired to use it to solve in real-time minimum-time optimal control laws. The (sub)-optimal cost function can be built through the algorithm described in Section 12.5.1. The optimal cost function for initial conditions at rest and initial conditions at the maximum velocity trim setting are reported in Figure 15.4. As can be seen from the figure, the optimal cost function is discontinuous; this is due to the quantization of trajectories. The discontinuities in the cost correspond to points at which the optimal coasting time τ becomes zero, thus triggering the selection of a different maneuver p; this means that the optimal trajectory, and hence the optimal cost, is not continuous in the initial conditions. This is especially evident close to the origin. In principle, the cost is zero only at $\psi = 0$ and is strongly positive everywhere else; this is due to the fact that if the initial condition is not at the desired target, at least one maneuver must be executed. Hence, the cost from any initial condition not at the target will be at least Γ_{min}—that is, the cost of the least expensive maneuvers. The adverse effect of this fact can be alleviated in practice by relaxing the final boundary condition to $\psi_f \in [-\epsilon, \epsilon]$, for some $\epsilon > 0$.

Based on the maneuver automaton structure outlined in this section, and the corresponding optimal cost just computed, a real-time controller was implemented on a three-degree-of-freedom helicopter. The software for this experiment was developed using Matlab/Simulink, with the addition of the Real-Time Workshop and the WinCon control shell. The computer used for the experiment is a Pentium II 300-MHz, running Windows 98. The conceptual implementation of the controller is quite straightforward, given the development in the preceding sections. The low-level controller, which runs at a fixed rate (100 Hz in the example), includes the details of the nominal evolution corresponding to each motion primitive (the information in Tables 15.1 and 15.2), together with an array of control laws.

The real-time reset of the motion primitive index was achieved in this simple example by polling the high-level controller at the same rate— that is, 100 MHz. More demanding applications can make full use of the partially synchronous nature of the maneuver automaton architecture by implementing more sophisticated interprocess communication schemes, ranging from signal passing and interrupt handling (i.e., to set the time at which the switch must occur) to the proxy message passing mechanism of QNX or the real-time FIFO queues in real-time Linux. More details on the real-time software implementations of maneuver automata will be given in future papers.

15.6.3. Experiment Results

The experiment is run as follows. First of all, a take-off phase is executed, in which the helicopter can acquire the desired state, hovering at zero pitch angle, and zero heading (that is on the vertical of the take-off position). At $t = 5$ s, a command is given to hover at $-90°$. At time $t = 12$ s, before the helicopter reaches the previously commanded position, a new command is issued, according to which the helicopter must hover at $+360°$. The reference trajectory is generated by the maneuver automaton, and a tracking controller ensures that the state of the helicopter stays "close" to the reference trajectory (more details on the design of feedback controllers to integrate with the maneuver automaton are given in references 12 and 30). The recorded data from the experiment are shown in Figure 15.5. As can be seen from the picture, the generated suboptimal trajectory is consistent with the system's dynamics and can be followed very accurately by a very simple controller. Very little knowledge of the system's dynamics (at least in an analytical model) was used in generating the trajectory, and only results from experiments formed the basis of the maneuver library.

In the picture on the right is reported the evolution of the discrete state $q \in Q_T \cup Q_M$; that is, we have a trace of the optimal sequence of motion primitives executed during the experiment. The indices of the motion primitives are given in Tables 15.1 and 15.2, and we use the convention that trim

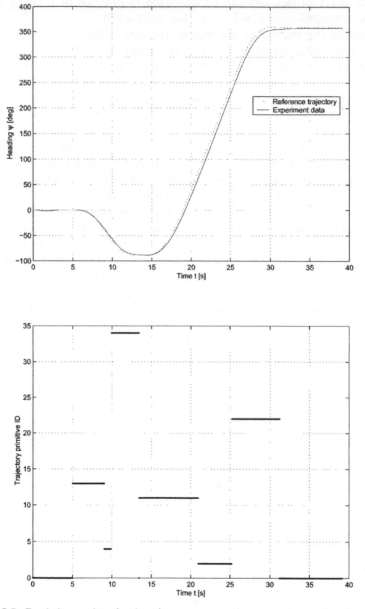

Figure 15.5. Real-time suboptimal trajectory generation experiment for the 3-DOF helicopter.

trajectories have indices smaller than 10, and maneuvers have indices 10 and higher.

15.7. CONCLUSION

In this chapter, we outlined a new approach to the development of integrated guidance and control systems for autonomous vehicles. Very promising results have been obtained through a process of *quantization* of the dynamics of a physical system, which leads to the construction of a language consisting of simple behaviors which can be composed to achieve complex tasks.

The ideas and the concepts outlined in this chapter can be further pursued to improve the methodology and extend the results; most importantly, it will be of primary interest to analyze the implications and the consequence of such a quantization approach on the development on high-confidence real-time software for dynamical systems.

Through the quantization of the continuous system's dynamics, it will be possible to develop a "programming language" which completely determines the physical state (or bounds its deviation from a reference state). The incorporation of such a programming language into more traditional software design frameworks will result in a new approach to the design and development of software-driven dynamical systems. In particular, we believe that this will make feasible for the analysis of the complete system otherwise poorly scalable techniques such as static analysis, abstract interpretation, and model checking.

Moreover, the model of computation that arises from such a quantization is inherently (partially) asynchronous, in the sense that decisions must be taken not at a fixed rate, but rather at a rate that depends on the behaviors being executed. The extreme compactness of the relevant information on the physical state of the system, and of its current behavior, has the potential to be particularly appropriate for distributed systems, which are subject to limits and constraints on the communication rate, or on the reliability of the communication channels. This is a consequence of the fact that the quantized state has a much higher semantic density than the traditional continuous state-space representation of the physical system.

Applications and extensions of the maneuver automaton approach in the above-mentioned directions are the object of current research by the authors.

REFERENCES

1. R. Alur, C. Courcoubetis, T. Henzinger, and P. H. Ho, Hybrid automata: An algorithmic approach to the specification and verification of hybrid systems, in R. L. Grossman and A. Nerode and A. P. Ravn and H. Rischel, editors, *Hybrid Systems*, *Lecture Notes in Computer Science*, Vol. 736, Springer-Verlag, New York, 1993, pp. 209–229.

2. M. S. Branicky, *Studies in Hybrid Systems: Modeling, Analysis, and Control*, Ph.D. thesis, Massachusetts Institute of Technology, Cambridge, MA, 1995.

3. N. Lynch, R. Segala, and F. Vandraager, Hybrid I/O automata revisited, in M. D. Di Benedetto and A. Sangiovanni-Vincentelli, editors, *Hybrid Systems IV: Computation and Control, Lecture Notes in Computer Science*, Vol. 2034, Springer-Verlag, New York, 2001, pp. 403–417.

4. T. Henzinger, P. H. Ho, and H. Wong-Toi, Algorithmic analysis of nonlinear hybrid systems, *IEEE Transactions on Automatic Control* **43**(4):540–554, 1998.

5. N. H. McClamrock, and I. Kolmanovsky, Hybrid switched mode control approach for V/STOL flight control problems, in *Proceedings of the IEEE Conference on Decision and Control*, Kobe, Japan, 1996.

6. M. J. van Nieuwstadt and R. M. Murray, Real-time trajectory generation for differentially flat systems, *International Journal of Robust and Nonlinear Control* **8**(11):995–1020, September 1998.

7. V. Manikonda. P. S. Krishnaprasad, and J. Hendler, A motion description language and a hybrid architecture for motion planning with nonholonomic robots, in *Proceedings of the IEEE Conference on Robotics and Automation*, 1995, pp. 2021–2028.

8. M. A. Dahleh and I. J. Diaz-Bobillo, *Control of Uncertain Systems: A Linear Programming Approach*, Prentice-Hall, Edgewood Cliffs, NJ, 1995.

9. K. Zhou and J. C. Doyle, *Robust and Optimal Control*, Prentice Hall, Englewood Cliffs, NJ, 1995.

10. S. Boyd, L. El Ghaoui, E. Feron, and V. Balakrishnan, *Linear Matrix Inequalities in System and Control Theory, Studies in Applied Mathematics*, Vol.15 SIAM, Philadelphia, PA, 1994.

11. E. Frazzoli, M. A. Dahleh, and E. Fero, A hybrid control architecture for aggressive maneuvering of autonomous helicopters, in *IEEE Conference on Decision and Control*, December 1999.

12. E. Frazzoli, *Robust Hybrid Control for Autonomous Vehicle Motion Planning*, Ph.D. thesis, Massachusetts Institute of Technology, Cambridge, MA, June 2001.

13. R. M. Murray and Z. Li and S. S. Sastry, *A Mathematical Introduction to Robotic Manipulation*, CRC Press, 1994.

14. V. I. Arnold, *Mathematical Methods of Classical Mechanics, GTM*, 2nd edition, Vol. 60, Springer-Verlag, New York, 1989.

15. P. Rouchon and J. Rudolph, Invariant tracking and stabilization: Problem formulation and examples, in D. Aeyels, F. Lamnabhi-Lagarrigue, and A. van der Schäft, editors, *Stability and Stabilization of Nonlinear Systems, Lecture Notes in Control and Information Sciences*, Vol. 246, Springer-Verlag London, London, 1999, pp. 261–273.

16. A. M. Bloch, P. S. Krishnaprasad, J. E. Marsden, and R. M. Murray, Nonholonomic mechanical systems with symmetry, *Archives of Rational Mechanical Analysis*, **136**:21–99, 1996.

17. F. Bullo, N. E. Leonard, and A. D. Lewis, Controllability and motion algorithms for underactuated Lagrangian systems on Lie groups, *IEEE Transactions on Automatic Control* **45**(8):1437–1454, August 2000.

18. E. Frazzoli, M. A. Dahleh, and E. Feron, Real-Time motion planning for agile autonomous vehicles, *AIAA Journal on Guidance, Control and Dynamics* **25**(1):116–129, 2002.

19. A. Marigo and A. Bicchi, Steering driftless nonholonomic systems by control quanta, in *Proceedings of IEEE Conference on Decision and Control*, 1998.

20. M. W. McConley, B. D. Appleby, M. A. Dahleh, and E. Feron, A computationally efficient Lyapunov-based scheduling procedure for control of nonlinear systems with stability guarantees, *IEEE Transactions on Automatic Control*, January 2000.

21. R. R. Burridge, A. A. Rizzi, and D. E. Koditscheck, Sequential decomposition of dynamically dexterous robot behaviors, *International Journal of Robotics Research*, **18**(6):534–555, June 1999.

22. J. S. Shamma and J. R. Cloutier, Gain-scheduled missile autopilot design using linear parameter varying transformations, *AIAA Journal on Guidance, Control and Dynamics* **16**(2):256–263, March–April 1993.

23. I. Kaminer, A. Pascoal, E. Hallberg, and C. Silvestre, Trajectory tracking for autonomous vehicles: An integrated approach to guidance and control, *AIAA Journal on Guidance, Control and Dynamics* **21**(1):29–38, January–February 1998.

24. G. J. Pappas, G. Lafferriere, and S. Sastry, Hierarchically consistent control systems, *IEEE Transactions on Automatic Control* **45**(6):1144–1160, June 2000.

25. V. Gavrilets, E. Frazzoli, B. Mettler, M. Piedmonte, and E. Feron, Aggressive maneuvering of small autonomous helicopters: A human-centered approach, *International Journal of Robotics Research* **20**(10):795–807, 2001.

26. A. E. Bryson and Y. C. Ho, *Applied Optimal Control*, Hemisphere Publishing, New York, 1975.

27. Z. Shiller and H. Chang, Trajectory preshaping for high-speed articulated systems, *ASME Journal of Dynamic Systems, Measurement and Control* **117**(3):304–310, 1995.

28. D. P. Bertsekas and J. T. Tsitsiklis, *Neuro-Dynamic Programming*, Athena Scientific, Belmont, MA, 1996.

29. N. Elia and S. K. Mitter, Stabilization of linear systems with limited information, *IEEE Transactions on Automatic Control* **46**(9):1400–1460, 2001.

30. E. Frazzoli, M. A. Dahleh, and E. Feron, Trajectory tracking control design for autonomous helicopters Using a backstepping algorithm, in *American Control Conference*, Chicago, IL, 2000.

CHAPTER 16

MULTIMODAL CONTROL OF CONSTRAINED NONLINEAR SYSTEMS

T. JOHN KOO, GEORGE J. PAPPAS, and SHANKAR SASTRY

Editors' Notes

This chapter is concerned with the design of multimode control systems. It assumes that output tracking controllers have already been developed for individual modes of a constrained nonlinear system. The problem addressed is the determination of a mode sequence along with reference trajectories for each mode in order for the system to reach a desired final mode.

An algorithm is presented for the construction of a control mode graph by abstracting the reachability properties of the given set of control modes. Therefore, the discrete and continuous-time aspects of the problem can be decoupled. The discrete aspect, in which the sequence of modes is determined, can be solved as a search over a finite graph which encodes the safe mode switches. Once the mode sequence is known, the reference trajectories can be selected based on desired criteria—the construction guarantees that reference trajectories for all intermode transitions along the sequence exist.

The effectiveness of the design framework is demonstrated using a nonlinear helicopter model, with hover, cruise, ascend, and ascend, and descend modes. Finally, the authors map their approach to a three-level implementation architecture and discuss connections with hybrid automata theory and embedded system implementations that combine multiple models of computation to promote formal methods for the design of high-confidence embedded control systems.

16.1. INTRODUCTION

Based on operational, financial, and environmental considerations, large-scale systems ranging from automated highway systems [1], air traffic management systems [2], unmanned aerial vehicle networks [3], communication networks and power distribution networks have been advocated to have higher levels

Software-Enabled Control. Edited by Tariq Samad and Gary Balas
ISBN 0-471-23436-2 © 2003 IEEE

of automation. Recent advances of embedded system technologies in sensing, computation and communication have enabled the rapid realization of sophisticated, high-performance embedded controllers. The control of large-scale systems is extremely challenging since by nature the systems are distributed and highly dynamic, the environments in which the systems reside in are usually rapidly evolving, and multiobjective design specifications intensify the complexity of system design. One natural way to manage the complexity of system design is by compositional methods. Of particular interest is a multimodal control paradigm in which control systems are designed by hierarchically nesting compositions of modes of operation such that each mode of operation is designed to cope with a designated scenario with respect to a design specification while the organization of modes of operation depends on the ordering of these specifications.

A multimodal control system can be modeled as a hierarchical nesting of parallel and serial compositions of discrete and continuous components. Furthermore, a model of computation (MOC) [4] governs the behaviors and interactions of components at each level of the hierarchy. Hybrid systems [5–7] are considered as formalisms used to describe a complex system as combinations of MOCs. This naturally leads to the generalization of the design problem for the control of large-scale systems as a problem of multimodal control synthesis/design in the modeling framework of hybrid systems. The high-profile and safety-critical nature of such applications has fostered a large and growing body of work on formal methods for hybrid systems: mathematical logics, computational models and methods, and automated reasoning tools supporting the formal specification and verification of performance requirements for hybrid systems, along with the design and synthesis of control programs for hybrid systems that are provably correct with respect to formal specifications.

A multimodal control paradigm, which assumes that a set of controllers of satisfactory performance have already been designed and must be used, is considered. Each controller may be designed for a different set of outputs in order to meet the given performance objectives and system constraints. When such a collection of *control modes* is available, an important problem is to be able to accomplish a variety of high-level tasks by appropriately switching between the low-level control modes. Multimodal control has been studied, especially in the context of stability and safety; see reference 8 for stability results of switching between stable linear time-invariant controllers, reference 9 for *safe* switching conditions for systems with pointwise-in-time constraints on state and control, and references 10 and 11 for controller designs and switching conditions for satisfying multiple objectives such as safety and optimal performance. In reference 12, we have proposed a framework for the synthesis of mode switching for reachability specifications. The problem of controller synthesis for preserving stability of a global equilibrium point has been studied in reference 8. Here, the stability of each control mode is assumed to be preserved with respect to the state variables of interest and taken care of by the control design. The switching condition between modes defined in reference 12 is a generalization of the result presented in reference 9 which requires the controllers to be designed in a total order with respect to safety specifications. The outcome of applying our algorithm results in a partial ordering of the given control modes that cannot only capture the outcome generated as shown in reference 9, but also is more expressive since it captures more possible switching combinations.

In this chapter, a *control mode* is defined as the operation of the system under a controller that is *guaranteed* to track a certain class of output trajectories while simultaneously avoiding violation of specified *safety* constraints related to the states and inputs. Given a set of control modes, the mode switching problem attempts to find a finite *sequence* of the control modes as well as *switching conditions* in order to satisfy the given reachability tasks. Hence, the mode switching problem can be posed as follows:

Problem 16.1. Given a control system and a finite set of control modes for the system, determine whether there exists a finite sequence of modes that will steer the system from an initial control mode to a desired final control mode. If such a sequence exists, then determine the switching conditions.

By considering the set of control modes, feedback greatly simplifies the continuous models in each discrete location since the complexity of the continuous behavior is now reduced to the complexity of the trajectories we design. Therefore, many reachability computations that are required in our approach can be greatly simplified by properly *designing* the desired trajectories. Even though feedback control simplifies the continuous complexity, the problem of having nested reachability computations is still present. In order to avoid such expensive computations, as shown in reference 12, we place a *consistency* condition in our mode switching logic which is reminiscent of the notion of *bisimulation* [13]. We propose an algorithm, which, given an initial set of control modes, constructs a *control mode graph* which refines the initial control modes but is consistent. Construction of the mode graph can be done off-line or every time a new control mode is designed, allowing the mode switching problem to be efficiently solved online, in real time. A helicopter-based unmanned aerial vehicle (UAV) is used as a design example to demonstrate the effectiveness of the proposed design paradigm for solving the multimodal control problem.

16.2. FORMULATION OF MULTIMODAL CONTROL PROBLEM

In this section, we introduce a concept of control mode, and precisely define Problem 16.1 as a *mode switching problem*. First, consider a nonlinear system modelled by differential equations of the form

$$\dot{x}(t) = f(x(t)) + g(x(t))u(t)$$

$$x(t_0) = x_0, \qquad t \geq t_0 \tag{16.1}$$

where $x \in \mathbb{R}^n$, $u \in \mathbb{R}^p$, $f(x): \mathbb{R}^n \to \mathbb{R}^n$, and $g(x): \mathbb{R}^n \to \mathbb{R}^n \times \mathbb{R}^p$. The system is assumed to be as smooth as needed. Each control mode corresponds to a output tracking controller applying to the nonlinear system (16.1). We now define a concept of control mode.

Definition 16.1 (Control Modes). *A control mode, labeled by q_i where $i \in \{1, \ldots, N\}$, is the operation of the nonlinear system (16.1) under a closed-loop feedback controller of the form*

$$u(t) = k_i(x(t), r_i(t)) \tag{16.2}$$

associated with an output $y_i(t) = h_i(x(t))$ such that $y_i(t)$ shall track $r_i(t)$ where $y_i(t)$, $r_i(t) \in \mathbb{R}^{m_i}$, $h_i: \mathbb{R}^n \to \mathbb{R}^{m_i}$, $k_i: \mathbb{R}^n \times \mathbb{R}^{m_i} \to \mathbb{R}^p$ for each $i \in \{1, \ldots, N\}$. We assume that $r_i \in \mathcal{R}_i$, the class of output trajectories associated with the control mode q_i. When the initial condition of the system (16.1) is in the set $S_i(r_i) \subseteq X_i$, output tracking is guaranteed and the state satisfies a set of state constraints $X_i \subseteq \mathbb{R}^n$.

The trajectory $r_i(t)$ is the desired output trajectory, and $y_i(t)$ is the output vector which shall track $r_i(t)$. Notice that in general the initial set may be a function of the trajectory r_i; thus we denote it as $S_i(r_i)$. This is because even though trajectory tracking controllers are guaranteed to converge for any initial condition, trajectory tracking in the presence of state constraints or input constraints can be guaranteed only if the initial tracking error is sufficiently small. State constraints are specified by X_i for $i = 1, \ldots, N$. The state constraints are introduced due to the physical limits of the system and the control design.

In this chapter we are interested in switching between controllers rather than the design of output tracking controllers. We therefore make the following assumption.

Assumption 16.1. *For each control mode q_i, $i \in \{1, \ldots, N\}$, we assume that a controller of the form (16.2) has been designed which achieves output tracking such that $y_i(t)$ shall track $r_i(t)$ where $r_i \in \mathcal{R}_i \neq \varnothing$, while the state satisfies the set of state constraints $x(t) \in X_i \subseteq \mathbb{R}^n$, when the initial condition of the system (16.1) is in the set $S_i(r_i) \subseteq X_i \subseteq \mathbb{R}^n$.*

The above assumption is justified given the maturity of output tracking controllers for large classes of linear and nonlinear systems [14, 15]. Based on different design methodologies, the notion of *output tracking* could be different (uniform, asymptotic, exponential, etc.).

Example 16.1 (Point Mass). Consider the dynamics of a point mass that can be modeled as a double chain of integrators,

$$\dot{x}_1 = x_2$$

$$\dot{x}_2 = u \tag{16.3}$$

where $x_1, x_2, u \in \mathbb{R}$ represent the position, velocity, and acceleration of the point mass, respectively. Define the state as $x = [x_1 \ x_2]^T \in \mathbb{R}^2$. Assume that there are two controllers designed for (16.3) and the corresponding control modes are specified as

Mode	Output	Reference	Constraint
q_1	$y_1 = x_1$	r_1	X_1
q_2	$y_2 = x_1$	r_2	X_2

where $X_2 \subset X_1 = \mathbb{R} \times (\underline{v}, \bar{v})$ with $\underline{v} < 0 < \bar{v}$. The given controllers are linear controllers, and the design is simply based on pole placement. In order to satisfy Assumption 16.1, the controller in control mode q_1 is designed such that $\mathcal{R}_1 \subseteq \mathbb{R}^n$ and $S_1(r_1) = B([r_1 \ 0]^T, \delta_1)$ with $\delta_1 < \min(\underline{v}, \bar{v})/M_1$ where M_1 is the overshoot constant which can easily be obtained by Lyapunov theorems. Control mode q_2 is similarly defined but the poles are placed differently. It results in faster response but larger overshoot; that is $M_2 > M_1$. Hence, $\mathcal{R}_2 \subseteq \mathbb{R}^n$ and $S_2(r_2) = B([r_2 \ 0]^T, \delta_2)$ with $\delta_2 < \min(\underline{v}, \bar{v})/M_2$. Therefore, we have $S_2(r_2) \subset S_1(r_1)$ if $r_1 = r_2$.

Given two control modes as shown in Example 16.1, one cannot simply switch from one control mode to another due to incompatible constraints and trajectories. One can easily complicate the situation by introducing many more control modes to serve different performance objectives. A natural question is then whether this mode reachability task as defined in Problem 16.1 can be achieved by a *finite sequence* of modes. Based on the discussion, we can now define the mode switching problem that we will address in this chapter.

Problem 16.2 (Mode Switching Problem). Given an initial control mode q_S with desired reference r_S, does there exist a sequence of control modes such that the system can reach a desired mode q_F with reference r_F? If so, then determine a mode sequence $q_S \to \cdots q_i \to q_j \cdots \to q_F$ along with trajectories r_i for each control mode q_i, as well as conditions for switching between the control modes.

For the control modes defined in Example 16.1, one can define a task of having mode q_1 as an initial mode and ask for a finite control mode sequence to reach mode q_2. Note that Problem 16.2 is a reachability problem. In this simple example, the problem can be solved by examining the mode switching condition between modes.

In the above mode switching problem, there is enough structure to take advantage of in order to simplify the complexity of the synthesis task. First of all, the continuous controllers are assumed to have been designed, and therefore we do not have to design the continuous part of the system, but simply determine the mode switching conditions. Furthermore, certain conditions on the allowable mode switches can allow us to reduce the complexity of the synthesis problem by *maximally decoupling the discrete and continuous aspects of the synthesis*.

16.3. A MODE SWITCHING CONDITION

Consider a mode switch from mode q_i to mode q_j. A mode switch from mode q_i to q_j could be allowed if during the operation of the system under mode q_i for some $r_i \in \mathcal{R}_i$, the state reaches the allowable set of initial conditions $S_j(r_j)$ for some $r_j \in \mathcal{R}_j$; that is, there exist $r_i \in \mathcal{R}_i$ and $r_j \in \mathcal{R}_j$ such that

$$\exists x_0 \in S_i(r_i) \exists t \geq 0 \exists x \in S_j(r_j) \qquad \text{s.t.} \quad x = \phi_i(t, r_i, x_0) \qquad (16.4)$$

where $\phi_i(t, r_i, x_0)$ denotes the *flow* of system (16.1) operating in mode q_i with the controller defined by (16.2) for initial condition x_0 and desired output trajectory r_i. If one allows this type of mode switching, then reachability critically depends on the particular choice of initial conditions since some initial conditions in $S_i(r_i)$ may reach the set

In general, nested reachability computations are required for solving such a reachability problem. Furthermore, since loops can be considered in feasible sequences, the number of possible mode sequences could be infinite. Therefore, for this type of mode switching, decidability becomes a central issue [16]. We now characterize the reachable set within each mode.

Definition 16.2 (Predecessor Set). *Given a set $P \subseteq X_i$, a trajectory $r_i \in \mathcal{R}_i$, the reach set $Pre_i(P, r_i)$ in mode q_i is defined by*

$$Pre_i(P, r_i) = \left\{ x_0 \in X_i | \exists t \geq 0 \exists x \in P \quad \text{s.t.} \quad x = \phi_i(t, r_i, x_0) \right\} \qquad (16.5)$$

$Pre_i(P, r_i)$ consists of all states that can reach the set P in mode q_i for a given output trajectory r_i, at *some* future time. Furthermore, because of Assumption 16.1, we have a guarantee that the state constraints are satisfied throughout the whole trajectory, that is $\phi_i(t, r_i, x_0) \in X_i$ for all $t \geq 0$. Hence,

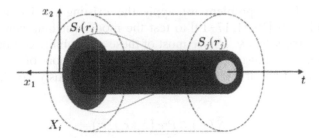

Figure 16.1. Consistent mode switching condition.

using (16.5), condition (16.4) can be rewritten as

$$S_i(r_i) \cap Pre_i\big(S_j(r_j), r_i\big) \neq \varnothing \qquad (16.6)$$

In order to avoid the nested computations mentioned above, as well as to break free of restricted decidability results, we constrain our allowable mode switches.

Definition 16.3 (Consistent Mode Switching). *Assume that control mode q_i satisfies Assumption 16.1, that is $\phi_i(t, r_i, x_0) \in X_i$ for all $t \geq 0$ with initial conditions starting from $S_i(r_i)$ where $r_i \in \mathcal{R}_i$. A transition from mode q_i to mode q_j is allowed only if there exist $r_i \in \mathcal{R}_i$ and $r_j \in \mathcal{R}_j$ such that*

$$S_i(r_i) \subseteq Pre_i\big(S_j(r_j), r_i\big) \qquad (16.7)$$

Therefore, if there exist trajectories r_i (in mode q_i) and r_j (in mode q_j) such that, if the system starts at *any* $x_0 \in S_i(r_i)$, then switching from mode q_i to q_j can occur at some time t such that $\phi_i(t, r_i, x_0) \in S_j(r_j)$. The consistent mode switching condition is shown in Figure 16.1. The condition expressed in Definition 16.3 is a consistency condition that guarantees that our ability to get from mode q_i to mode q_j for the particular trajectory pair (r_i, r_j) is independent of the choice of initial condition in $S_i(r_i)$. Hence, a mode switching from mode q_i to mode q_j is possible, if there exists a trajectory $r_i \in \mathcal{R}_i$ that will steer the system state to an initial set $S_j(r_j)$ with $r_j \in \mathcal{R}_j$ independently of where we start in $S_i(r_i)$. The collection of the trajectory pair (r_i, r_j) is specified by the following definition

$$\mathcal{R}^{ij} = \big\{(r_i, r_j) \in \mathcal{R}_i \times \mathcal{R}_j \big| \text{Condition (16.7) is satisfied}\big\} \qquad (16.8)$$

Therefore, *every* trajectory pair $(r_i, r_j) \in \mathcal{R}^{ij}$ will steer the system from mode q_i to mode q_j. For each $(r_i, r_j) \in \mathcal{R}^{ij}$, the only thing that depends on the initial condition is *when* the state will reach $S_j(r_j)$, but not *if* the state will reach $S_j(r_j)$. In our problem we apply the existing results in computing the reachable sets [10, 11, 17–19] to test the mode switching condition (16.7) and compute the sets $\mathcal{R}^{ij}$. Since most of these reachability computations are approximate, one must consider an overapproximation of $S_i(r_i)$ and an underapproximation of $Pre_i(S_j(r_j), r_i)$, in order to satisfy condition (16.7), that is

$$\overline{S}_i(r_i) \subseteq \underline{Pre}_i\big(\underline{S}_j(r_j), r_i\big) \tag{16.9}$$

where $\overline{P}$ and $\underline{P}$ denote the overapproximation and underapproximation of a set P, respectively.

16.4. MODE SEQUENCE SYNTHESIS

By focusing on the trajectory sets $\mathcal{R}^{ij}$ rather than the initial sets, the mode switching condition (16.7) makes the mode switching problem much more tractable. Furthermore, the construction presented in this section will abstract the mode switching logic into a purely discrete graph. Therefore one can first determine the sequence of modes using standard algorithms for discrete graph reachability, and then determine the continuous parameters r_i for each mode. This will decouple the discrete from the continuous aspects of the problem, allow continuous techniques for continuous problems, and allow discrete techniques for discrete problems.

Consider a collection of control modes $Q = \{q_1, \ldots, q_N\}$. If there exist trajectory pairs $(r_i, r_j) \in \mathcal{R}^{ij}$ that can transfer the system from mode q_i to q_j, there would be a transition $q_i \to q_j$. However, given $q_i \to q_j$ and $q_j \to q_k$, if $\mathcal{R}^{ij} \cap \mathcal{R}^{jk} = \varnothing$ there does not exist a trajectory r_j, which will take a point $x \in S_i(r_i)$ to $S_k(r_k)$ via $S_j(r_j)$. In order to construct a consistent *control mode graph* such that the high level mode switching logic is implementable at the lower level by the continuous controllers, transitivity should be preserved.

As shown in reference 12, each control mode q_i gets refined to $2N$ submodes, where N submodes stand for entering mode q_i from any other mode q_j, and N more copies for exiting mode q_i towards any other mode q_j. Therefore, this control mode graph has some discrete memory, in the sense that each state represents not only which mode the system is in, but also which mode will either precede it or has preceded it. If the set $\mathcal{R}^{ij}$ can be expressed as a decoupled product of the form $\mathcal{R}^{ij} = \mathcal{R}_i^{ij} \times \mathcal{R}_j^{ij}$ where $\mathcal{R}_i^{ij} = \{r_i \in \mathcal{R}_i | (r_i, r_j) \in \mathcal{R}^{ij}\}$ and $\mathcal{R}_j^{ij} = \{r_j \in \mathcal{R}_j | (r_i, r_j) \in \mathcal{R}^{ij}\}$, then the choice of trajectory $r_i \in \mathcal{R}_i^{ij}$ in mode q_i would work for any trajectory $r_j \in \mathcal{R}_j^{ij}$ in mode q_j, that is,

$$\forall r_i \in \mathcal{R}_i^{ij} \; \forall r_j \in \mathcal{R}_j^{ij} \qquad \text{condition (13.4) is satisfied.} \tag{16.10}$$

This decoupling allows us to consider switching via submodes. Within each mode, we can check for submode consistency by simply performing set intersections. Since there are maximally $2N$ submodes of N modes, a total of N^2 pairwise reachability computations and $N(N)^2 = N^3$ intersections must be computed.

Algorithm 16.1 (Consistent Control Mode Graph)

Input Control Modes $Q = \{q_1, \ldots, q_N\}$
Output Control Mode Graph $(Q_c, \to_c)$
Initialize $Q_c := \varnothing, \to_c := \varnothing$
Determine Mode Interconnections
for $i = 1 : N$; **for** $j = 1 : N$
  Compute sets $\mathcal{R}^{ij}$ using (3.4) and (3.5)
    if $\mathcal{R}^{ij} = \mathcal{R}^{ij}_i \times \mathcal{R}^{ij}_j$;
    $q_i^{ij} := q_i, q_j^{ij} := q_j,$
    $Q_c := Q_c \cup \{q_i^{ij}, q_j^{ij}\},$
    $\to_c := \to_c \cup \{(q_i^{ij}, q_j^{ij})\}$
    end if
end for; end for
Determine Submode Interconnections
for $j = 1 : N$
  $\check{Q} := \{q_j^{nj} \in Q_c \mid \exists n \ \ s.t. \ \left(q_n^{nj}, q_j^{nj}\right) \in \to_c\}$
  $\hat{Q} := \{q_j^{jm} \in Q_c \mid \exists m \ \ s.t. \ \left(q_j^{jm}, q_k^{jm}\right) \in \to_c\}$
  for all $q_j^{ij} \in \check{Q}$; **for all** $q_j^{jk} \in \hat{Q}$
    if $\mathcal{R}^{ij}_j \cap \mathcal{R}^{jk}_j \neq \varnothing$;
    $\to_c := \to_c \cup \{(q_j^{ij}, q_j^{jk})\}$
    end if
  end for; end for
end for

We now summarize the ideas and present an algorithm for constructing the consistent control mode graph. The algorithm starts with the pairwise reachability computations (16.7) and (16.8) and performs the submode interconnections. After applying the algorithm, we obtain a finite *control mode graph* $(Q_c, \to_c)$ which has been shown in reference 12 to be *consistent*. Without loss of generality, in the following discussion, we assume that the given initial and final control mode in Q can be represented by $q_0 \in Q_c$ and $q_F \in Q_c$, respectively. Given an initial control mode $q_0 \in Q_c$, the problem of whether we can reach control mode $q_F \in Q_c$ can be efficiently solved using standard reachability algorithms.

Furthermore, one can determine the shortest path (minimum number of mode switches) between mode q_S and q_F in the control mode graph. The structure that we have imposed on our control mode graph immediately results in the following solution to the mode switching problem.

Theorem 16.1 (Mode Switching Solution). *Given a collection of control modes Q, consider the mode switching Problem 16.2. Construct the consistent control mode graph $(Q_c, \to_c)$ as described in Algorithm 16.1. If there exists a path in the consistent control mode graph between q_S and q_F with feasible trajectories r_S and r_F, then Problem 16.2 is solvable.*

Having determined the sequence of modes that can steer our system from q_0 to q_F, we are left with the problem of determining the parameters r_i for each mode of the sequence. By construction, such parameters exist and may be selected from the computed sets. Furthermore, it is reasonable to pose the problem of choosing r_i within mode q_i as an optimization or an optimal control problem.

16.5. MULTIMODAL CONTROL OF A HELICOPTER-BASED UAV

The design framework presented in previous sections has been applied to the control of a helicopter model [20] as depicted in Figure 16.2. The derivation of the nonlinear model is based on Newton–Euler equations for rigid-body dynamics described in longitudinal and vertical axes and complex force and moment generation processes.

We first present the controller design to illustrate how to compute the reachable sets, then we show how to check the consistent mode switching condition between control modes. The x, z axes of the spatial frame are pointing in north and down directions. The body x axis is defined from the

Figure 16.2. Ursa Minor—a Berkeley UAV.

center of gravity to the nose of the helicopter, and the body z axis is pointing down from the center of gravity. The motion of the helicopter is controlled by the main rotor thrust, T_M and the longitudinal tilt path angle, a_M. The pitch angle is defined by θ. The equations of motion can be expressed as

$$\begin{bmatrix} \ddot{p}_x(t) \\ \ddot{p}_z(t) \end{bmatrix} = \frac{1}{m} \begin{bmatrix} \cos\theta(t) & \sin\theta(t) \\ -\sin\theta(t) & \cos\theta(t) \end{bmatrix} \begin{bmatrix} -T_M(t)\sin a_M(t) \\ -T_M(t)\cos a_M(t) \end{bmatrix} + \begin{bmatrix} 0 \\ g \end{bmatrix} \tag{16.11}$$

$$\ddot{\theta}(t) = \frac{1}{I_y}\left(M_M a_M(t) + h_M T_M(t)\sin a_M(t)\right) \tag{16.12}$$

The state vector and input vector are defined as $x = [p_x, \dot{p}_x, p_z, \dot{p}_z, \theta, \dot{\theta}]^T \in \mathbb{R}^6$ and $u = [T_M, a_M]^T \in \mathbb{R}^2$, respectively.

Control Mode	Output	Reference	Constraint
q_1: Hover	$y_1 = [p_x, p_z]^T$	r_1	X_1
q_2: Cruise	$y_2 = [\dot{p}_x, \dot{p}_z]^T$	r_2	X_2
q_3: Ascend	$y_3 = [\dot{p}_x, \dot{p}_z]^T$	r_3	X_3
q_4: Descend	$y_4 = [\dot{p}_x, \dot{p}_z]^T$	r_4	X_4

Define $X_1 = X_2 = \mathbb{R} \times (\underline{v}_x, \bar{v}_x) \times \mathbb{R} \times (\underline{v}_z, \bar{v}_z) \times (-\pi/2, \pi/2) \times \mathbb{R}$, $X_3 = \mathbb{R} \times (\underline{v}_x^{cr}, \bar{v}_x) \times \mathbb{R} \times (\underline{v}_z, \bar{v}_z^{as}) \times (-\pi/2, \pi/2) \times \mathbb{R}$, and $X_4 = \mathbb{R} \times (\underline{v}_x^{cr}, \bar{v}_x) \times \mathbb{R} \times (\underline{v}_z^{de}, \bar{v}_z) \times (-\pi/2, \pi/2) \times \mathbb{R}$ where $\underline{v}_x < 0 < \underline{v}_x^{cr} < \bar{v}_x$ and $\underline{v}_z < \underline{v}_z^{de} < 0 < \bar{v}_z^{as} < \bar{v}_z$. To satisfy Assumption 16.1, several control design methodologies can be used to design a controller for each discrete control mode q_i where $i \in \{1, 2, 3, 4\}$. Each controller implementation can be specified as $u = k_i(x, r_i)$ with $r_i \in \mathcal{R}_i$ where $\mathcal{R}_i$ defines the class of admissible output trajectories in mode q_i, and the performance of the closed-loop system can be specified by *initial set*, $S_i(r_i)$, and *flow*, $\phi_i(t, r_i, x_0)$ where $x_0 \in S_i(r_i)$. We assume that all output trajectories are *constant* trajectories, therefore, all controllers are setpoint regulators. Choosing computable classes of trajectories makes the reachability computations simpler.

Given the specifications for the control modes, a nonlinear control scheme [21] based on outer flatness is applied for the design of the controllers. For each mode, the closed-loop dynamics with states defined by $x_{ex} = [p_x, \dot{p}_x, p_z, \dot{p}_z, \theta, \dot{\theta}, T_M, a_M]^T \in \mathbb{R}^8$ can be decoupled into an inner system and two outer subsystems which specify the dynamics in x and z directions. In the following presentation, the Hover mode is presented to illustrate how the reachable set can be computed.

For q_1, the output tracking controller is designed such that $y_1(t)$ shall track $r_1 = [r_{1x}, r_{1z}]^T$ and the output tracking error is uniformly ultimately bounded. Furthermore, because of satisfying Assumption 16.1, the controller is designed with initial set* $S_1(r_1) = B([r_{1x}, 0]^T, \epsilon_{1x}) \times B([r_{1z}, 0]^T, \epsilon_{1z}) \times S_{in}$ where $r_1 \in \mathcal{R}_1 = \mathbb{R}^2$, $\epsilon_{1x}, \epsilon_{1z} > 0$, and $S_{in} \subseteq (-\pi/2, \pi/2) \times \mathbb{R}^3$ such that for $x(t_0) \in S_1(r_1)$, then

$$
\begin{aligned}
\|e_{1x}(t)\| &\leq M_{1x} \exp(-\alpha_{1x} t)(\|e_{1x}(t_0)\| + \delta_{1inx}) \\
\|e_{1z}(t)\| &\leq M_{1z} \exp(-\alpha_{1z} t)(\|e_{1z}(t_0)\| + \delta_{1inz}) \\
x_{in} &\in X_{in}, \forall t_0 \leq t < t_0 + T_1
\end{aligned}
\qquad
\begin{cases}
\|e_{1x}(t)\| \leq \delta_{1x} \\
\|e_{1z}(t)\| \leq \delta_{1z} \\
x_{in} \in S_{in}, \forall t \geq t_0 + T_1
\end{cases}
$$

$$(16.13)$$

for some T_1 M_{1x}, M_{1z}, α_{1x}, α_{1z}, δ_{1inx}, δ_{1inz}, δ_{1x}, $\delta_{1z} > 0$. In the above, $e_{1x} = [p_x - r_{1x}, \dot{p}_x]^T$, $e_{1z} = [p_z - r_{1z}, \dot{p}_z]^T$, and $x_{in} = [\theta, \dot{\theta}, T_M, a_M]^T$. Equation (16.13) explicitly overspecifies the reachable set of the mode q_1 by examining the stability property. For other modes, although the control designs are slightly modified for tracking different outputs, the reachable sets are similarly computed. In Figure 16.3, we show the initial sets of all the control modes by projecting them onto the $p_x - \dot{p}_x$ and $p_z - \dot{p}_z$ planes where the projection operator is defined as $\Pi_i : x_{ex} \mapsto (p_i, \dot{p}_i)$ for $i \in \{x, z\}$. In summary, the control modes can be specified by

Control Mode	Trajectory Set	Initial Set
q_1	$\mathcal{R}_1 = \mathbb{R}^2$	$S_1 = B(0,4) \times B(0,4) \times S_{in}$
q_2	$\mathcal{R}_2 = [-3,3] \times \{0\}$	$S_2 = \mathbb{R} \times (-3.5, 3.5) \times (-3.5, 3.5) \times S_{in}$
q_3	$\mathcal{R}_3 = [2,4] \times [-3,0]$	$S_3 = \mathbb{R} \times (1.5, 4.5) \times (-3.5, 0.5) \times S_{in}$
q_4	$\mathcal{R}_4 = [2,4] \times [0,3]$	$S_4 = \mathbb{R} \times (1.5, 4.5) \times (-0.5, 3.5) \times S_{in}$

where $S_{in} = B(0, 0.2)$, $X_{in} = (-\pi/2, \pi/2) \times \mathbb{R}^3$ and the associated parameters are defined as $\underline{v}_x = -6$, $\underline{v}^{cr} = 1$, $\bar{v}_x = 6$, $\underline{v}_z = -6$, $\bar{v}_z^{as} = 1$, $v_z^{de} = -1$, $\bar{v}_z = 6$.

Given the set of control modes, we generated the consistent control mode graph by applying Algorithm 16.1. In Figure 16.4, we illustrate the idea of computing the reachable sets on the $p_x - \dot{p}_x$ plan. One can easily see the advantage of using feedback, since it is straightforward not only to check the consistent mode switching condition but also to determine the feasible range of trajectories—that is, compute the sets $\mathcal{R}^{ij}$. In particular, consider the pair (q_1, q_2)—that is, the transition from hover to cruise. As can be seen from the left side of Figure 16.4, the consistency condition is trivially satisfied since the ball $S_1(r_1)$ will *eventually* shrink toward the setpoint $(r_{1x}, 0)$ and, as a result, will be totally contained inside $S_2(r_2)$ for any r_2. Therefore, in this case $\mathcal{R}^{12} = \mathcal{R}_1 \times \mathcal{R}_2$. Therefore, feedback allows us to check very easily the consistency condition and compute the sets $\mathcal{R}^{ij}$. The right side of Figure 16.4 shows the similar graphical computation for the mode transition (q_2, q_3),

*$B(r, \epsilon) = \{\eta | \|\eta - r\| < \epsilon\}$.

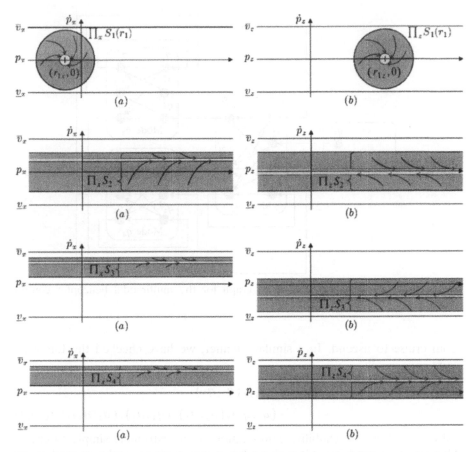

Figure 16.3. Projection of $S_1(r_1)$, $S_2(r_2)$, $S_3(r_3)$, and $S_4(r_4)$ onto (a) the $p_x - \dot{p}_x$ plane; (b) the $p_z - \dot{p}_z$ plane.

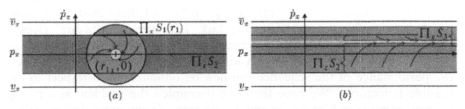

Figure 16.4. Graphical illustration of performing reachability computation for checking consistent mode switching condition on the $p_x - \dot{p}_x$ plane: (a) $q_1 \rightarrow q_2$; (b) $q_2 \rightarrow q_3$.

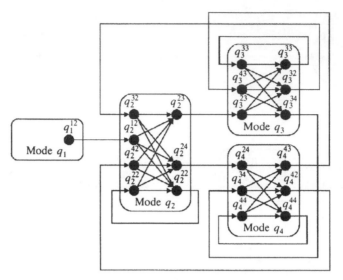

Figure 16.5. Consistent control mode graph for the multimodal helicopter control example.

from cruise to ascend. In a similar manner, we have checked the following pairs;

$$\{(q_1,q_2),(q_2,q_2),(q_2,q_3),(q_2,q_4),(q_3,q_2),$$

$$(q_3,q_3),(q_3,q_4),(q_4,q_2),(q_4,q_3),(q_4,q_4)\}$$

All of the above reachability computations were extremely simple to check. The result of applying Algorithm 16.1 is summarized in the control mode graph that is shown in Figure 16.5.

Recall the *high-altitude takeoff* task, which is the task of having the Hover mode q_1 as an initial mode and requiring a finite control mode sequence to reach the ascend mode q_3. We can now see from Figure 16.5 that q_1 has $\{q_1^{12}\}$ as a submode, and $q_3 = \{q_3^{23}, q_3^{43}, q_3^{32}, q_3^{34}\}$, and there exist many paths which are feasible for achieving the task. However, $q_1^{12} q_2^{12} q_2^{23} q_3^{23}$ gives a solution to the task with the minimum number of mode switches; that is, $q_1 \to q_2 \to q_3$. Given a cost function with respect to the continuous variables, the performance of the sequence can now be optimized with respect to the feasible trajectories. We have therefore decoupled the problem as a purely discrete graph search problems and a collection of continuous designs within each mode.

Simulation results of the controlled system based on the selected sequence are shown in Figure 16.6. In the simulation, we can choose $r_1 = [0\ 0]^T \in \mathcal{R}_1^{12}$, $r_2 = [2\ 0]^T \in \mathcal{R}_2^{12} \cap \mathcal{R}_2^{23}$, and $r_3 = [3\ -1]^T \in \mathcal{R}_3^{23}$. The initial conditions of the outer system are $p_x(0) = -2$, $\dot{p}_x(0) = -0.2$, $p_z(0) = 1$, $\dot{p}_z(0) = 0.5$. The initial condition of the inner system is $x_{in}(0) \in S_{in}$. Mode switchings occur at $t = 20$ for $q_1 \to q_2$ and at $t = 45$ for $q_2 \to q_3$.

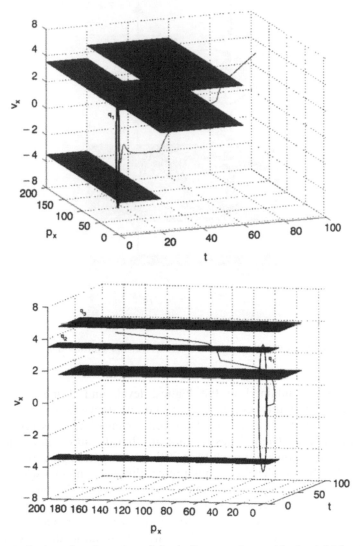

Figure 16.6. Projected trajectories of the helicopter along with the initial sets of the next control modes from different view angles are shown. Notice that an immediate transition $q_2 \to q_3$ after $q_1 \to q_2$ is not allowed until $x(t)$ enters the initial set $S_3(r_3)$.

16.6. HYBRID AND EMBEDDED SYSTEM MODELS

Our approach presented in previous sections consists of extracting a finite graph which refines the original collections of modes, but is consistent with the physical system, in the sense that the high level design has a feasible implementation. A three-level system architecture for implementing the approach is proposed and it is depicted in Figure 16.7. The consistent control

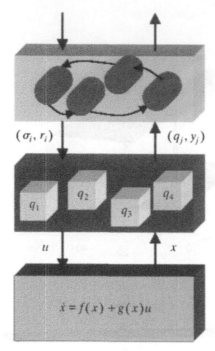

Figure 16.7. Hierarchical system architecture for the multimodal control system.

mode graph is contained in the highest level, and it interacts with the lower-level with output events (σ_i, r_i) and input events (q_j, y_j) where the symbol $\sigma_i \in \Sigma$ with $|\Sigma| = |Q|$ is introduced for enabling a discrete transition to control mode q_i. The indexes i, j indicate that the automaton is in discrete state q_j, and it is going to switch to q_i if the corresponding guard condition is satisfied. The bottom level contains the continuous dynamics which is represented by differential equations, and it interacts with the higher-level with the continuous state vector x and the continuous input vector u. If we combine the mid-level with the bottom-level, it can be modeled as a *hybrid automaton* [7] as shown in Figure 16.8. The hybrid automaton is essentially a finite-state machine with a closed-loop dynamics embedded in each discrete location. The continuous state in a discrete location evolves continuously according to differential equations, as long as the location's invariant remains true; then, when a transition guard becomes true, the discrete state proceeds to another discrete location.

There exist simulation and verification tools for hybrid systems which can be used to analyze the behaviors of multimodal systems. Ptolemy II [22] developed at UC Berkeley supports the use of well-defined models of computation (MOCs) that govern the interaction between components for the modeling, simulation, and design of concurrent, real-time, embedded systems. It provides a clean way to integrate different models by hierarchically nesting parallel and serial compositions of heterogeneous components.

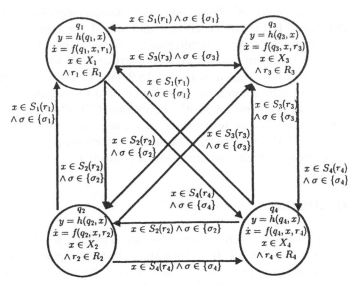

Figure 16.8. The multimodal system in the helicopter control example is modeled as a hybrid automaton.

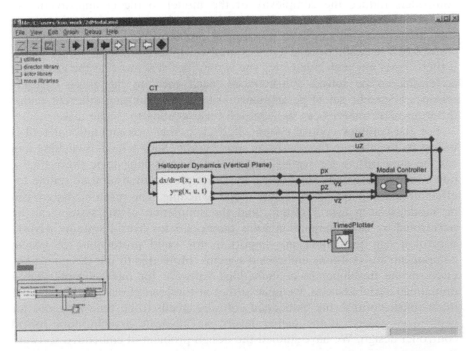

Figure 16.9. Models of the embedded controller and vehicle dynamics in Ptolemy II.

Depending on the level of abstraction and domain-specificity of design practice, different MOCs can be chosen and mixed for composing hybrid systems. The proposed system architecture has been implemented as an embedded system by mixing various MOCs. There are two levels of hierarchy introduced in the design, shown in Figure 16.9. At one level, the embedded controller is modeled as a composite component and the vehicle dynamics is modeled as a set of ordinary differential equations (ODEs). At another level, the embedded controller is refined by two components: the finite state machine (FSM) which governs the switching of control modes and the multimodal controller which contains a set of controllers.

16.7. CONCLUSION

In this chapter, we have focused on a formal approach for hybrid system design for multimodal control. In the multimodal control paradigm, a set of controllers of satisfactory performance have already been designed and must be used. Each controller may be designed for a different set of outputs in order to meet the given performance objectives and system constraints. Our approach consists of extracting a finite graph which refines the original collections of modes, but is consistent with the physical system, in the sense that the high-level design has a feasible implementation. Extracting a finite graph critically depends on the fact that the closed-loop, output tracking controllers reduce the complexity of the model to the complexity of the output trajectories. Therefore, by choosing simple trajectories, such as constant trajectories, we can solve meaningful problems, such as the multimodal control of a helicopter-based UAV. Given an initial mode and a final mode, if there exists any path connecting the two modes, it shows that the given task is feasible to be solved. Furthermore, each path on the graph basically encodes a specific set of performance criteria. By choosing different paths, different performances can be achieved for executing the same task.

For the complex control system, it is clear that a computational tool is necessary to generate the consistent mode graph. Such a mode switching tool can be used offline for synthesizing the mode switching logic every time a new mode is designed. The control mode graph can then be used online for efficient and dependable real-time mode switching. The resulting system can be modeled as a hybrid system, and the simulation of the system can be performed by using simulation tools developed for hybrid systems. Hybrid control design techniques are important for rapid prototyping of system components for real-time embedded systems. Motivated by our design experience on the development of embedded software for our helicopter-based unmanned aerial vehicles, we believe that at the level closest to the environment under control, the embedded software needs to be time-triggered for guaranteed safety; at the higher levels, we advocate asynchronous hybrid controller design. We have studied the design problem of embedded software

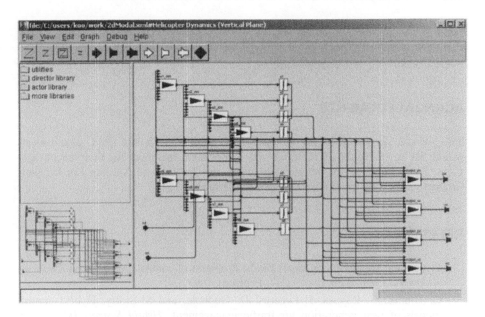

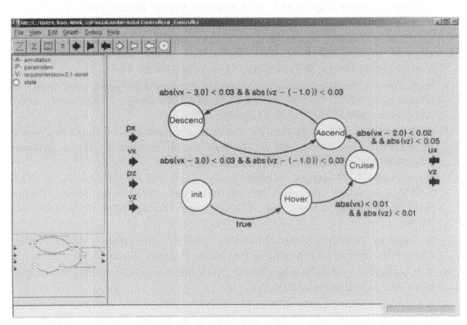

Figure 16.10. Models constructed in Ptolemy II. *Top*: The vehicle dynamics being modeled by a set of ordinary differential equations. *Bottom*: The finite-state machine governing the switching of control modes.

for multivehicle multimodal systems and a hierarchical architecture which promotes verification is presented in reference 23 for the construction of *high-confidence* embedded systems.

ACKNOWLEDGMENTS

This work is supported by the DARPA SEC grant F33615-98-C-3614. The authors would like to thank the Berkeley Aerial Robot team members for their encouragement, Paulo Tabuada for helpful discussion, and Jie Liu and Xiaojun Liu for their support on the construction of the Ptolemy II models.

REFERENCES

1. P. Varaiya, Smart cars on smart roads: Problems of control, *IEEE Transactions on Automatic Control* **38**(2):195–207, February 1993.
2. C. Tomlin, G. Pappas, J. Lygeros, D. Godbole, and S. Sastry, Hybrid control models of next generation air traffic management, *Hybrid Systems IV*, *Lecture Notes in Computer Science*, Vol. 1273, Springer-Verlag, Berlin, 1997, pp. 378–404.
3. T. J. Koo, B. Sinopoli, A. Sangiovanni-Vincentelli, and S. Sastry, A formal approach to reactive system design: A UAV flight management system design example, in *Proceedings of IEEE International symposium on Computer-Aided Control System Design*, Kohala Coast, Hawaii, September 1999.
4. E. A. Lee and A. Sangiovanni-Vincentelli, A framework for comparing models of computation, *IEEE Transactions on Computer-Aided Design of Integrated Circuits and Systems*, **17**(12):1217–1229, December 1998.
5. R. W. Brockett, Hybrid models for motion control systems. in *Essays in Control*: *Perspectives in the Theory and Its Applications*, H. Trentelman and J. Willems, editors, Birkhäuser, Boston, 1993, pp. 29–53.
6. R. Alur and T. Henzinger, Modularity for timed and hybrid systems, in *CONCUR97: Concurrency Theory*, A. Mazurkiewicz and J. Winkowski, editors, *Lecture Notes in Computer Science* Vol. 1243, Springer-Verlag, New York, 1997, pp. 74–88.
7. J. Lygeros, Hybrid systems: Modeling, analysis and control. Memorandum UCB/ERL M99/34, University of California, Berkeley, Spring 1999.
8. J. P. Hespanha and A. Stephen Morse, Switching between stabilizing controllers, submitted to *Auotmatica*, 2000.
9. I. Kolmanovsky and E. G. Gilbert, Multimode regulators for systems with state & control constraints and disturbance inputs, in *Lecture Notes in Control and Information Sciences* 222, *Control Using Logic-Based Switching*, A. Stephen Morse editor, pp. 104–117, Springer-Verlag, London, 1997.
10. J. Lygeros, C. Tomlin, and S. Sastry, Controllers for reachability specifications for hybrid systems *Automatica* **35**(3), March 1999.
11. E. Asarin, O. Bournez, T. Dang, O. Maler, and A. Pnueli, Effective synthesis of switching controllers for linear systems, *Proceedings of the IEEE* **88**(2):1011–1025.

12. T. J. Koo, G. J. Pappas, and S. Sastry, Mode switching synthesis for reachability specifications, in *Hybrid Systems: Computation and Control, Lecture Notes in Computer Science*, Vol. 2034, Springer-Verlag, New York, 2001.

13. R. Milner, *Communication and Concurrency*, Prentice-Hall, Englewood Cliffs, NJ, 1989.

14. A. Isidori and C. I. Byrnes, Output regulation of nonlinear systems, *IEEE Transactions on Automatic Control* **35**(2):131–140, February 1990.

15. S. Sastry, *Nonlinear Systems: Analysis, Stability, and Control*, Springer-Verlag, New York, 1999.

16. R. Alur, T. Henzinger, G. Lafferriere, and G. J. Pappas, Discrete abstractions of hybrid systems, *Proceedings of the IEEE* **88**(2):971–984, July 2000.

17. A. Chutinan and B. H. Krogh, Verification of polyhedral-invariant hybrid systems using polygonal flow pipe approximations, in *Hybrid Systems: Computation and Control, Lecture Notes in Computer Science*, Springer-Verlag, New York, 1999.

18. A. B. Kurzhanski and P. Varaiya, Ellipsoidal techniques for reachability analysis, in *Hybrid Systems:Computation and Control, Lecture Notes in Computer Science*, Springer-Verlag, New York, 2000.

19. G. Lafferriere, G. J. Pappas, and S. Yovine, Symbolic reachability computations for families of linear vector fields, *Journal of Symbolic Computation* **32**(3):231–253, September 2001.

20. J. Liu, X. Liu, T. J. Koo, B. Sinopoli, S. S. Sastry, and E. A. Lee, Hierarchical hybrid system simulation, in *Proceedings of the 38th Conference on Decision and Control*, Phoenix, Arizona, December 1999.

21. T. J. Koo and S. Sastry, Differential flatness based full authority helicopter control design, in *Proceedings of IEEE Conference on Decision and Control*, Phoenix, Arizona, December 1999, pp. 1982–1987.

22. E. A. Lee, Overview of the Ptolemy Project, *Technical Memorandum UCB/ERL*, M01/11, University of California, Berkeley, March 2001.

23. T. J. Koo, J. Liebman, C. Ma, and S. Sastry, Hierarchical approach for design of multi-vehicle multi-modal embedded software, in *Embedded Software, Lectures Notes in Computer Science*, Vol. 2211, Springer-Verlag, New York, 2001

24. G. J. Pappas, G. Lafferriere, and S. Sastry, Hierarchically consistent control systems, *IEEE Transactions on Automatic Control* **45**(6):1144–1160, June 2000.

CHAPTER 17

TOWARD FAULT-ADAPTIVE CONTROL OF COMPLEX DYNAMIC SYSTEMS

GABOR KARSAI, GAUTAM BISWAS, SRIRAM NARASIMHAN,
TAL PASTERNAK, SHERIF ABDELWAHED, TIVADAR SZEMETHY,
GABOR PECELI, GYULA SIMON, and TAMAS KOVACSHAZY

Editors' Notes

An important problem in the control of complex systems, and one that inherently exhibits hybrid dynamics, is fault management and recovery. If a component fails in a system, the continuous behavior will suddenly change. In an autonomous system, the controller must be adaptive to ensure continued useful operation when faults occur.

Fault-adaptive control encompasses a number of hard problems: the detection of a fault, its identification and assessment, the selection of a new control algorithm, the reconfiguration of the plant to "disconnect" the failed component, and the launching of the new control algorithm. The authors propose a model-based approach for designing control systems that are capable of accommodating faults. This chapter focuses on one aspect of this approach; however, an overview of the fault-adaptive control architecture is also given.

The authors model a plant as a hybrid bond graph. Bond graphs are energy-based models, and they are extended for hybrid dynamics by representing mode transitions as controlled junctions. A hybrid observer has been developed that uses this model to track the system behavior within and across modes.

Two complementary approaches to fault detection and isolation are outlined, one based on hybrid models, the other on discrete-event models. The former makes use of the hybrid observer, qualitative reasoning techniques, and real-time parameter estimation. The discrete approach uses failure propagation graphs. Controller reconfiguration relies on a previously developed controller library—an appropriate controller is selected based on current conditions. The authors discuss ideas for mitigating the large transients that can arise during controller switching. A two-tank system is used as an example throughout the chapter.

Software-Enabled Control. Edited by Tariq Samad and Gary Balas
ISBN 0-471-23436-2 © 2003 IEEE

17.1. INTRODUCTION

Today's complex systems, like high-performance aircraft, require sophisticated control techniques to support all aspects of operation: from flight controls through mission management to environmental controls, just to give a few examples. All this, of course, is done using a multitude of computer systems, all of which rely heavily on software technology. Software systems now play a dual role. Not only do they implement system functionalities, but they are also becoming the primary vehicle for system integration. One of the main goals of software is to implement control functions: open- and closed-loop control, from low-level regulation to high-level supervisory control. However, software enables new capabilities in control. It offers a framework that provides great flexibility for developing novel algorithms that significantly improve the performance of the system. Furthermore, brand new functionalities can be created that could not be implemented in any other way.

Any real-life system is prone to physical (hardware) and logical (software) failures. These systems also require a high degree of reliability and safety, therefore, the effects of these failures must be mitigated and control must be maintained under all fault scenarios. If systems are designed with redundancy, control decisions have to be made about when and how backup systems should be activated and how exactly the reconfiguration should be executed. For instance, aircraft often have redundant actuators for control surfaces. If one actuator fails, then the second actuator can still drive the control surface, although larger forces will be required. In order to manage the fault scenario described, we need to make a series of decisions and take control actions, such as (i) the fault has to be detected, (ii) the fault source—the actuator—has to be identified and the magnitude of failure estimated (e.g., is it a partial degradation or a total failure), (iii) depending on the nature of failure, a new control algorithm has to be selected that can compensate for the partial or complete loss of the actuator, (iv) the plant has to be reconfigured so that the faulty actuator can be moved "offline," and (v) the new control algorithm has to be brought up with the good actuator in a way that current operation is maintained. All these decisions must be made by a control system that incorporates not only simple regulatory loops and the supervisory control logic, but also a set of components that detect, isolate, and manage faults, in coordination with the control functions.

Traditional control theory gives very little guidance to the implementer of these systems. Mathematical models and formal analysis techniques have been developed for specific fault scenarios, but there is no general theory of control system design and analysis that encompasses all possible scenarios. Solutions applied to existing systems tend to take a pragmatic approach. Potential fault situations are pre-enumerated, and appropriate fault accommodation actions are built into the supervisory controller for each case. The approach works well for these cases, but may break down in unforeseen situations. Furthermore, most fault-adaptive control techniques are geared toward handling broken components. In many realistic situations, the system suffers only partial degradation and failures. If we can build online capabilities to detect and estimate these partial failures, more sophisticated control algorithms can be designed to keep the system operational under these conditions. Early discussions of these ideas can be found in references [1–3].

For the DARPA SEC project, we are developing a systematic model-based approach to the design and implementation of control systems that can accommodate faults. We call this approach fault-adaptive control technology (FACT). Developing fault-adaptive control requires us to solve a number of technical problems beyond the capabilities of traditional control approaches. First, faults must be detected while the system is in operation. System dynamics is complex, and sensors can be noisy, therefore, differentiating degraded faulty behavior from nominal behavior of the plant quickly is a nontrivial problem. Fault detection must be followed by rapid fault isolation and estimation of the fault magnitude. Then a decision has to be made online on how to reconfigure the control system to accommodate the fault. Many alternatives may have to be evaluated, and metrics will have to be defined that select an optimal configuration, if it can be computed in a feasible manner, or the best possible reconfiguration will have to be derived under given time and resource constraints. Finally, the reconfiguration must be executed, which means that set points and control parameters may have to be changed, or a different controller may have to be selected to continue system operation. The challenge is to bring together methodologies from fault diagnostics, control theory, signal processing, software engineering, and systems engineering to build the integrated online FACT system.

In this chapter, our focus is on the model-based fault isolation schemes. Section 17.2 discusses a reference architecture for FACT systems. Section 17.3 presents our scheme for modeling hybrid systems—that is, continuous physical systems with discrete supervisory controllers. Section 17.4 describes the hybrid observer scheme for tracking nominal system behavior. Section 17.5 discusses the fault isolation methodologies. Preliminary results that demonstrate the effectiveness of our approach are presented. Section 17.6 briefly discusses fault-adaptive control and controller reconfiguration. The summary and conclusions appear in Section 17.7 of the chapter. We illustrate the basic modeling concepts and our diagnosis algorithms using a two-tank system as the plant, with a supervisory controller. This system, while admittedly simple, has a relationship to real-life systems, such as aircraft fuel systems, which have a similar structure. Therefore this work may easily scale up to fault diagnosis and control of such real-life systems.

17.2. FACT ARCHITECTURE

Our overall approach, illustrated in Figure 17.1, is centered on model-based approaches for fault detection, fault isolation and estimation, and controller selection and reconfiguration for hybrid systems. The plant is "connected" to

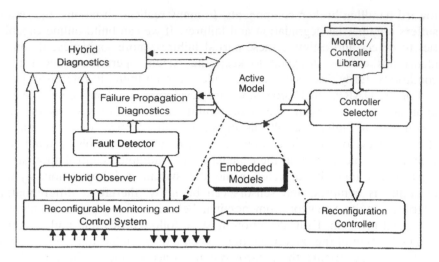

Figure 17.1. Fault adaptive control architecture.

the reconfigurable monitoring and control system block. We assume that the systems we deal with combine continuous dynamics of the plant and PID controllers with supervisory control implemented as computer programs. Hybrid models [4], derived from hybrid bond graphs [5], systematically integrate continuous and discrete system dynamics and discrete events to establish the core of the modeling framework. The supervisory controller, modeled as a generalized finite-state automaton, generates the discrete events that cause reconfigurations in the continuous energy-based bond graph models of the plant. Fault detection involves comparison of the expected behavior of the system generated from the hybrid models with actual system behavior, to determine when discrepancies occur. This requires the design and implementation of hybrid observers that estimate the continuous dynamic states of the system and detect mode transitions in the system operation. Sophisticated signal analysis and filtering methods linked to the hybrid observers are used for detecting deviations from nominal behavior and triggering the fault isolation schemes. Details of the signal analysis schemes are discussed elsewhere [6].

Our diagnostic schemes integrate the use of failure-propagation-graph-based techniques for discrete-event diagnosis [7] and combined qualitative reasoning and quantitative parameter estimation methods for parameterized fault isolation [8] of degraded components (sensors, actuators, and plant components). The dynamic system state accumulated from the observer (discrete system mode plus continuous state vector) and fault isolation units (status of faulty and degraded sensors, actuators, and plant components) define the active system state model. The tracking, fault detection, and fault isolation mechanisms, illustrated on the left of Figure 17.1, together constitute a bottom-up computational approach for estimating the dynamic system state (nominal or faulty) by monitoring plant and controller variables.

The reconfiguration controller uses this information to select from the controller library the controller that is most effective in maintaining desired system operation and performance. This requires the definition of metrics and decision criteria that govern the controller selection process. The selection and reconfiguration mechanisms operate in a top-down manner, using the dynamic state information to effect changes in supervisory control mechanisms, such as selection (not synthesis) of feedback control mechanisms, and retuning of low level regulators, such as PID or model-based controllers. The overall computational architecture combines the bottom-up and top-down computational schemes in a seamless manner, via the shared active model.

The implementation and support for the online FACT architecture is based on our model-integrated computing paradigm [9]. To achieve this, we have created (1) a graphical modeling environment that facilitates building hybrid models of the plant and controllers and (2) a set of run-time components that can execute the code synthesized from the models. This code, when integrated with the generic FACT run-time components, instantiate the architecture for a specific application domain.

17.3. MODELING HYBRID SYSTEMS AND CONTROLLERS

We assume that the plant is made up of components, such as tanks and pipes, that exhibit continuous behaviors. Other components, such as valves and switches, can be turned off and on at rates much faster than the normal dynamics of the plant. There are also components, such as pumps and motors, that exhibit continuous behaviors, interspersed with more discrete on/off transitions. Plants that exhibit these mixed continuous/discrete behaviors are modeled as *hybrid systems*. Some of the discrete changes can be attributed to changes in system configuration imposed by the supervisory controller, and other changes can be attributed to abstracting complex nonlinear system behaviors into piecewise linear behaviors [10]. In this section, we describe our hybrid bond graph modeling approach and present the scheme for modeling the supervisory controller.

17.3.1. Hybrid Bond Graphs

We use bond graphs as the modeling paradigm in the continuous domain [11]. Bond graphs represent energy-based models of the system in terms of the effort and flow variables of the system. Bonds specify interconnections between elements that exchange energy, which is given by the rate of flow of energy, *power = effort × flow*. Bond graphs represent a generic modeling language that can be applied to a multitude of physical systems, such as electrical, fluid, mechanical, and thermal systems. There exist standard

techniques to build bond graph models of systems based on physical principles. State equations can be systematically derived from the bond graph
representation of the system. We can systematically derive temporal causal
graphs, the models for qualitative diagnostic analysis, from bond graphs [12].

We use an enhanced form of bond graphs, called hybrid bond graphs
(HBGs) [5] that include controlled junctions to facilitate the modeling of
discrete mode transitions in system behavior. Consider the example of a
two-tank system shown in Figure 17.2. The system consists of tanks, flow
sources (simplified models of pumps), connecting pipes, and outlet pipes.
Some of the pipes contain valves that can be opened and closed by the
supervisory controller. The flow source can also be turned on and off by the
controller. Two types of discrete events (or "jumps") can occur in the system:

1. *Controlled Events*: These are external controller actions that cause
   changes in the configuration of the system. Opening or closing of the
   valve on the flow source, Sf1, pipe is determined by supervisory controller signals. This represents a controlled event.

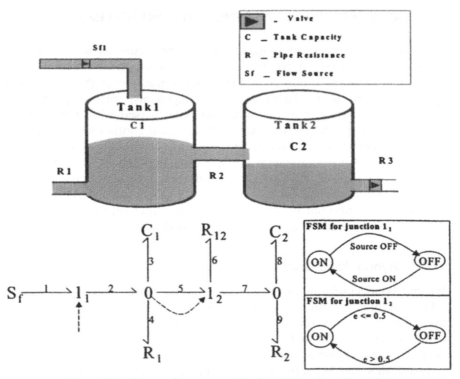

Figure 17.2. Two-tank system and its hybrid bond graph model.

2. *Autonomous Events*: These are changes that occur in the configuration of the system when its internal variables go above and below prespecified values. When the level of fluid in tank 1 reaches the height of pipe R2, fluid flow commences through pipe R2. This represents an autonomous event, since the presence or absence of flow is determined by internal variables (heights of tanks 1 and 2) and not by the controller.

The hybrid bond graph model of the two-tank system of Figure 17.2 is illustrated on the bottom part of the figure. The two tanks are modeled as capacitors. The pipes in the system are modeled as simple resistances. Fluid flow through the pipes are assumed to be at low velocities, therefore, the inertia of the fluid is ignored. The inlet pipe is modeled as an idealized flow source, Sf1, with constant inflow rate. The 0- and 1-junctions are analogous to parallel and series junctions in electrical circuits. Fluid flow behavior is defined in terms of the effort variable, pressure, and the flow variable, fluid flow rate. The switching 0- and 1-junctions represent idealized discrete switching elements that can turn the corresponding energy connection on and off. The physical on/off state for each of these controlled junctions is determined by external control signals and continuous variables crossing prespecified thresholds. These can be specified as finite state sequential automata. The sequential automata that control the on/off states for all the controlled junctions are shown in Figure 17.2. There are four possible modes of continuous operation of the two-tank system. The valve connecting the source to tank 1 may be ON or OFF. This would correspond to the switching junction 1_1 in the HBG. The flow between the tanks is a function of the level of fluid in the tanks. This is captured by switching junction 1_2. Since each junction can be *on* or *off*, the system can have four distinct modes of operation.

17.3.2. Controller Models

In the FACT architecture, the reconfigurable monitoring and control component represents all the traditional monitoring and control functions in an application. We envision that this component is implemented mainly in software, although some components might utilize dedicated hardware components. This component is also "reconfigurable": Its subcomponents, their parameters, and their interconnection can be changed during system operation.

To represent this reconfigurable monitoring and control component, we have developed a modeling language, called controller modeling language (CML). The approach followed here is that of model-integrated computing [9]. CML represents controllers on two levels:

- On the *regulatory* level, it represents controllers using computational blocks that form a signal flow diagram. The signal flow diagram has

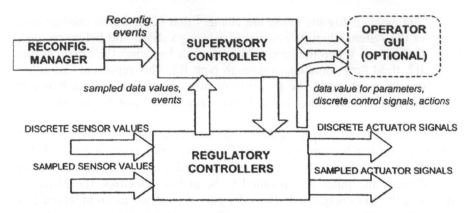

Figure 17.3. Relationship between the supervisory and regulatory controllers.

process-network semantics: Each block is a process that is scheduled for execution upon arrival of data on its inputs. Then the process performs some calculations and may generate output data that are sent to downstream blocks. After finishing processing, the process terminates and waits for the next triggering data.

- On the *supervisory* level, it represents controllers using a hierarchical finite-state machine approach, in the style of Statecharts [13]. State machines are triggered by events, may perform state transitions triggered by events or timers, and generate other events and actions.

The two levels are in a master–slave relationship. CML can also represent reconfigurable controllers: Actions executed by the supervisory controller may result in changes in the regulatory controllers. Furthermore, CML has provisions for interfacing the supervisory control logic with the output of the fault diagnostics system.

The relationship between the two controller layers; supervisory and regulatory, is shown in Figure 17.3. The regulatory layer operates in a discrete-time fashion; that is, it receives discrete (sporadic) and sampled data from the plant, and it generates discrete (sporadic) and sampled data for the actuators. On the other hand, the supervisory controller operates in a discrete-event mode, that is, it has no explicit notion of time. It receives sampled data values and discrete events generated in the regulatory layer, sends new data values for parameters, and sends events in the form of discrete control signals to the regulatory layer. The supervisory controller can also trigger the execution of reconfiguration actions. As mentioned above, during reconfiguration the design procedures associated with the regulatory blocks will be triggered to recalculate parameter values.

17.4. THE HYBRID OBSERVER

The hybrid observer tracks the system behavior across different modes of operation. This involves two steps:

- Tracking continuous system behavior in individual modes of operation.
- Identifying and executing all mode changes including controlled and autonomous jumps. Transitioning from one mode to the other involves (i) switching the state equation model that defines continuous behavior in a mode and (ii) applying the reset function to derive the initial state in the new mode.

The observer uses the state equations models—derived by symbolic analysis from the hybrid bond graph model—for tracking the continuous behavior in a particular mode of operation. The analysis also derives the controlled and autonomous events that define mode transition conditions as the system behavior evolves in time. Solving for the mode transitions requires access to controller signals for controlled jumps and requires predictions of state variable values for autonomous jumps. We rewrite all autonomous jump conditions in terms of the state variables of the system. The state variable estimates are obtained from the hybrid observer, and these values are used to determine if autonomous jumps have occurred. If a mode change occurs in the system, the observer switches the tracking model (to a different set of state space equations), initializes the state variables in the new mode (using a "reset" function, again derived, from the hybrid bond graph model), and continues to track system behavior with the new model [14]. Figure 17.4 illustrates our approach to building a hybrid observer.

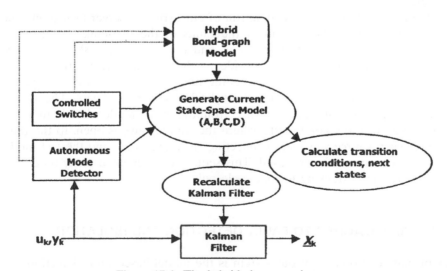

Figure 17.4. The hybrid observer scheme.

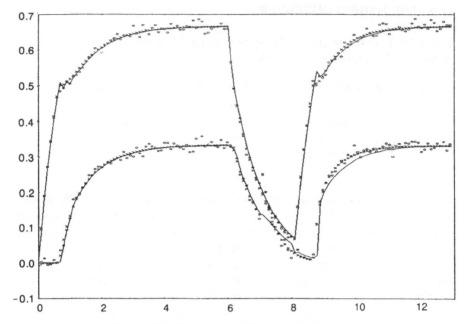

Figure 17.5. Sample run of the hybrid observer.

Since the input and output of the system may be affected by processor disturbances and measurement noise, we use a Kalman filter [15] to track system behavior in a single mode of operation. For a given state space model the Kalman gain matrix can be computed from the covariance matrices, as usual.

Figure 17.5 illustrates a sample run of our hybrid observer as it tracks the pressure values at the bottom of the tanks in the two-tank system. (The horizontal axis shows time in seconds, and the vertical axis shows pressure values.) The crosses (×) show the ideal behavior, the circles are the noisy measurement values obtained by adding noise to the ideal values, and the continuous black line represents the hybrid observer's estimate. The system goes through a set of modes, starting from the initial mode when both valves are closed, to the second mode, where the source valve is open, to the third mode, when source valve and connecting valve are open, and the last mode, where the source valve is closed. This sequence is repeated many times, but the figure shows tracking for one cycle only.

17.5. APPROACHES TO FAULT DETECTION AND ISOLATION

A primary component of our system is the model-based fault detection and isolation (FDI) subsystem that can deal with sensor, actuator, and parametric faults in the system. Traditional FDI methods [16–19] are primarily directed toward additive faults that include failures in sensors and actuators. Isolation of parametric component faults, which are multiplicative, requires the use of sophisticated parameter estimation techniques [19]. Numerical techniques for state and parameter estimation often face convergence and accuracy problems when dealing with high-order models that may contain nonlinearities [19, 20]. Parameter estimation techniques are often biased by measurement noise, and they may need specialized approaches to compensate for these situations [19, 21]. Accurate parameter estimation also requires persistent excitation of the input, and this may not always be true during system operation. Furthermore, these schemes are applicable in continuous real-valued spaces, and they do not easily extend to situations where mode transitions cause discontinuous changes in the system models and system variables. Discrete-event based diagnosis techniques have been proposed, but they require the pre-compiling of the fault models and fault trajectories into finite-state machines (FSMs) for tracking nominal and faulty system behavior [22, 23]. In the section below, we will show how an alternative representation form can be used which does not require the explicit construction of FSMs.

When one deals with hybrid systems that include discrete transitions, extending these continuous methodologies becomes intractable, because the residual transformation functions have to be precomputed for all modes of operation. Furthermore, when faults occur, predicting the true system mode in itself becomes a challenging task. The fault isolation problem becomes even more complex when the fault occurs in an earlier mode, but is detected in a later mode of operation. The predicted mode sequence may no longer be the true mode sequence the system goes through after the occurrence of the fault. Additional methods have to be introduced for detecting mode transitions, switching the system model when such transitions occur, and correctly initializing the system state, so that the fault observers perform correctly.

Typically, mode changes introduce discrete effects that cause transients, and it may be difficult to separate the fault transients from the transients caused by mode changes. Therefore, extending continuous FDI schemes to hybrid systems is a nontrivial task.

We use two approaches to the FDI problem that generalize traditional approaches: (i) the use of a robust qualitative fault isolation scheme based on tracking fault transients combined with a parameter estimation scheme for refining fault hypotheses and (ii) fault diagnostics based on discrete event models represented as fault propagation graphs. We discuss each of these methodologies in greater detail next.

17.5.1. Diagnosis Using Hybrid Models

Our diagnosis methodology consists of three main steps, (i) using a hybrid observer to track system behavior, (ii) detecting fault occurrences, and (iii) isolating faults in the system. The hybrid observer, discussed in the previous section, uses the models of the system to track system behavior. The fault detection schemes that compare the measurements made on the system and

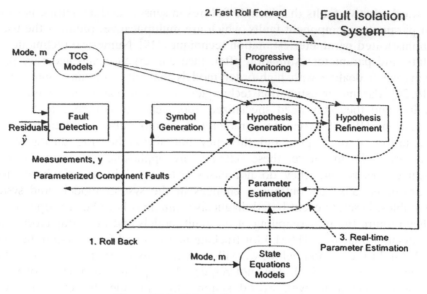

Figure 17.6. Hybrid diagnosis scheme.

the predictions from the observer to look for significant deviations in the observed signals are discussed elsewhere [16]. Our fault detectors for continuous systems have to be modified to signal faults only when abrupt changes cannot be attributed to mode changes [14, 24].

The overall scheme for hybrid diagnosis is illustrated in Figure 17.6. We overcome limitations of quantitative schemes by combining robust qualitative reasoning mechanisms with quantitative parameter estimation schemes for parametric fault isolation [8]. Hybrid bond graph models discussed in Section 17.3 form the basis for generating parameterized timed causal graphs (TCGs), a representation that captures system dynamics as causal links between system variables, annotated by temporal relations, such as instantaneous effects and integral relationships [23]. The bond graph representation explicitly includes component parameters that govern system dynamics as resistive, capacitive, inertial, transformer, and signal propagation elements. The TCG representation makes explicit the effect of changes in parameter values on the dynamics of system variables. The fault isolation methodology for hybrid systems is broken down into three steps, as follows:

1. A fast *roll-back process* using qualitative reasoning techniques to generate possible fault hypotheses. Since the fault could have occurred in a mode earlier than the current mode, fault hypotheses need to be characterized as a two-tuple ⟨*mode, fault parameter*⟩, where mode indicates the mode in which the fault occurs, and fault parameter is a parameter of the implicated component whose deviation possibly explains the observed discrepancies in behavior.

2. A quick *roll-forward process* using progressive monitoring techniques to refine the possible fault candidates. The goal is to retain only those candidates whose fault signatures are consistent with the current sequence of measurements. After the occurrence of a fault, the observers predictions of autonomous mode transitions may no longer be correct, therefore, determining the consistency of fault hypotheses also requires the fault isolation unit to roll forward to the correct current mode of system operation.

3. A *real-time parameter estimation* process using quantitative parameter estimation schemes. The qualitative reasoning schemes are inherently imprecise. As a result a number of fault hypotheses may still be active after Step 2. We employ a least squares estimation technique on the input–output form of the system model to estimate values of the fault parameter that are consistent with the sequence of measurements made on the system.

Each of these steps is explained in greater detail next.

Fast Roll-Back Process Using Back-Propagation. We use a qualitative reasoning mechanism to identify an initial candidate set to explain the discrepancy in the observations and predictions. This is achieved by back-propagating the qualitative value of the discrepancy $(-, 0$ or $+)$ through the temporal causal graph of the system. Back-propagation for initial hypotheses generation has to be performed across modes the system has traversed through because the fault may have occurred in a previous mode but the manifestations are not seen until a later mode. For example, this happens when none of the observed variables are affected in the mode in which the fault occurs. The problem is that once a fault occurs, the predicted mode sequence of the observer may no longer be correct, and a worst-case analysis may require considering all possible modes of the system in generating fault hypotheses. However, we make the assumption that the controller model is correct; therefore, *the observer must have predicted the correct mode sequence until the fault occurred. As a consequence, the mode in which the fault occurred must be in the predicted trajectory of the observer.*

The back-propagation algorithm determines all possible component parameter deviations that are consistent with observed discrepancies in each of the modes in the mode trajectory up to the current mode. This ensures that the true fault hypothesis, which includes the fault and the mode in which the fault occurred ($\langle mode, fault\ parameter \rangle$), will be included in our initial hypothesis set. As a reasonable heuristic, we limit our search by looking back only k modes. k is determined by measurement selection analyses [25], which verifies that the effects of any fault must manifest within k mode changes.

As an example scenario, we introduce a R1+ fault (this corresponds to a partial block in outlet pipe1, causing an increase in its resistance) into the system in mode 10, when the level of fluid in the tank 1 is very close to the height of the connecting pipe. This causes the autonomous transition, where flow occurs in the connecting pipe but the observer does not predict this transition. However, the fault causes the level of fluid in tank 2 to start increasing from a 0 value. The observer still works with the model where the second tank is isolated; therefore, the predicted pressure value in tank 2 is 0. At some point this leads to Tank2Pressure + (above nominal) discrepancy. The back-propagation algorithm triggered by this discrepancy generates candidates, in what the algorithm thinks is the current mode (source flow on, tanks disconnected), but the actual mode is source flow on and tanks connected. The mode produces C1− and R1+ as candidates. The system also backtracks to the previous mode (source off and tanks disconnected), but that mode does not generate any new candidates.

Quick Roll-Forward Process Using Progressive Monitoring. The quick roll-forward process is initiated by applying a progressive monitoring scheme [12] to compare signatures for the hypothesized faults against observations within a mode of system behavior. Typically, a mismatch would indicate that the hypothesized fault is not consistent with the observations, and the hypothesis would be dropped. In the hybrid system case, it may also imply that the hypothesized mode is not the current mode the system is operating in. In that case, we hypothesize potential autonomous mode changes that could have taken place but were not hypothesized by the observer after the occurrence of the fault. Therefore, the goal of the quick roll-forward process is to arrive at situations where the system measurements are consistent with the hypothesized fault and predicted mode.

We can qualitatively predict future behavior of the system under each of the hypothesized fault conditions by forward-propagating through the TCG [8, 12]. These predictions include magnitude and higher-order derivatives of the variables of the system in the mode that the fault occurred. If the qualitative predictions do not match the observations in the mode, we consider possible autonomous mode changes from the current mode using the model of the controller plus additional information from the system to limit the number of possible mode transitions. For each of the hypothesized modes, the new TCG is used to derive a new set of qualitative predictions. If the predictions still do not match the observations, then the candidate mode is dropped. For a fault hypothesis, if all mode candidates are dropped, then the fault candidate is dropped since none of the possible mode change sequences could make the predictions for the fault candidate and the observations consistent. To limit the search in the quick roll-forward phase, the parameter k is again invoked to limit the number of autonomous mode transitions that have to be considered.

Going back to our example fault scenario, the signature for C1− implies that there should be a discontinuous change in Tank1Pressure. Since a discontinuous change in pressure was not observed, we eliminate C1− as a candidate. The signature for R1+ in the current mode indicates that the

Tank2Pressure is unaffected. This does not agree with the observations, so we hypothesize a possible autonomous transition, apply this transition, and generate fault signatures for candidate R1+ in this new mode. In this mode, the signatures for Tank1Pressure and Tank2Pressure match the actual observations and so R1+ is still considered to be a possible candidate.

Real-Time Parameter Estimation. For the quantitative analysis, we estimate the deviated parameter values for each of the remaining fault candidates. To do this, we rewrite the state space equations in terms of the parameter associated with the fault candidate and use least-squares estimation techniques to derive the faulty parameter value [8]. First, the input−output transfer function model of the system is derived as follows:

The input−output model is expressed in terms of the component parameters by applying the Mason's gain rule

$$y(t) = \frac{g(q^{-1})}{h(q^{-1})} u(t)$$

to the TCG model of the system. Note that the above formula works only for linear systems. For nonlinear systems (which are approximated by piecewise linear models), a separate formula has to be developed for each region, and the parameter estimation procedure has to take these regions into account. For each hypothesis, we start a new parameter estimator. To make the computation simpler and achieve convergence faster, the parameter estimator estimates only those g and h coefficients that include the hypothesized fault parameter. All other coefficients are fixed using nominal values of the remaining parameters. If the computed g and h coefficients converge, we can invert the relation between the coefficients and the parameter to estimate the parameter value. If a controlled transition occurs during the estimation process, we calculate the new input−output model and continue the estimation process using the new model. We have demonstrated by empirical analysis that the estimator corresponding to the true fault parameter converges, whereas the estimators for the other parameters diverge.

In our example system, we note that the R1 fault parameter affects only two of the h coefficients. Computing the R1 value from these two coefficients, we get R1 = 9.87 and R1 = 10.08, respectively. Since these values are close, we make the determination that this is the true fault parameter and use a simple average to estimate the faulty value of R1 to be 9.98.

17.5.2. The Discrete Approach

The discrete approach uses a discrete model of the dynamic system. Discrete models can offer an effective abstraction for diagnosis of large systems [22]. We employ a graphic modeling paradigm based on failure propagation

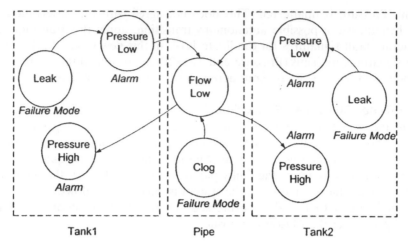

Figure 17.7. Example of a failure propagation graph for the two-tank system.

graphs (FPGs), a set-relational model that can easily represent nondeterminism. Nondeterminism is a characteristic of discrete event models that are derived as an abstraction of deterministic hybrid dynamical systems. We have implemented the discrete diagnosis algorithm using ordered binary decision diagrams [27] (OBDDs), which are used to manipulate sets and relations symbolically, thus enabling the representation of very large state-spaces.

Failure Propagation Model. Failure propagation graphs [7] capture the relation between *failure modes* of physical components and *discrepancies* (functional failures) that can be attributed to these failures. Discrepancies may have associated alarms that indicate the presence or absence of the discrepancy. Failures propagate to discrepancies and discrepancies can cascade as well. The result is a directed graph, which leads from failure modes to observed alarms in the system. Figure 17.7 illustrates a failure propagation graph for the two-tank system.

The failure propagation graph represents trajectories that link the set of failures that can occur in the system to possible alarms. This modeling approach has the advantage of being intuitive, and it captures knowledge about the system at a high level. At the same time, it is important to assign precise meaning to the graph with relation to discrete abstractions of a hybrid dynamical system.

Starting from a closed-loop model of a hybrid dynamical system, a discrete-event model of the closed-loop system can be obtained by defining events that correspond to region crossings defined by hypersurfaces in the hybrid state-space [23, 26]. The set of discrepancies in the failure propagation graph are interpreted as a distinguished set of events that correspond to the occurrence of faults in system components. The failure propagation graph describes the order of discrepancies in time, following the failure occurrence.

The failure propagation graph is expressed as a six-tuple: (V, E, A, F, Q, T), where (V, E) is the graph (vertices and edges), A is a set of alarms, F is a set of failure modes, Q is a relation $Q \subseteq V \times A$, and T is a relation $T \subseteq V \times F$. To fully represent nondeterminism in the model, which allows for diagnosis of multiple simultaneous faults, it is useful to define Q and T as relations between the nodes of the graph and the powerset of the alarms and failures, respectively. This approach is detailed in reference 27. For computational efficiency, the FPG model can be represented symbolically using OBDDs.

Diagnosis Algorithm for Single Fault Scenario. The diagnosis algorithm, which makes the single-fault assumption, starts by parsing the FPG model into two relations, *Ancestors* $\subseteq A \times A$ and *Descendants* $\subseteq F \times A$, representing the relation between failures and alarms. The diagnosis module is initialized with three sets: *Hypothesis* $\subseteq F$, *AlreadyRinging* $\subseteq A$, and *MissingUpstream* $\subseteq A$, all initialized to empty sets. A hypothesis data structure is maintained, which includes possible failure modes matched with a ranking,

```
global set Hypothesis, AlreadyRinging, MissingUpstream;
global const set Ancestors, Descendants;

procedure RefineHypothesis (in set Alarms)
local set NewFailureModes, NewMissingUpstream, MissingAncestor,
        PromotedNewFailureModes;
begin
(* Calculate new failure modes that have descendants in the current set
   of alarms but are not in the hypothesis set yet*)
NewFailureModes :=  relationalproduct(Descendants,Alarms) intersect
                    complement(Hypothesis);
(* Add new failure modes to the hypothesis set *)
Hypothesis := hypothesis union NewFailureModes;
(* Calculate missing ancestor alarms: the ancestors of the current
   alarms, which are not in the set of missing upstream alarms and are
   not ringing yet *)
MissingAncecstors := relationalproduct(Ancestors,Alarms) intersect
                     complement(MissingUpstream) intersect
                     complement(AlreadyRinging);
(* Calculate new missing upstream alarms: all the descendant alarms of
   the current hypothesis set, which are also in the missing ancestor
   alarm set *)
NewMissingUpstream := relationalproduct(Descendants,Hypothesis)
                      intersect MissingAncestors;
(* Update set of missing upstream alarms *)
MissingUpstream := MissingUpstream union NewMissingUpstream;
(* Update set of already ringing alarms *)
AlreadyRinging := AlreadyRinging union Alarms;
(* Increment rank of faults which have new supporting alarms and no new
   missing upstream alarms *)
PromotedNewFailureModes := relationalproduct(DescendantAlarms,Alarms)
                           intersect
                           complement(relationaproduct(Descendants,
                                      NewMissingUpstream));
end
```

Figure 17.8. Discrete diagnosis hypothesis refinement algorithm.

designating their plausibility. The plausibility is a function of the alarms that occur *downstream* from a failure mode and are considered *supporting* alarms versus alarms that occur upstream from supporting alarms and are considered to be *missing* alarms. Whenever new alarms are reported, the hypothesis set is refined based on the set-relational computation algorithm that appears in Figure 17.8.

Example: Two-Tank System. The two-tank system is operated with fluid flowing from Tank 1 to Tank 2 through the connecting pipe. For a fault scenario in which a decrease in pressure in Tank 2 is detected, followed by an increase in pressure in Tank 1, the results of the diagnosis using the algorithm in Figure 17.8 are: "Pipe.Clog" with rank 2 and "Tank1.Leak" with rank 1 and "Tank2.Leak" with rank 1.

17.6. CONTROLLER SELECTION

Our approach is to develop a library of controllers, which is indexed by sets of characteristics. The goal is to use the information about current system state—that is, the current mode of operation and system state vector along with failed and degraded states of components and subsystems to select a controller that best suits current and long-term performance objectives.

We address the controller reconfiguration task on two levels. At the supervisory (discrete) level, reconfiguration implies modification of high-level control actions. This can take the form of replacing a current action sequence by a new sequence, or altering the sequence of actions in the current set. This type of reconfiguration requires that the supervisory control logic be explicitly represented as a data structure. Our challenge is to (a) adopt model-based approaches to representing supervisory control programs and (b) develop reconfiguration procedures governed by different kinds of fault conditions. At the lower (continuous) level of control, the system relies on regulators, which can range from simple switching controllers, to PID mechanisms, and then model-based controllers. Reconfiguration at this level can take on three different forms:

1. Set point changes for handling simple fault situations, such as a partially degraded component.
2. Controller tuning for handling cases where the fault changes the plant dynamics (e.g., changes in the capacitive and inertial parameters in the plant) and retuning of the controller is a viable solution.
3. Structural changes (i.e., rewiring or replacing the regulators) may compensate for complex faults where the current controller architecture is unable to maintain the desired control because of a significant fault (e.g., sensor faults, actuator faults, and major structural changes in the plant, such as pump failures or valves stuck at closed).

All of the above cases may lead to the introduction of large switching transients into the system. We are investigating a number of ways to manage the reconfiguration. Two examples are as follows:

- *The Blender Approach.* In this technique, a "new" controller gradually replaces the "old" controller. The reconfiguration starts with the old and new controllers connected to a tapered switch. Initially, the output of the "old" controller is fully connected to the plant. As time proceeds, the switch setting is gradually moved from the old to the new controller setting. At the end of the process, the new controller completely replaces the old one. Interesting research issues that we have to deal with include design of the blending function for control signals at intermediate stages of the tapered switching process and the speed (and thus the dynamics) at which the switching is accomplished to avoid unnecessary transients in the plant dynamics during the reconfiguration process.
- *The State Initialization Approach.* If rapid reconfiguration is required, the tapered switch approach may not be fast enough. In this case, the new controller should replace the old one, possibly within the sampling interval set for the system. To avoid large bumps, the internal state of the new controller should be initialized in such a way that it generates a control signal after reconfiguration that is minimally different from the last signal generated by the old controller.

There is an interesting and highly relevant aspect of controller reconfiguration that is also being addressed: the explicit management of reconfiguration transients. Early results [17, 18] show that there are a number of techniques available for mitigating reconfiguration transients in control systems. If the selected approach of controller reinitialization and/or blending does not meet the requirements for the reconfiguration dynamics, then other, more explicit transient suppression techniques can be applied to mitigate the effects of switching.

17.7. CONCLUSION AND FUTURE WORK

We have applied our continuous and discrete FDI methodology to diagnosing faults in a two-tank system with a number of valves. A simple supervisory controller model took the system through a number of filling, emptying, and mixing cycles. We were successful in tracking continuous system behavior through discrete mode changes, and isolating faults when they occurred, with the discrete and continuous diagnostics algorithms. As a next step, we would like to extend the two diagnostic algorithms to work in a more cohesive fashion, and inform each other as they come up with fault hypotheses. Once this step is completed, we will introduce the controller selection mechanisms to have a comprehensive implementation of the FACT architecture that has been presented in this chapter.

We are also looking at applying this technology to more real-world applications, such as the fuel transfer system in modern aircraft. The physical components of the fuel system include a number of tanks, interconnecting pipes, valves, and pumps. In addition, the system is equipped with sophisticated controls to support reliable and robust fuel delivery under a variety of flight conditions, at the same time ensuring that the aircraft center of gravity is not compromised. In addition, the controllers have to deal with a number of fault scenarios, such as pump failure and pipe leaks. The goal under such conditions is not to compromise aircraft safety, but to save as much fuel as one can to continue the current mission. To achieve this, the system should utilize built-in redundancy mechanisms to compensate for the failure and, at the same time, maintain control. We are currently developing models of a generic aircraft fuel system, and we are testing and validating the FACT tools and techniques on a number of example scenarios that have been generated using a high-fidelity simulator.

ACKNOWLEDGMENTS

The DARPA/ITO SEC program (F33615-99-C-3611), and The Boeing Company have supported the activities described in this chapter. We would like to thank Dr. Kirby Keller and Mr. Mark Kay for their help.

REFERENCES

1. P. J. Antsaklis and K. M. Passino, Introduction to intelligent control systems with high degree of autonomy, *Introduction to Intelligent and Autonomous Control*, P. J. Antsaklis and K. M. Passino, editors, Chapter 1, Kluwer Academic Publishers, 1993, pp. 1–26.

2. P. J. Antsaklis, Defining intelligent control, Report of the Task Force on Intelligent Control, P. J Antsaklis, Chair, in *IEEE Control Systems Magazine*, pp. 4–5 and 58–66, June 1994.

3. P. J. Antsaklis, K. M. Passino and S. J. Wang, An introduction to autonomous control systems, *IEEE Control Systems* 11(4):5–13, June 1991.

4. M. S. Branicky, V. Borkar, and S. Mitter, A unified framework for hybrid control: Background, model, and theory, in *Proceedings of the 33rd IEEE Conference on Decision and Control*, Lake Buena Vista, FL, Paper No. LIDS-P-2239, 1994.

5. P. J. Mosterman and G. Biswas, A theory of discontinuities in physical system models, *Journal of the Franklin Institute* 335B:401–439, 1998.

6. E. J. Manders, P. J. Mosterman, and G. Biswas, Signal to symbol transformation techniques for robust diagnosis in TRANSCEND, in *Tenth International Workshop on Principles of Diagnosis*, Loch Awc, Scotland, 1999, pp. 155–165.

7. A. Misra, J. Sztipanovits, and J. Carnes, Robust diagnostics: Structural redundancy approach, Knowledge Based Artificial Intelligence Systems in Aerospace and Industry, SPIE's Symposium on Intelligent Systems, Orlando, 1994.

8. E. J. Manders, S. Narasimhan, G. Biswas, and P. J. Mosterman, A combined qualitative/quantitative approach for efficient fault isolation in complex dynamic systems, *4th Symposium on Fault Detection, Supervision and Safety Processes*, 2000, pp. 512–517.

9. J. Sztipanovits, G. Karsai, Model-integrated computing, *IEEE Computer*, **April**:110–112, 1997.

10. P. J. Mosterman and G. Biswas, Towards procedures for systematically deriving hybrid models of complex systems, in *Hybrid Systems: Computation and Control, Third International Workshop, Lecture Notes in Computer Science*, Vol. 1790, N. Lynch and B. Krogh, editors., Springer-Verlag, Berlin, March 2000, pp. 324–337.

11. R. C. Rosenberg, and D. C. Karnopp, *Introduction to Physical System Dynamics*, McGraw Hill, 1983.

12. P. J. Mosterman and G. Biswas, Diagnosis of continuous valued systems in transient operating regions, *IEEE Transactions on Systems, Man and Cybernetics* **29**:554–565, 1999.

13. D. Harel and M. Politi, *Modeling Reactive Systems with Statecharts: The Statemate Approach*, McGraw-Hill, New York, 1998.

14. S. Narasimhan, G. Biswas, G. Karsai, T. Pasternak, and F. Zhao, Building observers to handle fault isolation and control problems in hybrid systems, in *Proceedings of the 2000 IEEE International Conference on Systems, Man, and Cybernetics*, Nashville, TN, 2000, pp. 2393–2398.

15. A. Gelb, *Applied Optimal Estimation*, MIT Press, Cambridge, MA, 1979.

16. R. J. Patton, P. M. Frank, and R. N. Clark (editors), *Issues of Fault Diagnosis for Dynamic Systems*, Springer-Verlag, London, 2000.

17. H. L. Jones, Fault Detection in Linear Systems, Ph.D. thesis, Massachusetts Institute of Technology, 1973.

18. R. Mangoubi, *Robust Estimation and Failure Detection, A Concise Treatment*, Springer Verlag, New York, 1998.

19. J. J. Gertler, *Fault Detection and Diagnosis in Engineering Systems*, Marcel Dekker, New York, 1998.

20. J. Chen and R. J.Patton, *Robust Model-Based Fault Diagnosis for Dynamic Systems*, Kluwer Academic, Boston, MA, 1999.

21. L. Ljung, *System Identification: Theory for the User*, Prentice-Hall, Englewood Cliffs, NJ, 1987.

22. M. Sampath et al., Fault Diagnosis using discrete-event models, *IEEE Transactions on Control Systems Technology* **4**(2):105–124, 1996.

23. J. Lunze, A timed discrete event abstraction of continuous dynamic systems, *International Journal of Control* **72**:1147–1164, 1999.

24. S. Narasimhan and G. Biswas, Using supervisory controller models for more efficient diagnosis of hybrid systems, *Hybrid Systems: Control and Computation, International Workshop*, Rome, Italy, 2000, submitted.

25. S. Narasimhan, P. J. Mosterman, and G. Biswas, A systematic analysis of measurement selection algorithms for fault isolation in dynamic systems, *9th International Workshop on Principles of Diagnosis*, Cape Cod, MA, May 1998, pp. 94–101.

26. X. D. Koutsoukos, P. J. Antsaklis, J. A. Stiver, and M. D. Lemmon, Supervisory Control of Hybrid Systems, in *Proceedings of the IEEE: Special Issue on Hybrid Systems*, P. J. Antsaklis, editor, 2000, pp. 1026–1049.

27. T. Pasternak, Extended Relational Models for Diagnosis, Masters thesis, Vanderbilt University, August 2000.

28. R. Alur et al., Hybrid automata: An algorithmic approach to the specification and verification of hybrid systems, in R. L. Grossman et al., editors, *Lecture Notes in Computer Science*, Vol. 736, Springer, Berlin, 1999, pp. 209–229.

29. J. Lunze, Diagnosis of quantized systems by means of timed discrete-event representation, in *Proceedings of Third International Workshop on Hybrid Systems, Computation and Control, Lecture Notes in Computer Science*, Vol. 1790, March 2000, pp. 258–271.

30. G. Simon, T. Kovcshzy, and G. Péceli, Transients in reconfigurable control loops, *IEEE Instrumentation and Measurement Technology Conference, IMTC/2000*, Baltimore, Maryland, 2000.

31. G. Simon, T. Kovcshzy, and G. Péceli, Transient management in reconfigurable systems, *International Workshop on Self Adaptive Software*, Oxford University, England, 2000.

32. C. S. Pierce, Note B: The logic of relatives, in *Studies in Logic by Members of the Johns Hopkins University*, Little Brown and Co., Boston, 1883

33. A. Ledeczi, A. Bakay, and M. Maroti, Model-integrated embedded systems, in *Self Adaptive Software, Lecture Notes in Computer Science*, Vol. 1936, Springer-Verlag, New York, 2001.

CHAPTER 18

COMPUTATIONAL TOOLS FOR THE VERIFICATION OF HYBRID SYSTEMS

CLAIRE J. TOMLIN, STEPHEN P. BOYD, IAN MITCHELL, ALEXANDRE BAYEN, MIKAEL JOHANSSON, and LIN XIAO

Editors' Notes

The hybrid systems framework provides an appealing means for verifying the safety of dynamical systems. This chapter addresses safety verification as a reachability problem: Given an unsafe subset of the system state space and an initial state, is the former reachable from the latter? If it is reachable despite any controllable actions that we can take then the system is provably unsafe.

The continuous-time nonlinear dynamics of the system need to be considered in assessing reachability. The authors' formulation requires the solution of a Hamilton-Jacobi partial differential equation, and a grid-based numerical solution approach based on level set methods is used for this purpose. Simulation examples are presented for three flight management applications: two-aircraft collision avoidance, the related problem of conflict resolution, and ensuring safety during final landing approach.

The exact reachability computation is prey to the curse of dimensionality: its computational complexity is exponential with respect to the continuous dimension. This chapter also presents an alternative approach, which is based on over-approximating the reachable set of states with a polyhedron. This is also computationally intractable since the propagation of the system's dynamics will result in a potentially unlimited number of constraints (faces of the polyhedron), but the authors have developed a novel technique for identifying and pruning redundant and irrelevant constraints. This technique promises to be computationally feasible for very high dimensional problems. The basis of the approach is the computation of the maximum volume ellipsoid contained in a polyhedron, a computation that can be formulated as a convex optimization problem (for which global and efficient algorithms are available).

Note: Research supported by the DARPA Software Enabled Control (SEC) Program administered by AFRL under contract F33615-99-C-3014.

Software-Enabled Control. Edited by Tariq Samad and Gary Balas
ISBN 0-471-23436-2 © 2003 IEEE

18.1. INTRODUCTION

For about the past 10 years, researchers in the traditionally distinct fields of control theory and computer science verification have proposed models, along with verification and controller synthesis techniques for complex, safety critical systems. The area of *hybrid systems theory* studies systems that involve the interaction of discrete event and continuous-time dynamics, and it provides an effective modeling framework for complex continuous systems with large numbers of operating modes. Examples include continuous systems controlled by a discrete logic such as the autopilot modes for controlling an aircraft, systems of many interacting processes such as air or ground transportation systems in which discrete dynamics are used to model the coordination protocols among processes, or continuous systems that have a phased operation, such as biological cell growth and division.

The current and potential impact of hybrid systems lies in the confluence of computational methods from control theory and from formal methods in computer science verification. In the examples mentioned above, the system dynamics are complex enough that traditional analysis and control methods based solely on differential equations are not computationally feasible; analysis based solely on discrete event dynamics ignores critical system behavior. Our interest lies in developing computational tools for analyzing and controlling the behavior of hybrid systems; our eventual goal is to develop a real-time tool to provide online verification that a hybrid system satisfies its specified behavior. This goal addresses one of the central themes of software-enabled control, in which we seek a systematic methodology for the design, validation, and implementation of control software.

In our work to date, the system specification that we are most interested in is that of system safety, which asks the question, Is a potentially unsafe configuration of the system reachable from an initial configuration? More important for control theory, given a set of desired configurations, it is crucial to be able to design the hybrid system control inputs to achieve these configurations. Previous work in this area had focused on hybrid systems with very simple continuous dynamic equations (such as clocks, or linear decoupled systems). Our research has three components: the problem formulation for computing *exact reachable sets* of hybrid systems and the design of a software tool to perform such calculations; the development of an algorithm for computing overapproximations of reachable sets which works efficiently in high dimensions; and the design of a real-time aircraft testbed for these algorithms. We will focus on the first component in this chapter and briefly outline our work on the second.

18.2. HYBRID SYSTEM MODEL

The main difference between our hybrid system model and leading models in the literature (see, for example, the work of Alur and Henzinger [1] on linear hybrid automata) is that we include accurate, nonlinear models of the hybrid system continuous dynamics. The model and algorithm have been developed jointly with Lygeros and Sastry; full details are presented in reference 2.

A hybrid automaton is a finite-state machine with discrete states $\{q_1, q_2, \ldots, q_k\}$, in which each discrete state has associated continuous dynamics $\dot{x} = f(q_i, x, v)$, with $x \in \Re^n$, and continuous inputs and disturbances $v = (u, d)$, where $u \in U$ and $d \in D$.

Definition 18.1 (Hybrid Automaton). *A hybrid automaton H is a collection*

$$H = (Q, X, \Sigma, V, \text{Init}, f, \text{Inv}, R)$$

where

- *$Q \cup X$ is the set of state variables, with Q a finite set of discrete states, and $X = \Re^n$;*
- *$\Sigma = \Sigma_1 \cup \Sigma_2$ is a finite collection of discrete input variables, where Σ_1 is the set of discrete control inputs, and Σ_2 is the set of discrete disturbance inputs;*
- *$V = U \cup D$ is the set of continuous input variables, where U is the set of continuous control inputs, and D is the set of continuous disturbance or uncontrollable inputs; we denote the spaces of input (disturbance) trajectories as the sets of piecewise continuous functions $\mathcal{U}$ ($\mathcal{D}$ respectively), which take values in U (D respectively);*
- *Init $\subseteq Q \times X$ is a set of initial states;*
- *$f : Q \times X \times V \to TX$ is a vector field describing the evolution of x for each $q \in Q$; f is assumed to be globally Lipschitz in X (for fixed $q \in Q$) and continuous in V;*
- *Inv $\subseteq Q \times X \times \Sigma \times V$ is called an invariant, and defines combinations of states and inputs for which continuous evolution is allowed;*
- *$R : Q \times X \times \Sigma \times V \to 2^{Q \times X}$ is a reset relation, which encodes the discrete transitions of the hybrid automaton.*

We refer to $(q, x) \in Q \times X$ as the *state* of H and to $((\sigma_1, \sigma_2), (u, d)) \in \Sigma \times V$ as the *input* of H. The control actions (σ_1, u) model those inputs over which the designer has control, and the disturbance actions (σ_2, d) model inputs over which the designer has no control, such as uncertainties in the actions or behaviors of the other aircraft in the system. We assume that the designer has complete knowledge of the *bounds* on these disturbance actions (Σ_2, D).

Associated to the hybrid automaton H is a *specification* which describes the condition that one would like the system to satisfy. Our work has been motivated by verification and synthesis for safety critical applications, and as such we have been primarily interested in safety specifications. These specifications are encoded as subsets of the state space of the hybrid system: The set $G_0 \subseteq Q \times X$ is that subset which is defined *a priori* to be unsafe. The problem specification is therefore to keep the system state outside of the unsafe set.

18.3. EXACT REACH SET COMPUTATION USING LEVEL SETS

In this section, we describe the development of a general-purpose tool for exact reachable set computation—the core of which is a new variant of a "local level set" algorithm that efficiently computes an accurate representation of the reachable set boundary. We demonstrate in an example the numerical convergence of our computation by analyzing the results as the continuous state space grid is made finer, a standard method of validation for scientific computing codes. In this way, we show that high accuracy can be achieved at the cost of increased computational time and space. We illustrate our tool in a single-mode aircraft conflict resolution example [3, 4] and in a three-mode conflict resolution example, as well as in an example of a six-mode commercial aircraft auto-lander [5], which exhibits nondeterminism and cycles in its discrete behavior.

Our motivation for this component of the research stems from the belief that for many applications of hybrid systems, it is important to be able to accurately represent the reachable set. We have dealt primarily in the safety verification of avionic systems, where accurate representation of the safe region of operation translates into the ability to operate the system closer to the boundaries of that region, at a higher performance level than previously allowed. For very high-dimensional state spaces, additional logic (such as projection operators) or new techniques (such as the convex overapproximations of the next section) will be needed; however, our results in this chapter show that it is feasible to do exacting computation for hybrid systems with nonlinear continuous dynamics in three continuous state dimensions and six discrete modes, and we believe that it will be feasible to extend this up to five continuous dimensions and large numbers of discrete modes.

18.3.1. Reachability for Hybrid Systems

In this section we summarize the general framework for handling the interaction between discrete and continuous dynamics (following reference 2 and 5).

Fundamentally, reachability analysis in discrete, continuous, or hybrid systems seeks to partition states into two categories: those that can reach a given target set and those that cannot. We will label these two sets of states G and $E = G^c$, respectively (the symbol G has been used traditionally to represent unsafe sets, E is used as a symbol for "escape" sets). Any inputs to the hybrid automata are assumed to lie in bounded sets and to have the goal of locally maximizing or minimizing the reachable set: At each iteration, the reachability algorithm chooses values for inputs ξ_G that maximize the size of

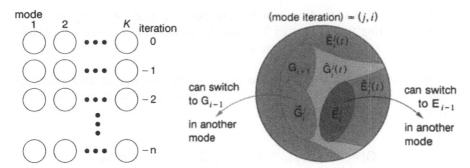

Figure 18.1. Iterative reachability algorithm: Showing detail of iteration for discrete mode j at iteration i.

G and values for inputs ξ_E that minimize the size of G (and hence maximize the size of E). Any nondeterminism in the transition relation is also utilized to consistently maximize or minimize G, depending on the goal of the reachability computation. For hybrid automata, the discrete inputs σ and continuous inputs v can be assigned to the two categories $\xi_G = (\sigma_G, v_G)$ and $\xi_E = (\sigma_E, v_E)$ according to whether they seek to maximize or minimize G. For example, for our hybrid system model with control actions (σ_1, u) and disturbance actions (σ_2, d), we would have that $\xi_G = (\sigma_G, v_G) = (\sigma_2, d)$, and $\xi_E = (\sigma_E, v_E) = (\sigma_1, u)$.

The reachability computation follows an iterative, two-stage algorithm shown graphically in Figure 18.1. The outer iteration computes reachability over the discrete switches, producing iterates G_i and E_i at iteration $i = 0, -1, -2, \ldots$, where a negative index is used to indicate that the algorithm is initialized with a target set and run backward to compute those states that can reach the target set. The inner iteration runs a separate continuous reachability problem in each of the discrete modes $j = 1, 2, \ldots, K$ to compute the estimates G_i^j and E_i^j. We define the "switch" sets:

- $\vec{G}_i^j$ contains all states in mode j from which a discrete transition to a state in G_{i-1} (typically a state in another mode) can be forced to occur through the application of a discrete input σ_G; these states will be defined by the invariant of mode j and the guards of the transitions from mode j.

- $\vec{E}_i^j$ contains all states from which a discrete transition to a state in E_{i-1} can be forced to occur through the application of a discrete input σ_E; these states are also defined by the invariant of mode j and the guards of transitions from mode j.

Then the goal of the continuous reachability tool is to identify the "flow" sets:

- $\tilde{G}_i^j(t)$ contains states from which for all v_E there exists v_G that will force the resulting trajectory to flow into $G_{i-1}^j \cup \vec{G}_i^j$ within time t.

- $\tilde{\mathsf{E}}_i^j(t)$ contains states from which there exists ν_E that for all ν_G will force the resulting trajectories to flow into $\vec{\mathsf{E}}_i^j$ within time t or to stay outside of $\mathsf{G}_{i-1}^j \cup \tilde{\mathsf{G}}_i^j$ for at least time t.

Note that in some problems the order of the existential and universal quantifiers in the definition above must be reversed. Given these sets, we obtain

$$\mathsf{G}_i^j = \lim_{t \to \infty} \tilde{\mathsf{G}}_i^j(t) \qquad \mathsf{G}_i = \bigcup_{j=1}^K \mathsf{G}_i^j, \qquad \mathsf{G} = \lim_{t \to \infty} \mathsf{G}_i$$

$$\mathsf{E}_i^j = \lim_{t \to \infty} \tilde{\mathsf{E}}_i^j(t), \qquad \mathsf{E}_i = \bigcup_{j=1}^K \mathsf{E}_i^j, \qquad \mathsf{E} = \lim_{t \to \infty} \mathsf{E}_i$$

where G_0^j is the set of initial conditions of the reachability problem and $\mathsf{E}_0^j = (\mathsf{G}_0^j)^c$. Simple modifications of this algorithm suffice to solve finite-time-reachability problems.

The procedure described above, developed in references 2 and 5, was motivated by the work of references 6 and 7 for reachability computation and controller synthesis on timed automata, and that of reference 8 for controller synthesis on linear hybrid automata. In that development the reachability problem's objective was to determine E—the largest controllable invariant subset of the state space—by computing the set of states G that were reachable in backwards time from the set of predefined unsafe states. In terms of the definitions above, control inputs from this problem lie in ξ_E and disturbance inputs in ξ_G. For safety, any model nondeterminism would be used to maximize the unsafe set G.

18.3.2. Continuous Reachability Using Level Sets

While practical algorithms for computing discrete reachability over many thousands of states have been designed and implemented, determination of continuous reachability for even low-dimensional systems is still an open problem. The continuous portion of a hybrid reachability problem requires methods of performing four key operations on sets: unions, intersections, tests of equality, and evolution according to the discrete mode's continuous flow field. The choice of representation for sets dictates the complexity and accuracy of these operations; consequently, continuous reachability algorithms can be classified according to how they represent sets.

For our exact representation scheme, we characterize the set being tracked implicitly by defining a "level set function" $J(x, t)$ throughout the continuous state space which is negative inside the set, zero on its boundary, and positive outside, and which encodes the initial data in $J(x, 0)$. The intersection of two such sets is simply the maximum of their level set functions at each point in state space, and the union is the minimum; a variety of easily implemented equality tests are possible. Evolution of a level set under a nonlinear flow field is governed by the Hamilton–Jacobi (HJ) partial differential equation (PDE) (see, for example, reference 3)

$$-\frac{\partial J(x,t)}{\partial t} = \min\left\{0, \max_{\nu_{\min}} \max_{\nu_{\max}} f(x, \nu_{\min}, \nu_{\max})^T \nabla J(x,t)\right\}$$

$$= \min\{0, H(x, \nabla J(x,t))\} \tag{18.1}$$

where $\nu_{\min}$ are those continuous inputs trying to minimize the size of the set being tracked, and $\nu_{\max}$ are those inputs trying to maximize its size. The order of the optimization must be chosen appropriately for the situation. The first minimization on the right-hand side of Eq. (18.1) ensures that the set can only grow (states that are unsafe cannot become safe at a future time). That the subzero level sets of the solution of (18.1) yields the correct reachable set is proved in reference 9. The implicit representation has a number of advantages when compared with the explicit representations that other researchers are pursuing, including a conceptually simple representation of very general sets and a size that is independent of the complexity of the set (although it grows exponentially with dimension). In addition, a set of sophisticated numerical techniques to accurately solve PDEs may be drawn upon for computation. In the remainder of this section, we focus on the representation (18.1) and assume that the modeler can compute the appropriate optimization over inputs in (18.1) if given x and $\nabla J(x,t)$.

The HJ PDE (18.1) is well known to have complex behavior. Even with smooth initial data $J(x,0)$ and continuous Hamiltonian $H(x, \nabla J)$, the solution $J(x,t)$ can develop discontinuous derivatives in finite time; consequently, classical infinite time solutions to the PDE are generally not possible. In the quest for a unique weak solution, Crandall and Lions introduced the concept of the viscosity solution [10]. For most problems of interest, finding the analytic viscosity solution is not possible (see reference 11 for cases in which it is possible), and so we seek a numerical solution. We approximate the solution of (18.1) on a Cartesian grid of nodes. Three terms in the equation must be approximated at each node, based on the values of the level set function at that node and its neighbors: the gradient ∇J, the Hamiltonian H, and the time derivative $\partial J(x,t)/\partial t$. We discuss each of these separately.

In each dimension at each grid point there exist both left and right approximations of the gradient ∇J, depending on which neighboring grid points' values are used in the finite-difference calculation. We label the vector of left approximations ∇J^-, and label the vector of right approximations ∇J^+, and will see below that ∇J^-, ∇J^+ or some combination of the two will be used to compute the numerical Hamiltonian $\hat{H}$. The accuracy of a derivative approximation is measured in terms of the order of its local truncation error; an order p method has error $\| \nabla J - \nabla J^{\pm} \| = \mathcal{O}(\Delta x^p)$. At the current time, we have implemented the basic first-order accurate approximation for speed and a weighted, essentially nonoscillatory fifth-order accurate approximation for high fidelity. We have chosen to use the well-studied Lax–Friedrichs numerical Hamiltonian approximation $\hat{H}$ [12]

$$\hat{H}(x, \nabla J^-, \nabla J^+) = H\left(x, \frac{\nabla J^- + \nabla J^+}{2}\right) - \tfrac{1}{2}\alpha^T(\nabla J^+ - \nabla J^-) \quad (18.2)$$

where $H(x, \nabla J)$ is given by Eq. (18.1) and the term containing the vector coefficient α is a high-order numerical dissipation added to damp out spurious oscillations in the solution. The time derivative of the PDE is handled by the method of lines: The value of the level set function J at each node is treated as an ODE $dJ/dt = \hat{H}$, with $\hat{H}$ given by Eq. (18.2). General ODE solvers, such as Runge–Kutta (RK) schemes, can then be applied.

The Hamilton–Jacobi equation (18.1) describes the evolution of the level set function over all of space. But we are only interested in its zero level set; thus, we can restrict our computational updates to nodes near the boundary between positive and negative $J(x, t)$—an idea variously called "local level sets" [13] or "narrowbanding" [14]. We have implemented a new variant of this method in our code. Because the boundary is of one dimension less than the state space, considerable savings are available for two- and three-dimensional problems. If the number of nodes in each dimension is n (proportional to Δx^{-1}) and the dimension is d, the total number of nodes is $\mathcal{O}(n^d)$; with the restriction on the timestep necessary for numerical stability, the total computational cost is $\mathcal{O}(n^{d+1})$. With local level sets, we reduce computational costs back down to $\mathcal{O}(n^d)$.

18.3.3. Examples

In this section we present three examples: a validation of our numerical implementation on a single mode, three-dimensional aircraft collision avoidance example (see [4, 3] for details), and two multimodal examples. Once a method of determining continuous reachability is available, the discrete iteration of the algorithm described in Section 18.3.1 is relatively straightforward. In fact, for discrete transition graphs with no cycles, it is possible to order the continuous reachability problems such that no discrete iteration is required (the three-mode example below). In order to examine the complications induced by discrete cycles—such as how to avoid zenoness, in what order to execute the continuous reachability problems, and how to determine which switchs are active—we present an example representing the landing of a civilian airliner.

Numerical Validation of Aircraft Collision Avoidance. This example features a control aircraft trying to avoid collision with a disturbance aircraft, where both aircraft have fixed and equal altitude, speed, and turning radius

—they may only choose which direction they will turn:

$$\dot{x}_r = -v_u + v_d \cos \psi_r + u y_r, \qquad \dot{y}_r = v_d \sin \psi_r - u x_r, \qquad \dot{\psi}_r = d - u$$

where $v_u = v_d = 5$ are the aircraft speeds, x_r and y_r are the relative planar location of the aircraft, and ψ_r is their relative heading. The inputs $|u| \leq 1$ and $|d| \leq 1$ are the control's and disturbance's respective turn rates. The initial unsafe set $J(x, 0)$ is the interior of the radius 5 cylinder centered on the ψ_r axis. Choosing optimal inputs according to Eq. (18.1) with $v_G = v_{max} = d$ and $v_E = v_{min} = u$, we get the optimal Hamiltonian:

$$H(x, p) = -p_1 v_u + p_1 v_d \cos \psi_r + p_2 v_d \sin \psi_r + |p_1 y_r - p_2 x_r - p_3| - |p_3|$$

Using our C++ implementation, grid sizes corresponding to 50, 70, 100, 140, and 200 nodes in each dimension were tried with a low-order accurate scheme (first-order space and time, hereafter referred to as the "(1, 1)" scheme) and a high-resolution scheme (fifth-order space and second-order time, hereafter the "(5, 2)" scheme). On the eight-million-node finest grid—only around 10% of which is being actively updated on any one timestep by the local level set algorithm—execution time for the (5, 2) scheme was about 18 hours on a Sun UltraSparc II with lots of memory. Reducing the grid size in half results in the expected eightfold savings in memory and time; hence, the coarsest grid takes only 15 minutes with the (5, 2) scheme.

Results are visualized[1] by the zero-level isosurface of the unsafe reachable set G, shown in Figure 18.2. On the left is a head-on view of the (5, 2) solution. On the right is a zoomed overhead view of the point of the bulge computed by the (1, 1) scheme for several grid sizes.

We compare solutions on the four coarser grids to the solution on the finest grid, using linear interpolation on the finest grid if necessary. Figure 18.3 demonstrates that the scheme is converging to the finest grid's solution of Eq. (18.1) at approximately a linear rate in both average error and pointwise maximum error. We cannot expect to show a higher-order convergence rate because of the linear interpolation used to evaluate the error.

Two conclusions can be drawn from Figures 18.2 and 18.3. First, for this example, low-order schemes are not at all competitive in terms of accuracy with the (5, 2) scheme: Our (5, 2) implementation can produce more accurate results in about 15 minutes using only the coarsest grid. Second, the pointwise maximum error of the (5, 2) scheme is always less than the grid spacing, so if a $50^{-1} = 2\%$ error is tolerable for this application, only this fastest, coarsest grid need ever be run.

---

[1] Figure 18.2 and Figure 18.8 visualize some level set surfaces as triangular meshes; these are not the meshes on which the Hamilton–Jacobi PDE was solved, but rather an artifact of three-dimensional Matlab visualization techniques.

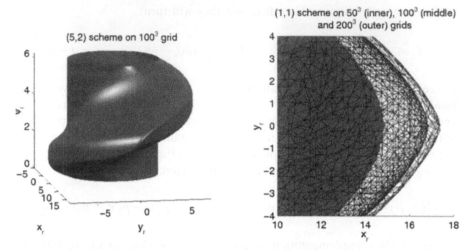

Figure 18.2. Reachable set for aircraft collision avoidance example.

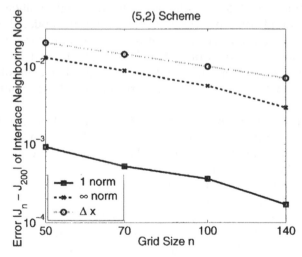

Figure 18.3. Convergence of $(5, 2)$ scheme to finest grid solution (J_n is the solution $J(x, t)$ on a grid size of n).

Three-Mode Conflict Resolution. Consider the following conflict resolution problem between two aircraft ($i \in \{1, 2\}$), which is presented in reference 15. Let $(x_r, y_r, \psi_r) \in \Re^2 \times [-\pi, \pi)$ represent the relative position and orientation of aircraft 2 with respect to aircraft 1. In terms of the absolute positions and orientations of the two aircraft, (x_i, y_i, ψ_i) for $i = 1, 2$, it may be verified that $x_r = \cos \psi_1 (x_2 - x_1) + \sin \psi_1 (y_2 - y_1)$, $y_r = -\sin \psi_1 (x_2 - x_1) +$

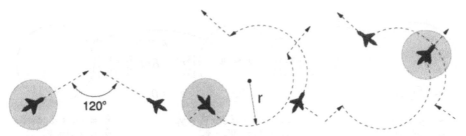

Figure 18.4. Two aircraft in two modes of operation: In the first and third figures, the aircraft follow a constant heading and velocity (in *Mode* 1); in the second figure, the aircraft follow a half circle (in *Mode* 2).

$\cos \psi_1(y_2 - y_1)$, $\psi_r = \psi_2 - \psi_1$, and thus

$$\dot{x}_r = -v_1 + v_2 \cos \psi_r + \omega_1 y_r$$

$$\dot{y}_r = v_2 \sin \psi_r - \omega_1 x_r \qquad (18.3)$$

$$\dot{\psi}_r = \omega_2 - \omega_1$$

where v_i is the velocity along the body axis of aircraft i, and ω_i is the rate of change of heading. A conflict occurs when two aircraft incur a lateral spacing of less than 5 nautical miles; for safety of the maneuver, the relative position (x_r, y_r) must remain outside of the interior of the 5-mile-radius disk centered at the origin $\{(x_r, y_r, \psi_r): x_r^2 + y_r^2 \leq 5^2\}$.

Consider the maneuver illustrated in Figure 18.4, with the protocol of the maneuver defined as follows: The aircraft are nominally in *Mode 1*, in which the aircraft fly straight and level with given constant velocities v_i and constant heading ψ_i; at a certain relative separation distance each aircraft turns to its right, follows an arc of a circle in *Mode 2*, in which the velocity is held at a prescribed constant value v_i, until it intersects its original trajectory, then turns to its right and returns to its desired trajectory.

The model allows instantaneous changes in heading, which is an obvious abstraction of the true aircraft dynamics and is treated here in order to clearly illustrate the algorithm; a model with heading capture dynamics is presented in reference 15. In each mode, the continuous dynamics may be expressed in terms of the relative motion of the two aircraft (18.3): In *Mode 1*, $\omega_i = 0$ for $i = 1, 2$; in *Mode 2*, $\omega_i = 1$ for $i = 1, 2$. We assume that both aircraft switch modes simultaneously, so that the relative orientation ψ_r is constant. This assumption simply allows us to display the state space in two dimensions, making the results easier to present. The problem statement is therefore the following: Generate the relative distance between aircraft at which the aircraft may switch safely from *Mode 1* to *Mode 2*, and a turning radius r in *Mode 2*, to ensure that the 5 nautical mile separation is maintained.

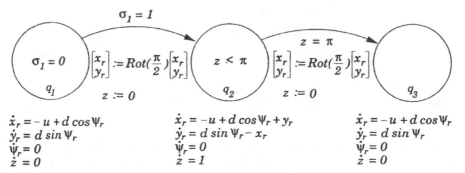

Figure 18.5. In q_1 both aircraft follow a straight course, in q_2 they follow a half circle, and in q_3 they return to a straight course.

The dynamics of the maneuver can be encoded by the hybrid automaton of Figure 18.5, where q_1 corresponds to *Mode 1* before the maneuver, q_2 corresponds to *Mode 2*, an "avoid mode," and q_3 corresponds to *Mode 1* after the avoid maneuver has been completed. There is one discrete control input σ_1, such that the switch from $\sigma_1 = 0$ to $\sigma_1 = 1$ triggers the transition from q_1 to q_2. The transition from q_2 to q_3 is required to take place after the aircraft have completed a half circle: Note that with $\omega_i = 1$, for $i = 1, 2$, it takes π time units to complete a half circle. The continuous state space is augmented with a timer $z \in \Re^+$ to force this transition. Let $x = (x_r, y_r, \psi_r, z)^T$. At each transition, both aircraft change heading instantaneously by $\pi/2$ radians; we represent this with the standard rotation matrix $Rot(\frac{\pi}{2})$. As discussed in the previous section, we assume that v_1 is controllable, and v_2 is the disturbance input with known bounds. Safety is defined in terms of the unsafe set:

$$G_0 = \{q_1, q_2, q_3\} \times \{x \in X : x_r^2 + y_r^2 \le 5^2\}$$

which encodes the fact that the two aircraft should never come within 5 nautical miles of each other.

The solution of the three-mode conflict resolution example is presented in Figure 18.6. Each of the 12 graphs in Figure 18.6 is a plot in the (x_r, y_r) axes, with each of the four rows representing an iteration of the algorithm, and the three columns corresponding to discrete states q_1, q_2, and q_3. The computation is initialized with the set G_0, which is shown as the dark-shaded disks. A fixed point is reached after three iterations. The dark-shaded region in the fourth row, first column of Figure 18.6 is the most interesting: It is a plot of the unsafe set of states in q_1. As long as the relative position of aircraft 2 with respect to aircraft 1 is not inside this region, then there exists a control policy such that the conflict is resolved. If the relative position is inside the

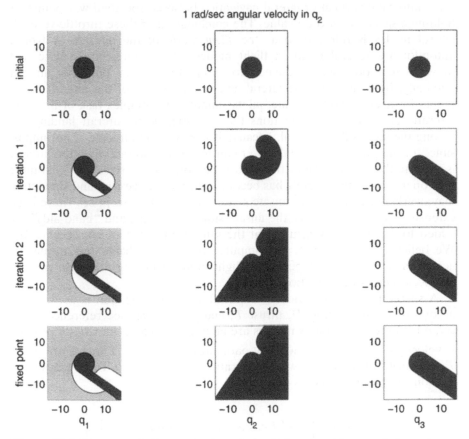

Figure 18.6. The rows indicate the iterations of the algorithm, and the columns indicate the discrete state. Each figure displays a projection of part of the unsafe set onto the (x_r, y_r) axes. A fixed point is reached after three iterations. Angular velocity $\omega_i = 1$ in q_2.

unshaded (white) region, then switching instantaneously will lead to a conflict; the aircraft must remain in q_1 until the relative state enters the light-shaded region, then either switch to q_2 or remain in q_1.

Aircraft Landing Example. The autopilots of modern jets are highly automated systems that assist the pilot in constructing and flying four-dimensional trajectories, as well as altering these trajectories online in response to air traffic control directives. The autopilot typically controls the throttle input and the vertical and lateral trajectories of the aircraft to automatically perform such functions as: acquiring a specified altitude and then leveling, holding a specified altitude, acquiring a specified vertical climb or descend rate, automatic vertical or lateral navigation between specified way points, or holding a specified throttle value. The combination of these throttle-vertical-lateral modes is referred to as the *flight mode* of the aircraft. A typical autopilot has several hundred flight modes—it is interesting to note that these flight modes were designed to automate the way pilots fly aircraft manually: By controlling the lateral and vertical states of the aircraft to set points for fixed periods of time, pilots simplify the complex task of flying an aircraft. Those autopilot functions that are specific to aircraft landing are among the most safety critical, because it is extremely difficult, if not impossible, for the pilot to safely take over control of the aircraft when the aircraft is close to the ground. Thus, the need for automation designs that guarantee safe operation of the aircraft has become paramount. Testing and simulation may overlook trajectories to unsafe states: "Automation surprises" have been extensively studied [16] *after* the unsafe situation occurs, and "band-aids" are added to the design to ensure that the same problem does not occur again. We believe that the ability to compute accurate reachable sets inside the aerodynamic flight envelope under flight mode switching may help prevent the occurrence of automation surprises.

A simple point mass model for aircraft vertical navigation is used, which accounts for lift L, drag D, thrust T, and gravity mg (see reference 4 and references therein). State variables are aircraft altitude z, horizontal position x, velocity $V = \sqrt{\dot{x}^2 + \dot{z}^2}$, and flight path angle $\gamma = \tan^{-1}(\frac{\dot{z}}{\dot{x}})$. Inputs are thrust T and angle of attack α, where aircraft pitch $\theta = \gamma + \alpha$. The equations of motion can be expressed as follows:

$$\frac{d}{dt}\begin{bmatrix} V \\ \gamma \\ x \\ z \end{bmatrix} = \begin{bmatrix} \frac{1}{m}[T\cos\alpha - D(\alpha,V) - mg\sin\gamma] \\ \frac{1}{mV}[T\sin\alpha + L(\alpha,V) - mg\cos\gamma] \\ V\cos\gamma \\ V\sin\gamma \end{bmatrix} \qquad (18.4)$$

The functions $L(\alpha,V)$ and $D(\alpha,V)$ are modeled based on empirical data [17] and Prandtl's lifting line theory [18]:

$$L(\alpha,V) = \tfrac{1}{2}\rho SV^2 C_L(\alpha), \qquad D(\alpha,V) = \tfrac{1}{2}\rho SV^2 C_D(\alpha)$$

where ρ is the density of air, S is wing area, and $C_L(\alpha)$ and $C_D(\alpha)$ are the dimensionless lift and drag coefficients.

In determining $C_L(\alpha)$ we will follow standard auto-lander design and assume that the aircraft switches between three fixed flap deflections $\delta = 0°$, $\delta = 25°$, and $\delta = 50°$ (with slats either extended or retracted), thus constituting a hybrid system with different nonlinear dynamics in each mode. This

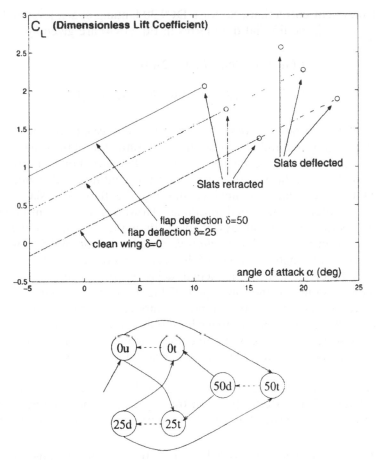

Figure 18.7. *Upper*: Lift coefficient $C_L(\alpha)$ model for the DC9-30 [17]. Circles located at $(\alpha_\delta^{max}, C_L(\alpha_\delta^{max}))$ indicate the stall angle and the corresponding lift coefficient in each mode. *Lower*: Discrete transition graph of slat and flap settings. Solid lines are controlled switches (σ_E in this version of the reachability problem), and dashed lines are uncontrolled switches (σ_G).

model is representative of current aircraft technology; for example, in Airbus cockpits the pilot uses a lever to select among four predefined flap deflection settings. We assume a linear form for the lift coefficient $C_L(\alpha) = h_\delta + 4.2\,\alpha$, where parameters $h_{0°} = 0.2$, $h_{25°} = 0.8$, and $h_{50°} = 1.2$ are determined from experimental data for a DC9-30 [17]. The value of α at which the vehicle stalls decreases with increasing flap deflection: $\alpha_{0°}^{max} = 16°$, $\alpha_{25°}^{max} = 13°$, $\alpha_{50°}^{max} = 11°$; slat deflection adds 7° to the α^{max} in each mode. The upper plot of Figure 18.7 gives a graphical summary of the possible configurations. The drag coefficient is computed from the lift coefficient as [18] $C_D(\alpha) = 0.041 + 0.045 C_L^2(\alpha)$ and includes flap deflection, slat extension, and gear deployment corrections. So for a DC9-30 landing at sea level and for all $\alpha \in [-5°, \alpha_\delta^{max}]$, the lift and drag terms in Eq. (18.4) are given by

$$L(\alpha, V) = 68.6(h_\delta + 4.2\alpha)V^2,$$

$$D(\alpha, V) = \left(2.81 + 3.09(h_\delta + 4.2\alpha)^2\right)V^2$$

Flap Deflection Dynamics Model. In reality, the decision to move from one deflection setting to another can occur at any time, but approximately 10 s is required for a 25°-degree change in flap deflection. For our preliminary implementation, we have chosen to ignore the continuous dynamics associated with discrete mode switching, allowing the flaps and slats to move instantly to their commanded positions. However, if such instantaneous controlled switches were always enabled, then the system would be zeno; therefore, we introduce delay, or *transition* modes $0t$, $25t$, and $50t$, which use the envelopes and flight dynamics of the regular modes $0u$, $25d$, and $50d$ and include a timer to model the actual delay in flap change (the discrete automaton is shown in the lower plot of Figure 18.7). A regular mode may make a controlled switch to a transition mode, so flight dynamics can be changed instantly. Transition modes have only a timed switch at $t = t_{delay}$, so controlled switches will be separated by at least t_{delay} time units and the system is nonzeno. For the executions shown below, $t_{delay} = 0.5$ s.

Landing. The aircraft enters its final stage of landing close to 50 feet above ground level [17, 19]. Restrictions on the flight path angle, aircraft velocity, and touchdown (TD) speed are used to determine the initial safe set E_0:

$$
\begin{cases}
z \leq 0 & \text{landing or has landed} \\
V > V_\delta^{stall} & \text{faster than stall speed} \\
V < V^{max} & \text{slower than limit speed} \\
V \sin\gamma \geq \dot{z}_0 & \text{limited TD speed} \\
\gamma \leq 0 & \text{monotonic descent}
\end{cases}
\cup
\begin{cases}
z > 0 & \text{aircraft in the air} \\
V > V_\delta^{stall} & \text{faster than stall speed} \\
V < V^{max} & \text{slower than limit speed} \\
\gamma > -3° & \text{limited descent flight path} \\
\gamma \leq 0 & \text{monotonic descent}
\end{cases}
$$

$$(18.5)$$

We again draw on numerical values for a DC9-30 [17]: stall speeds $V_{0u}^{stall} = 78$ m/s, $V_{25d}^{stall} = 61$ m/s, $V_{50d}^{stall} = 58$ m/s, maximal touchdown speed $\dot{z}_0 = 0.9144$ m/s, and maximal velocity $V^{max} = 83$ m/s. For passenger comfort, the aircraft's input range is restricted to $T \in [0\ \text{kN}, 160\ \text{kN}]$ and $\alpha \in [0°, 10°]$.

The interior of the surface shown in the first row of Figure 18.8 represents E_0 for each of the $0u$, $25d$, and $50d$ modes. The second row of the figure shows the safe envelope E when there is no mode switching. Portions of E_0 are excluded from E for two reasons. States near $z = 0$ correspond to low altitudes and are too close to the ground at steep flight path angles to allow

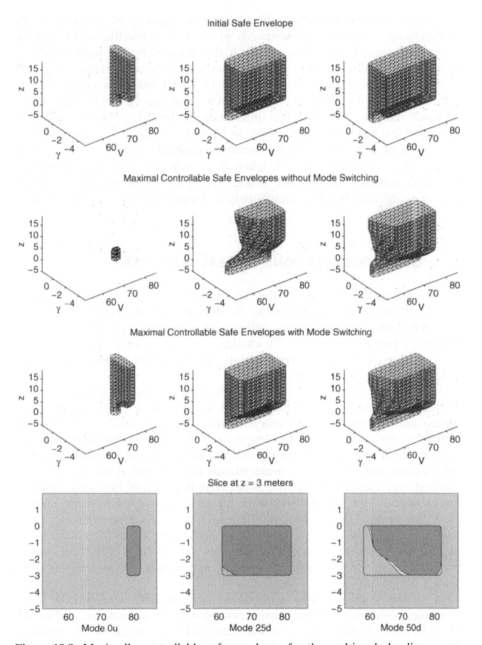

Figure 18.8. Maximally controllable safe envelopes for the multimode landing example. From left to right the columns represent modes $0u$, $25d$, and $50d$.

control inputs time to prevent the plane from crashing. States close to the stall velocity correspond to low speeds where there is insufficient lift and the flight path angle becomes steeper than that allowed by the flight envelope. This latter condition holds throughout the very narrow range of speeds allowed in mode $0u$, with the result that only post-touchdown states ($z \leq 0$) are controllable in this mode. The third row shows how E can be increased if switches are permitted (for example, mode $0u$ becomes almost completely controllable). Mode $50d$ is the best to be in for landing, and there is little difference in E with or without switching enabled. The fourth row shows slices of the set in the third row, taken at $z = 3$ m. The light gray regions are unsafe G, and the dark gray are safe E. The figure shows that modes $0u$ and $25d$ are safe only because there exists a discrete switch to a safe state in another mode.

18.4. OVERAPPROXIMATIONS OF REACHABLE SETS

In the reachable set computation of the previous section, computational complexity introduced by increasing the dimension of the continuous state is a limiting factor: The technique suffers from exponential growth with respect to the continuous dimension. To challenge this curse of dimensionality, we are investigating techniques for computing overapproximations to the reachable set of states.

In reference 20, we devise and implement a method based on projecting the reachable set of a high-dimensional system into a collection of lower-dimensional subspaces where computational costs are much lower. We formulate a method to evolve the lower-dimensional reachable sets such that they are each an overapproximation of the full reachable set, and thus their intersection will also be an overapproximation of the reachable set: The method uses a set of lower-dimensional Hamilton–Jacobi PDEs for which, in any projection, the set of disturbance inputs is augmented with the unmodeled dimensions. For the three-dimensional aircraft collision avoidance problem of the previous section, the result of performing the computation in the (x_r, y_r) space, and then back-projecting this set to form a cylindrical overapproximation of the reachable set in three dimensions, is computed in about 40 s.

A second technique is to overapproximate nonlinear dynamics by linear dynamics with bounded disturbance and to overapproximate reachable sets by polyhedra. Propagation of sets is then simply described by the propagation, through linear dynamics, of linear inequalities describing the polyhedral set. With this technique, however, comes the problem of the potentially unlimited number of inequalities needed to describe the polyhedral estimate. This problem has been addressed in the literature, yet previous solutions fall short in several ways: Exact pruning of polytopes [21] requires complex data structures to be maintained, and the number of "active" hyperplanes could still be unlimited; the use of simple polyhedral shapes for bounding the polyhedral set [22] is itself computationally intensive; the ellipsoidal overapproximation to the reachable set [23–25], while an elegant and efficient representation of the set, is generally too conservative an overapproximation.

We have developed a technique that combines the efficiency of the ellipsoidal data structure with the tightness of polyhedral overapproximation. The heart of our technique is a novel pruning procedure based on the maximum volume ellipsoid (MVE) contained in the polyhedron. The basic idea of the method is to rank the constraints by their distance (measured in the norm induced by the ellipsoid) to the center of the MVE and delete the ones that appear to be least relevant. Conceptually, constraints that are far away from the center are less relevant than those that are close. This approach will allow us to perform guaranteed pruning (where only provably redundant constraints are deleted).

For example, consider a discrete-time linear system (which represents a linear overapproximation of the nonlinear systems of the previous section):

$$x(t+1) = Ax(t) + Bw(t)$$

where $x(t) \in \Re^n$ is the state vector, $w(t) \in \Re^d$ is a bounded disturbance, and A and B are matrices with consistent dimensions. We assume that $w(t) \in \mathcal{W}(t)$ for some bounded convex set $\mathcal{W}(t)$. Assume that $x(t)$ is known to lie in a polyhedron $\mathcal{P}(t)$, where $\mathcal{P} \subseteq \mathbb{R}^n$ is the intersection of a finite number of half-spaces,

$$\mathcal{P} = \bigcap_{i=1}^{m} \mathcal{H}_i$$

Using the parameterization of half-spaces introduced above, we can represent $\mathcal{P}$ in the form

$$\mathcal{P} = \left\{ x \,\middle|\, f_i^T x \le g_i, \, i = 1, \ldots, m \right\} \equiv \left\{ x \,\middle|\, Fx \preccurlyeq g \right\} \tag{18.6}$$

where $\preccurlyeq$ denotes componentwise inequality. Now, an ellipsoid is defined as the image of a unit ball under an affine transformation

$$\mathcal{E} = \left\{ Pu + q \,\middle|\, \|u\| \le 1 \right\} \tag{18.7}$$

We assume that A is invertible, which is always the case if the dynamics is discretized from some continuous-time system, and we propagate the set $\mathcal{P}(t)$ through the dynamic matrix A:

$$A\mathcal{P}(t) = \left\{ x \,\middle|\, f_i(t)^T A^{-1} x \le g_i(t), \, i = 1, \ldots, m(t) \right\}$$

$$= \left\{ x \,\middle|\, F(t) A^{-1} x \le g(t) \right\}$$

To account for the bounded disturbance, we need to find the polytope $A\,\mathcal{P}(t) + B\mathcal{W}(t)$. An easily computed outer approximation of it is given by

$$\tilde{\mathcal{P}}_{t+1} = \left\{ z \middle| f_i(t)^T A^{-1} z \le g_i(t) + \left\| f_i(t)^T A^{-1} B \right\|_1, \, i = 1, \ldots, m(t) \right\}$$

$$= \left\{ z \middle| \tilde{F}(t) z \le \tilde{g}(t) \right\}$$

As $\mathcal{P}$ evolves through hybrid dynamics, its representation grows in dimension. To limit the computational requirements for storing and operating on $\mathcal{P}$, it is therefore necessary to occasionally simplify the description of $\mathcal{P}$ by dropping, or discarding, constraints. We will call this procedure polyhedral pruning.

The Maximum Volume Inscribed Ellipsoid (MVE). We are interested in finding the maximum volume ellipsoid that is contained in a polyhedron. Let $\mathcal{E}$ be an ellipsoid of the form (18.7), and let $\mathcal{P}$ be a polyhedron parameterized as in Eq. (18.6). Then, $\mathcal{E} \subset \mathcal{P}$ if and only if

$$\sup_{x \in \mathcal{E}} f_i^T x \le g_i, \quad i = 1, \ldots, m$$

Using the parameterization of the ellipsoid, this condition can be expressed as

$$\sup_{\|u\|_2 \le 1} f_i^T P u + f_i^T q \le g_i, \quad i = 1, \ldots, m$$

Maximization of the left-hand-side expression gives the equivalent containment condition

$$\left\| P f_i \right\|_2 + f_i^T q \le g_i, \quad i = 1, \ldots, m$$

Hence, we can find the MVE inside a polyhedron by solving the convex optimization problem

$$
\begin{aligned}
\text{maximize} \quad & \log \det P \\
\text{subject to} \quad & P = P^T \succeq 0 \\
& \left\| P f_i \right\|_2 + f_i^T q \le g_i, \quad i = 1, \ldots, m
\end{aligned}
$$

This problem follows in the general class of maxdet problems, which can be solved globally and efficiently using the algorithms described in reference 26. Moreover, specialized algorithms for computing the MVE have been proposed in references 27 and 28. The MVE gives an optimal (in terms of volume) ellipsoidal approximation of the polyhedron. Clearly, the MVE also provides an inner bound on the polyhedron, since $\mathcal{E}_{\text{mve}} \subseteq \mathcal{P}$. More surprisingly, perhaps, is that the MVE also provides an *outer* bound of the polyhedron: The ellipsoid obtained by scaling the MVE a factor n around its center is guaranteed to contain the polyhedron:

$$\mathcal{E}_{\text{mve}} \subseteq \mathcal{P} \subseteq q_{\text{mve}} + n\left(\mathcal{E}_{\text{mve}} - q_{\text{mve}}\right)$$

This is a dual result to the celebrated Löwner–John ellipsoid [29, 30] (which treats the minimum volume ellipsoid that contains a convex set). Thus, the MVE gives a good approximation of a polyhedron and provides simultaneous inner and outer bounds.

Polyhedral Pruning Using the MVE. The MVE also gives a natural measure for ranking the relevance of the constraints that define $\mathcal{P}$. For a half-space $\mathcal{H}_i = \{x \mid f_i^T x \le g_i\}$, we define the *relevance ranking* η_i as the largest factor by which the ellipsoid can be enlarged and yet still lie within the half-space:

$$\eta_i = \max\left\{\alpha \mid q_{\text{mve}} + \alpha\left(\mathcal{E}_{\text{mve}} - q_{\text{mve}}\right) \subseteq \mathcal{H}_i\right\}$$

It is easy to verify that the relevance ranking is given by

$$\eta_i = \left(g_i - f_i^T q\right) / \|Pf_i\|_2$$

This is, in fact, nothing more than the minimum distance from the half-space to the center of the ellipsoid in the P^2 metric. The relevance ranking has the following properties:

1. $\eta_i \ge 1$ for all i
2. $\eta_i = 1 \Rightarrow f_i^T x \le g_i$ active
3. $\eta_i \ge n \Rightarrow f_i^T x \le g_i$ redundant

We define a constraint to be *redundant* if the polyhedron $\mathcal{P}$ does not change when the constraint is removed. We will say that the polyhedron is *minimal* if it does not contain any redundant constraints, and we will say that a constraint is *active* if $f_i^T x^* = g_i$ for some $x^* \in \mathcal{P}$. A constraint is (strongly) redundant if and only if it is inactive for all $x \in \mathcal{P}$. The last property allows us to discard redundant constraints. There is a gray zone $\eta_i \in (1, n)$ for which the relevance ranking neither proves that a constraint is active nor proves that it is redundant. In practice, guaranteed pruning can often be too conservative in that it fails to discard many redundant constraints. We therefore use the ranking in a heuristic pruning procedure: Given $\mathcal{P} = \{x \mid f_i^T x \le g_i \ \ i = 1, \ldots, m\}$, we compute $\mathcal{E}_{\text{mve}}$, sort the inequalities by their relevance ranking, η_i, and keep the k most relevant (the ones with smallest η_i). This yields an approximate simplified polyhedron $\hat{\mathcal{P}}$, with $\mathcal{P} \subseteq \hat{\mathcal{P}}$. We believe this algorithm to be an effective method for computing fast polyhedral overapproximations of reachable sets, in very high dimensions.

18.5. SUMMARY

We have presented and numerically validated a tool for determining accurate approximations of reachable sets for hybrid systems with nonlinear continuous dynamics and adversarial continuous and discrete inputs. By developing convergent approximations of such complex systems, we will be better able to synthesize aggressive but safe controllers. As an example, the six-mode auto-lander shows that for envelope protection purposes the safest control decisions are to switch directly to full flap deflection, but to maintain airspeed until touchdown. With the summary data from the reachability analysis, such decisions can be made based on local state information; without it the auto-lander may not detect that low speeds—while still within the flight envelope—lead inevitably to unsafe flight path angles. In this arena, we have made real progress on problems in a few dimensions, where we have close to exact solutions. We show that the computation may be made faster as we loosen the numerical approximation, however, even with such approximation, these methods will not scale beyond five or six continuous dimensions.

In response to this problem, we are developing methods that gracefully extend to problems with a much higher dimension of the continuous state (these methods have been shown to work well in dimension up to 100). While these are approximate methods, they are, however, nonheuristic: when they determine that a reachable set does not intersect some other set, the result is certain.

Finally, in order to test our reachability tools, we are designing and building a system of three unmanned aerial vehicle (UAVs), with on-board sensing, navigation, and control systems enabling automatic flight. The Stanford DragonFly UAV Project currently consists of two 10-foot wingspan, fixed wing, unmanned aircraft, each equipped with GPS, inertial sensors, and on-board computers running QNX Neutrino.

REFERENCES

1. R. Alur, C. Courcoubetis, T. A. Henzinger, and P.-H. Ho, Hybrid automata: An algorithmic approach to the specification and verification of hybrid systems, in *Hybrid Systems*, R. L. Grossman, A. Nerode, A. P. Ravn, and H. Rischel, editors, *Lecture Notes in Computer Science*, Springer-Verlag, New York, 1993, pp. 366–392.

2. C. Tomlin, J. Lygeros, and S. Sastry, A game theoretic approach to controller design for hybrid systems, *Proceedings of the IEEE* **88**(7):949–970, July 2000.

3. I. Mitchell and C. Tomlin, Level set methods for computation in hybrid systems, in *Hybrid Systems: Computation and Control*, B. Krogh and N. Lynch, editors, *Lecture Notes in Computer Science*, Vol. 1790, Springer-Verlag, New York, 2000, pp. 310–323.

4. C. J. Tomlin, Hybrid Control of Air Traffic Management Systems, Ph.D. thesis, Department of Electrical Engineering, University of California, Berkeley, 1998.

5. I. Mitchell, A. Bayen, and C. J. Tomlin, Validating a Hamilton–Jacobi approximation to hybrid system reachable sets, in *Hybrid Systems: Computation and Control*, M. D. Di Benedetto and A. Sangiovanni-Vincentelli, editors, *Lecture Notes in Computer Science*, Vol. 2034, Springer-Verlag, New York, 2001, pp. 418–432.

6. O. Maler, A. Pnueli, and J. Sifakis, On the synthesis of discrete controllers for timed systems, in *STACS 95: Theoretical Aspects of Computer Science*, E. W. Mayr and C. Puech, editors, *Lecture Notes in Computer Science*, Vol. 909, Springer Verlag, Munich, 1995, pp. 229–242.

7. E. Asarin, O. Maler, and A. Pnueli, Symbolic controller synthesis for discrete and timed systems, in *Proceedings of Hybrid Systems II*, P. Antsaklis, W. Kohn, A. Nerode, and S. Sastry, editors, *Lecture Notes in Computer Science*, Vol. 999, Springer-Verlag, New York, 1995, pp. 1–20.

8. H. Wong-Toi, The synthesis of controllers for linear hybrid automata, in *Proceedings of the IEEE Conference on Decision and Control*, San Diego, CA, 1997.

9. I. Mitchell, A. M. Bayen, and C. J. Tomlin, Computing reachable sets for continuous dynamic games using level set methods, *IEEE Transactions on Automatic Control*, December 2001, submitted.

10. M. G. Crandall, L. C. Evans, and P.-L. Lions, Some properties of viscosity solutions of Hamilton–Jacobi equations, *Transactions of the American Mathematical Society* **282**(2):487–502, 1984.

11. A. M. Bayen and C. J. Tomlin, A construction procedure using characteristics for viscosity solutions of the Hamilton–Jacobi equation, in *Proceedings of the IEEE Conference on Decision and Control*, Orlando, FL, 2001.

12. M. G. Crandall and P.-L. Lions, Two approximations of solutions of Hamilton–Jacobi equations, *Mathematics of Computation* **43**(167):1–19, 1984.

13. D. Peng, B. Merriman, S. Osher, H. Zhao, and M. Kang, A PDE based fast local level set method, *Journal of Computational Physics* **165**:410–438, 1999.

14. J. A. Sethian, *Level Set Methods and Fast Marching Methods*, Cambridge University Press, New York, 1999.

15. C. J. Tomlin, I. Mitchell, and R. Ghosh, Safety verification of conflict resolution maneuvers, *IEEE Transactions on Intelligent Transportation Systems*, **2**(2):110–120, 2001, June.

16. A. Degani, Modeling Human-Machine Systems: On Modes, Error, and Patterns of Interaction, Ph.D. thesis, Department of Industrial and Systems Engineering, Georgia Institute of Technology, 1996.

17. I. M. Kroo, *Aircraft Design: Synthesis and Analysis*, Desktop Aeronautics Inc., Stanford, CA, 1999.

18. J. Anderson, *Fundamentals of Aerodynamics*, McGraw-Hill, New York, 1991.

19. United States Federal Aviation Administration, *Federal Aviation Regulations*, 1990, Section 25.125 (landing).

20. I. Mitchell and C. J. Tomlin, Overapproximating reachable sets by Hamilton–Jacobi projections, *Journal of Scientific Computing*. (To appear)

21. N. A. Bruinsma and M. Steinbuch, A fast algorithm to compute the $\mathbf{H}_\infty$-norm of a transfer function matrix, *Systems and Control Letters* **14**(5):287–293, 1990.

22. M. Milanese and G. Belforte, Estimation theory and uncertainty intervals evaluation in presence of unknown but bounded errors—linear families of models and estimators, *IEEE Transactions on Automatic Control* **27**(2):408–414, 1982.

23. F. C. Schweppe, Recursive state estimation: Unknown but bounded errors and system inputs, *IEEE Transactions on Automatic Control*, **AC-13**(1):22–28, 1968.

24. H. S. Witsenhausen, Sets of possible states of linear systems given perturbed observations, *IEEE Transactions on Automatic Control* **13**:556–558, 1968.

25. N. Shishido and C. J. Tomlin, Ellipsoidal approximations of reachable sets for linear games, in *Proceedings of the IEEE Conference on Decision and Control*, (Sydney, Australia), 2000.

26. S.-P. Wu and S. Boyd, SDPSOL: *A Parser/Solver for Semidefinite Programming and Determinant Maximization Problems with Matrix Structure. User's Guide, Version Beta*, Stanford University, June 1996.

27. Y. Nesterov and A. Nemirovsky, *Interior-Point Polynomial Methods in Convex Programming of Studies in Applied Mathematics*, SIAM, Philadelphia, PA, 1994.

28. Y. Zhang, An interior-point algorithm for the maximum-volume ellipsoid problem, Technical Report TR98-15, Rice University, Houston, Texas, 1998.

29. F. John, Extremum problems with inequalities as subsidiary conditions, in *Fritz John, Collected Papers* J. Moser, editor, Birkhauser, Boston, MA, 1985, pp. 543–560.

30. M. Grötschel, L. Lovász, and A. Schrijver, *Geometric Algorithms and Combinatorial Optimization. Algorithms and Combinatorics*, Springer-Verlag, New York, 1988.

PART V

CONCLUSIONS

CONCLUSIONS

CHAPTER 19

THE OUTLOOK FOR SOFTWARE-ENABLED CONTROL

TARIQ SAMAD and GARY BALAS

Control engineering has been ahead of its time over the last few decades. Our journals and conference proceedings are replete with sophisticated theories and algorithms, with analysis and simulations testifying to their potential for dramatically improving the dynamical behavior of complex physical systems. Yet, too often, these results have remained on the bookshelf. The new ideas are just too complicated, we've been told, for implementation in production avionics and other real-time automation and control systems. Indeed, in aerospace and many other high-technology domains, even the transition from analog to digital control or from mechanical to electronic systems has a relatively recent provenance (e.g., fly-by-wire didn't appear in commercial aircraft until the late 1980s).

Developments in software and computing are rapidly changing the automation landscape, and societal and economic pressures are dictating research into new, "software-enabled," capabilities. In the military aerospace arena, the new watchword is *autonomy*. Unpiloted aircraft have existed for decades, but they have been limited to preprogrammed or remotely controlled operation. Today, we seek to imbue them with intelligence. With little or no communication with or instructions from humans, an uninhabited autonomous vehicle (UAV) should be able to execute high-performance maneuvers, dynamically adapt its route, detect and evade threats, identify and compensate for faults, undertake takeoffs and landings, tolerate extreme environmental variations, and perform other functions that have hitherto always been assumed to require manual supervision, if not manual operation. Although much previous research in control and related fields has addressed such functions, in the absence of a sufficiently powerful automation and

Software-Enabled Control. Edited by Tariq Samad and Gary Balas
ISBN 0-471-23436-2 © 2003 IEEE

control infrastructure this work has had a theoretical rather than a practical orientation.

The DARPA-sponsored "Software Enabled Control" program represents possibly the first large-scale, targeted effort toward realizing the synergy outlined above. The research contributions that are resulting from this program—discussed in the previous chapters in this volume—herald a new generation of sophisticated, high-performance control systems. The UAV domain provides a common application focus to this research, although most of the developments are relevant for other domains of importance to industry and society.

In this concluding chapter, we survey some of what's been accomplished and discuss topics for ongoing and future work. We have elected not to mirror either the tripartite structure of the book or the categories used by DARPA in establishing the program, but to structure the discussion below along research themes that, in our view, are particularly important for future research in software-enabled control.

19.1. NEXT-GENERATION COMPUTING PLATFORMS FOR REAL-TIME CONTROL

In terms of computing capability, real-time execution platforms for control systems have always lagged behind, say, office automation, to the point where differences in processor speeds between a flight control system and a desktop PC can reach orders of magnitude. Several reasons can be cited for the relative computational impoverishment of real-time platforms. These include (a) the fact that their useful deployed lifetimes tend to be in the decades and (b) the fact that they have generally not been based on commercial off-the-shelf (COTS) technology, the leading edge of the IT revolution.

The SEC program is addressing the challenge of integrating advanced control with the state-of-the-art in computing through the development of an open control platform (OCP). Based largely on COTS components, the OCP is intended to enable a next generation of real-time system that can take advantage of the latest developments in computing and that can ensure that the former benefits from the evolution in the latter. Furthermore, as an "open" platform, the OCP will also facilitate the process of taking research results and deploying them in real applications. Proprietary interfaces and protocols have been a near-impenetrable barrier for third-party applications; the OCP promises to change this situation.

Progress in real-time control requires more than CPU throughput. Another important need is for a real-time infrastructure that supports more complex computing models. In current avionics systems (and all other real-time systems for that matter), designers must specify in advance exactly how much computing resource must be dedicated to each computational task in any given situation. New "modes" cannot be configured online, tradeoffs between competing objectives cannot be made dynamically, and adaptation is anathema. The predetermination of processing profiles wasn't a problem as long as computing tasks were computationally and algorithmically simple, but it has precluded the use of sophisticated, adaptive algorithms.

The capabilities the OCP is being designed to support include both dynamic reconfiguration and dynamic resource management. Initial demonstrations of these advances have already been conducted, and once the implementations are complete a significant obstacle to high-performance control will be cleared. We can look forward to running iterative, "anytime" algorithms, to dynamically swapping computing tasks as mission objectives and system state dictate, and to rapidly integrating new functions and software implementations within avionics systems.

A primary benefit of the middleware-based architecture developed in the OCP is the isolation of the application components from the underlying platform. This allows for the platform-independent development of application software and should also enable software behavior to be predictable across platforms. The current instance of the OCP is an event-driven, asynchronous architecture, and issues related to the asynchronous nature remain to be addressed. In particular, the dataflow for asynchronous systems can be nondeterministic and unpredictable. This is not the case with offline simulation software, which defines a dataflow specification of the system prior to each simulation (which proceeds synchronously and is hence repeatable). Asynchronous components, on the other hand, become active after all the required inputs are available. Execution of components within the current instance of the OCP will therefore likely be different from the original simulation trace and will likely change as a function of the system state. The difference and variation will make it difficult to validate software components or overall system behavior even when all the software components are implemented in the real-time operating system (RTOS) on the flight hardware. Nondeterminism of execution time and ordering would have to be addressed with either the addition of a time-triggered middleware mode or a well-defined protocol for component execution that is defined prior to real-time execution.

Even topics such as dynamic resource management and reconfigurable control are broader in their scope than current work is considering. The general problems—for example, of accommodating multiple variable-processing-need computing tasks and determining how models of computation can be used to allocate resources in unpredicted situations—still await fundamental contributions.

19.2. INCREASING AUTONOMY AND PERFORMANCE

With the OCP promising a suitably capable system infrastructure, other SEC projects are developing new algorithmic techniques that can exploit the OCP to enable greater UAV capability. Perhaps the most obvious advance from an algorithmic perspective is online optimization. A substantial majority of modern techniques used in control, signal processing, estimation, and related fields have, at their core, an optimization engine (or solver). Generally, the optimizers are model-based, which means that a model of the relevant aspects of the system must be executed every time (and often several times) for each invocation of the optimizer. The online use of solvers has been computationally prohibitive for real-time applications in systems with fast dynamics such as aircraft—the role of optimization for such application domains has largely been limited to design activities. (Exceptions exist, usually in those instances where the model of the system can be simplified to, for example, linear dynamics.)

The technique of model predictive control (MPC) exemplifies these general remarks. In MPC, an online optimization is undertaken in which a dynamic model of the system is exercised to determine the optimal control sequence over a future horizon. Usually the first control action identified in this sequence is used and the entire calculation is repeated at the next sample instant over an updated ("receding") horizon. MPC has had a revolutionary impact in chemical process control, but its application has been possible for two reasons, neither of which applies in the aerospace arena. First, the dynamics of most chemical processes are slow: Seconds and even minutes may elapse between sensor readings and corresponding control actuation. Second, the simplifying assumption of linear dynamics is usually justified in this domain, permitting a closed-form solution to what would otherwise be an iterative calculation. (It should also be noted that most MPC applications in the process industries execute on supervisory platforms.)

With its optimality and constraint-handling properties, MPC is widely recognized as a powerful technique for aerospace control as well, but it has been considered computationally intractable for such applications. This is the case no longer. Testbed applications have now been demonstrated in the SEC program. The demonstrations, on laboratory platforms and high-fidelity simulations, have highlighted the capabilities that nonlinear MPC can provide for UAV operation: high-performance maneuvers, accommodation to extreme weather conditions, and fault tolerance.

Software-enabled control also suggests a new take on the robustness-versus-performance tradeoff that remains central in control engineering. Since the online control law was perforce a simple one, there was no possibility of adapting it in-flight based on the predictive accuracy of the model from which it was derived. Now, computing resources are available to perform complex state and model estimation. We can manage uncertainty dynamically; and we can effect time-varying tradeoffs between robustness and performance, even as models are adapted online.

Another compute-intensive prospect that is now realizable onboard a vehicle is fault detection and identification. Again, with advances in real-time and onboard computing, we can now deploy more sophisticated and accurate techniques than conventional approaches such as simple limit checks. Both model-based methods, in which faults are identified and diagnosed by comparing the observed system output with the output of a model, and model-free methods, such as matching a possibly abnormal sensor pattern with sensor patterns from known failures, are being used. With good progress in fault detection and identification having being made, the area of fault accommodation is now attracting more interest in the research community. Here, we seek to automatically adapt control laws based on detected and identified faults to ensure the continued safety and performance of the vehicle.

Information technology is critical to making these attacks broadly successful for complex systems. Increasing processing power is just one facet. We can borrow methods from software architecture and software engineering to develop powerful integrated solutions in which estimators, solvers, fault detection and identification components, online identifiers, and other components are plug-and-play compatible and readily updated and maintained. The adaptive nature of some of these components raises the question of validation and verification of the software. Answers are being suggested under another SEC theme, discussed next.

19.3. HIGH-CONFIDENCE CONTROL

As remarked earlier, research in control engineering regularly produces new theories and algorithms that promise substantial improvement over the state of the practice. But only a small fraction of this research ultimately sees practical application. We often bemoan the "theory/practice gap" in control, and it is clear that there is no simple solution to the problem. PID and PID-like controllers, products of 1930s control engineering, continue to be the control approach of choice in a variety of industries, from process control to aerospace. Even when more advanced control algorithms are used on real-time platforms, such as model predictive control or feedback linearization, these algorithms are generally simplified to PID-like forms before they are implemented on the real-time system.

A major reason for the limited real-world impact of advanced control research has to do with difficulties related to *reliable implementation* on real-time computing platforms. Our current emphasis on "high performance" needs to be complemented with equal attention to "high confidence." The need for real-time computing distinguishes control engineering from many other algorithmic research fields: Real-time control is not office automation. It is not sufficient that a controller implementation give a suitable answer; it must provide the answer within a mandated deadline. Lack of predictability in response times or other computational characteristics is merely an inconvenience for, say, a data mining tool for financial forecasting. In control systems, on the other hand, the cost of a calculation regularly not completing within the sampling interval can be catastrophic for human lives, environmental safety, and economic returns.

Most algorithms for advanced control and autonomous operation are characterized by properties that render high confidence hard to achieve. Computational requirements can vary substantially from invocation to invocation, exhaustive testing can be an impossibility, analytical or theoretical results cannot vouch for 100% safety, and so on. Verification, validation, safety assurance, and certification—among the central topics in high-confidence control research—are established areas, but they take on an added dimension for autonomous systems. Characteristics of conventional approaches such as exhaustive analysis, deterministic processing, and a human operator as a fall back option are no longer applicable.

The practical impossibility of exhaustive analysis of complex algorithms and their software implementations is a particularly daunting challenge, but probabilistic techniques suggest a way out of the impasse. What is critical is that new techniques be rigorous and provide some (not necessarily complete) certitude that is meaningfully quantified. Hundred percent safety for any component of a system has been a convenient fiction, but the search for it is imbued in the assurance culture for life- and mission-critical systems. There may be no alternative to probabilistic methods for software components that represent complex functionality, and only after the technical problems are resolved can the culture change be promoted.

High confidence is an overlay on other topical areas in software-enabled control as well as a topic in its own right. For example, the widespread use of optimization algorithms can raise reliability concerns since a novel solution may be generated and deployed without a manual check. A promising approach to deal with this problem is the development of solvers that can assure that hard safety constraints are satisfied even as a performance criterion is maximized.

Recent additions to the SEC research portfolio have specifically targeted high confidence, and we foresee this issue emerging as a new research focus in control engineering generally.

19.4. MULTIVEHICLE COORDINATION AND COOPERATION

As research in automation and control targets ever-larger-scale systems, what counts as a system is no longer limited to a unitary "plant." The implication of this for research in UAVs is that not just single aircraft operation are of interest, but also, and increasingly, UAV formations. Just as we seek to automate how a pilot follows a flight plan, avoids pop-up threats, engages time-critical targets, and handles disturbances and failures, similarly we need to automate how pilots fly in formation, coordinating with each other and with manned aircraft, and how a formation as a whole can undertake control actions at different decision-making levels to maximize likelihood of mission success.

This does not mean that the performance of UAV formations is upper bounded by their manned counterparts. UAVs can execute higher-performance maneuvers—accelerations can be limited by airframe and components, not by human factors. Analogously, in principle UAV formations could be much tighter than piloted ones since human reaction time and latency constraints do not apply and the information exchanged between aircraft need not be limited to visual cues or verbal messages.

Tight formations have several advantages over dispersed multiaircraft operations. Significant drag reduction is possible with close-formation flight (as demonstrated by the V-formations of geese), reduced overall footprints can aid in stealth operations, and intervehicle transmissions require lower power. The technical challenges, however, are significant. Algorithms for maintaining formation geometry in maneuvers, subject to disturbances and uncertainties, need to be developed. Formation stability must be maintained; for example, tracking errors in a leading aircraft should not be amplified through the formation. Solutions need to ensure that a formation and its mission execution capability can tolerate aircraft faults and dropouts where feasible. In addition to the regulatory and tracking control requirements, modifications of the formation geometry as a result of faults will be needed.

There's also the interesting question of what information should be communicated between aircraft for formation control. Under which conditions is state information sufficient? If "preview" or intent communication is required, what should be its horizon and resolution? In general, what is the relationship between formation geometry, intervehicle communication requirements, and the desired formation performance and robustness? In addition, approaches to deal with the complex problem of the forming and breakup of formations need to be developed; the problem remains open even for the case of a single aircraft joining or leaving a tight formation.

Communication is central to coordination, and it takes on added requirements as the extent of automation and autonomy increases. An autonomous multivehicle formation must be able to communicate plans and replans, degradation and failures, pop-up threats and targets, and much other information. Sometimes radio communication must be limited by stealth considerations, occasionally to the point where complete radio silence is required. Human pilots deal with such situations by relying on visual signals—a shallow rocking of wings is a sign to attract attention, for example, and visual codes have been devised for many other messages. An outstanding need for autonomous formations is a multimodel language for coordination, encompassing syntax and semantics and ultimately pragmatics as well.

Finally, autonomous formation control and cooperative maneuvers will be greatly facilitated if the entire formation can be thought of as, and can perform as, one integrated computing platform. Different vehicles may have different resources (computing, sensing, communication, etc.) onboard, and a resource should be used in the way that is best suited for the mission and the formation as a whole rather than as "allocated" to one vehicle.

19.5. INTEGRATION OF PLANNING AND CONTROL

The decision making required of a fully autonomous vehicle encompasses all the functions involved in the successful execution of a mission. Most research in automation today focuses on one task or another, generally addressing it independently. Algorithms and representations are developed for mission replanning, route optimization, threat avoidance, outer-loop and inner-loop control, and fault detection and identification. Each problem is enough of a challenge that a near-exclusive focus is understandable, yet true autonomy is more than the sum of its constituent functions.

We often assume that a strictly defined hierarchical architecture will enable separately developed pieces to be effectively integrated, A mission plan may coarsely indicate times and points of mode switching; a route optimizer may use this information to generate waypoints; the waypoints would be input to an outer-loop module for trajectory generation, and the generated trajectory may serve as the reference command for inner-loop control. We currently consider all of these aspects as decoupled, and this decoupling is enforced by time-scale separation, so that outer-loop control and inner-loop control, say, do not interfere with each other. This structure simplifies design as well as computation, but it exacts a price in performance. In particular, response times suffer; the time-scale separation that is needed to make the hierarchical structure work limits the speed of reaction that is achievable. Yet few alternative architectures have been explored in any depth. This is an important topic in our view, with a potential impact that goes well beyond the UAV domain.

In a related vein, a common complexity management method in automation is mode-based design. For example, flight control laws for a vehicle may be designed separately for takeoff, landing, cruising at different altitudes, and different maneuvers. Mode-partitioned automation is a universal practice for complex engineering systems, but it is a design approach, not a design requirement, and incurs a performance penalty for mode transitions in particular. Experienced Air Force pilots will maintain that they are not following a strict mode-based decision-making process when they execute evasive trajectories or undertake other high-performance maneuvers, and that their "seat of the pants" style blends together modes into one seamless process. Readers may pause to consider whether they operate their vehicles (e.g., automobiles) in consciousness of multiple modes. The extensive use of mode partitioning may be avoidable with more sophisticated algorithms and the computing wherewithal to execute them onboard.

The examples suggest that one source for insight into alternative architectures for autonomy is biology. The fluidity and flexibility exhibited by natural organisms in their interactions with an uncertain, noisy environment remain unparalleled in engineered systems. Cognitive deliberation may (or may not) be the exclusive preserve of humankind—and thus lessons to be drawn for planning systems, in the most abstract sense of planning, may need to be drawn from ourselves—but animals across the spectrum of species complexity engage in coordinated, integrated, adaptive behaviors for hunting, foraging, and other activities necessary to their survival.

19.6. DESIGN AND DEPLOYMENT TOOLS

Progress in the areas of software architecture, computing capabilities, and real-time systems has significantly expanded opportunities to implement advanced planning, control, fault detection, identification, and estimation algorithms in future UAVs. Dynamic reconfiguration, dynamic resource management, and publish and subscribe protocols will allow highly integrated applications that represent a favorable alternative to current systems that enforce rigid compartmentalization and decoupling of software modules as discussed above.

An important component to the overall success of the software-enabled control paradigm is the tight coupling of the application design and development tools with the RTOS software architecture. Algorithm developers currently work in a design environment that is completely separated from the RTOS used for implementation. A goal of the open control platform is to merge these environments by providing the same interface and characteristics during real-time and non-real-time operations independent of the platform. Hence, system-level services such as dynamic allocation of CPU time, acquiring and broadcast of data, and system and world state information must be easily available to application engineers who often do not have the software engineering expertise of an RTOS programmer.

A step toward the seamless integration of the open control platform and application software tools is the development of application programming interfaces (APIs). An OCP controls API has been developed to aid users in creating OCP components from Simulink diagrams. The API handles the input/output interfaces of components within the OCP framework, though the coding of the internal behavior of each component is the user's responsibility. Generating the internal, application-specific code is still not trivial for many designers because it requires knowledge at the level of the operating system. To overcome this significant hurdle, transparent and seamless interfaces must be provided between software tools and the OCP middleware. The controls API is just one example of an interface that is needed. Similar APIs are needed for online optimization, route planning, trajectory generation, reconfiguration, mode switching, and communication, among others.

Current practice allows software design and development tools to focus exclusively on isolated, decoupled subsets of the overall system since individual software modules are integrated separately. For example, the flight stability augmentation system is integrated as a subroutine in the overall system. It is given a predetermined amount of computation time, and it must operate properly within that time slice. The stability augmentation system is designed independently from the fault detection and isolation subsystem, the trajectory generator, mission level management, and redundancy management. This environment simplifies design and development since multiple, diverse functions do not need to be integrated within a common framework. Advances in software-enabled control dramatically alter the landscape of system architecture and software design tools. Instead of being restricted by an enforced hierarchy within the software architecture, the designers have the freedom to customize architecture. Some see this freedom as anarchy, but to others it presents an opportunity for extracting more performance and flexibility from given computing resources. However, a new generation of design and development tools are needed before the opportunity is realized. The tools need to provide seamless integration of varied system components and help the designer manage the additional complexity that comes with the added capability afforded. Specific features that are desired include some degree of self-documentation within the design environment, reliable auto-code generation with identical performance across platforms, inclusion of deterministic and probabilistic uncertainty in system models, management of hierarchies of system (and subsystem) models, integrated verification and validation methods, and seamless linkages to other software tools. Tight connections with the RTOS and middleware (such as the OCP) are also necessary, and advances in these areas will require corresponding upgrades to design and development software environments.

19.7. FINAL WORDS

The theory/practice gap yawns wider for control than for most other disciplines. Even as the IT revolution has had a game-changing effect in many fields, dramatically improving the performance/cost ratio and spurring innovation and technology transfer, the impact on avionics and flight control has been considerably more modest. This is partly inevitable—the complexities of dynamics and mission-criticality make any attempt at fundamental transformation that much more formidable—but also partly a consequence of limited interdisciplinary collaboration. It is only through a synthesis of control engineering and computer science that the potential benefits of advances in computing and software technologies for complex engineering systems such as UAVs will be realized. Attaining this goal requires multidisciplinary research initiatives.

The DARPA Software Enabled Control program is such an initiative. The research being supported in this program encompasses over a dozen projects, which together address topics central to enabling high performance, reliability, and autonomy in UAVs and other systems, and each of which, in its own way, demonstrates the impact that software and computing technologies can have in control engineering. The DARPA-sponsored research initiative continues: As we write this conclusion, plans are being finalized for the flight testing of much of the work presented in this volume. At the same time, new programs and efforts that are offshoots (at least in an intellectual sense) of the SEC vision are being established.

This volume thus represents more of a progress report than the culmination of a research initiative. Yet the influence of the SEC program on research in control and real-time systems generally is already evident, and the gap between research and practical application has perceptibly narrowed. The goal remains a pursuit, but we're headed in the right direction.

INDEX

ABOUT THE EDITORS

Tariq Samad received a B.S. degree in Engineering and Applied Science from Yale University and M.S. and Ph.D. degrees in Electrical and Computer Engineering from Carnegie Mellon University. He has been employed by various Honeywell research and development organizations since 1986 and is currently a Corporate Fellow associated with the corporation's Automation and Control Solutions business group. Dr. Samad is Editor-in-Chief of *IEEE Control Systems Magazine* and his recent publications include the edited volumes *Perspectives in Control Engineering: Technologies, Applications, New Directions* (IEEE Press) and *Automation, Control, and Complexity: An Integrated View* (Wiley). His research interests include intelligent control, autonomous systems, and statistical verification of real-time systems.

Gary Balas received B.S. and M.S. degrees in civil and electrical engineering from UC Irvine and a Ph.D. in Aeronautics and Astronautics from the California Institute of Technology in 1990. He is a professor in the Department of Aerospace Engineering and Mechanics at the University of Minnesota and is codirector of the Control Science and Dynamical Systems Department. He is a co-organizer and developer of the MUSYN Robust Control Short Course and the μ-Analysis and Synthesis Toolbox used with MATLAB, and president of MUSYN, Inc. In 1999, he received the Outstanding Young Investigator Award from the ASME Dynamic Systems and Control Division and in 1993 was awarded a McKnight-Land Grant Professorship at the University of Minnesota. He is an Associate Fellow of the AIAA and a Senior Member of the IEEE. Professor Balas' research interests include robust and distributed control, integration of control and real-time software services, and application of advanced control to aerospace vehicles.

Printed and bound by CPI Group (UK) Ltd, Croydon, CR0 4YY

27/10/2024

14580331-0002